THE ART OF THE NOVELLA

THE ART OF THE NOVELLA

Eight Short Novels

EDITED BY

Arnold B. Sklare

C. W. POST COLLEGE
OF LONG ISLAND UNIVERSITY

THE MACMILLAN COMPANY

NEW YORK

Second Printing, 1965

Library of Congress catalog card number: 65–11876

The Macmillan Company, New York
Collier-Macmillan Canada, Ltd., Toronto, Ontario

PRINTED IN THE UNITED STATES OF AMERICA

For I. S. S.

PREFACE

This anthology comes of a belief that most college students gain an insufficient introduction to long works of prose fiction. Courses that introduce students to genres of serious literature usually do not allow time for a novel or, at most, permit the reading of one or two. After the usual haphazard sampling of short stories, students often find only small pleasure in large works.

Because of its usually greater complexity and greater range of possible effects, the novella can offer a deeper insight into the art of fiction than that afforded by the short story. Readers who enjoy good short novels in college classes may well wish to take the upward step to full-length novels.

The short novels here are arranged in order of complexity of theme and style, and they are presented with little paraphernalia, so that there are no barriers between the student and the examples. A note on each selection gives the reader a suggestion of what to watch for; a few pertinent biographical and bibliographical facts concerning each writer are gathered at the end of the volume. Appreciation of the works will, in most cases, come largely through classroom analysis and written assignments.

By design, six of the eight stories in the collection are by authors who wrote in English. The American college reader needs to develop whatever knowledge he can of the power of his own language as an instrument of art. But he should also begin to have a perspective of other literatures. Two short novels from other languages in the best translations available will help point the way.

ARNOLD B. SKLARE

CONTENTS

THE ART OF THE NOVELLA

INTRODUCTION

A novella is a short novel. The term is properly used today to describe a prose fiction piece that is too short to be a novel and too long to be a short story. Yet it is not useful to dogmatize on the question of length of a novel, a novella, or a short story.

Novels may be very long—more than a thousand pages, or short—less than three hundred pages; but we know what a novel is nonetheless. A novel is a fictional prose narrative of considerable length, in which characters and actions are portrayed in a plot of some complexity. A novella is the same thing, except that it is of less than considerable length. People who dare to deal in numbers in such matters say that "considerable length" may be fifty thousand words or more. No story in this book is that long.

Surely, however, length in itself is not the only criterion for categorizing a work of literary art as a short story, short novel, or novel. The crucial issue, in addition to the length of a novella, is its complexity balanced against its unity of effect. The problem has to do with the nature of the beautiful, its essential characteristics, and the tests by which it may be judged and related to the human mind. The nature of the novella is difficult to define in isolation, though an analysis of the relationship between form and content in particular examples provides a constructive exercise. After similar studies have been made of short stories and novels, results from each genre may profitably be compared.

In any event, the question of novella-genre should not be made kin to that of how many angels can stand on the head of a pin. Few authorities today would be willing to cavil; a useful descriptive tag is needed for a not-short story, not-novel piece, and novella *is often chosen.*

Certain sophisticated readers will prefer the word novelette. *On occasion, the French word* nouvelle *is also used. But* novella *appears to be liked by many other English-speaking literary people because it has the weight of tradition. The term was first used to describe a type of narrative, current in the sixteenth century, consisting of a number of short stories, many of them older* fabliaux *in a new dress, told by characters in the main history. Boccaccio's* Decameron *is an example, and Chaucer's* Canterbury Tales *represents an English variety of this class of literature.*

There are, indeed, great short novels in all the modern European languages. While six of the eight stories in the present collection are by English and American authors, two are translated from the French and the Russian. So suited is the novella-form to the creation of masterly fiction that many excellent continental writers of the last hundred years were drawn to it.

With the exception of Katherine Mansfield, the writers represented in this collection also wrote full-length novels, and all of them wrote other short novels or short stories. The reader who finds these novelle *interesting may thus range widely among novels and short stories by these writers as well as others.*

A student may not experience the full intellectual excitement of the stories in this book without an awareness of their art. The appreciation of prose fiction is not possible without an understanding of such essentials as character, plot, setting, narrative technique, and theme. Analysis of these components, in proportion to the needs of each story and each writer, helps to produce the greatest of all intellectual satisfactions that a reader can enjoy—understanding based on adequate knowledge.

LEO TOLSTOY

Tolstoy became a legend in his own time. The drama of his tumultuous, prolonged religious conversion was carried out in the full view of a world which, during his lifetime, hungrily devoured everything he wrote. His prose fiction, in our day, of course stands on its own merits. But even at this distance, the reader comes to know that a story such as The Death of Ivan Ilych *is a personal confession, a personal expiation, a personal hymn of thanks. In this blazing footnote to Tolstoy's spiritual autobiography, art and life become one magnificent whole.*

To say that a short novel about death reverentially celebrates life is not to subject Tolstoy to the outrageous indignity of a glib phrase. Nor is there, in thus recognizing the author's almost unbearable energy for living, any desire to gloss over his equally dynamic apperception of dying. Ilych's death agony is Tolstoy's lament for the quality of Ivan's life. With great brilliance, Tolstoy penetrates to the core of an ordinary man's existence: Ivan was spiritually dead long before his physical decay.

*In these terms, the student knows unmistakably that he is confronted by a serious moral tract. Ivan is Everyman—an average nineteenth-century bureaucratic functionary, a bourgeois, a middle-class citizen—*un homme moyen sensuel. *Americans may think of him as a general equivalent of the man in the gray flannel suit. The cancer which consumes Ivan's flesh is a horror, but so is the cancer of his spirit. Only in his last hours does knowledge penetrate to Ivan's excited brain. Salvation is gained. Tolstoy, the man and the artist, gives thanks that he himself did not have to undergo Ivan's awesome mortification.*

Questions may legitimately arise as to whether Tolstoy found his own preaching so compelling that he allowed it to overrule his art. Is it aesthetically necessary, for example, to document so minutely the final stages of Ivan's demise? What is thereby gained or lost? If Ivan's life and death are central, is interest in them deflected or heightened by the revelations made about his family, friends, and doctors? Is

Ivan completely individualized, or does he remain a prototype throughout? Has this novella gone too far in making its point? Can life and art successfully come together as completely as they do here?

It has often been said that art improves the quality of life by exposing the labyrinthine possibilities of inner and outer human experience. Tolstoy anatomizes in cruel detail a man frozen into the semblance of life by observing only its forms while his essential and vital impulses die within him. Truthful emotions and honest responses are unlearned and unremembered; there is attrition of the soul. Swooning toward his ultimate release, Ivan Ilych speaks to his spirit and asks, "But what is the right thing?"

The Death of Ivan Ilych

I

DURING an interval in the Melvinski trial in the large building of the Law Courts, the members and public prosecutor met in Ivan Egorovich Shebek's private room, where the conversation turned on the celebrated Krasovski case. Fëdor Vasilievich warmly maintained that it was not subject to their jurisdiction, Ivan Egorovich maintained the contrary, while Peter Ivanovich, not having entered into the discussion at the start, took no part in it but looked through the *Gazette* which had just been handed in.

"Gentlemen," he said, "Ivan Ilych has died!"

"You don't say so!"

"Here, read it yourself," replied Peter Ivanovich, handing Fëdor Vasilievich the paper still damp from the press. Surrounded by a black border were the words: "Praskovya Fëdorovna Golovina, with profound sorrow, informs relatives and friends of the demise of her beloved husband Ivan Ilych Golovin, Member of the Court of Justice, which occurred on February the 4th of this year 1882. The funeral will take place on Friday at one o'clock in the afternoon."

From *Ivan Ilych and Hadji Murad and Other Stories* by Leo Tolstoy, translated by Louise and Aylmer Maude. Oxford University Press, 1935. Reprinted by permission.

Ivan Ilych had been a colleague of the gentlemen present and was liked by them all. He had been ill for some weeks with an illness said to be incurable. His post had been kept open for him, but there had been conjectures that in case of his death Alexeev might receive his appointment, and that either Vinnikov or Shtabel would succeed Alexeev. So on receiving the news of Ivan Ilych's death the first thought of each of the gentlemen in that private room was of the changes and promotions it might occasion among themselves or their acquaintances.

"I shall be sure to get Shtabel's place or Vinnikov's," thought Fëdor Vasilievich. "I was promised that long ago, and the promotion means an extra eight hundred rubles a year for me besides the allowance."

"Now I must apply for my brother-in-law's transfer from Kaluga," thought Peter Ivanovich. "My wife will be very glad, and then she won't be able to say that I never do anything for her relations."

"I thought he would never leave his bed again," said Peter Ivanovich aloud. "It's very sad."

"But what really was the matter with him?"

"The doctors couldn't say—at least they could, but each of them said something different. When last I saw him I thought he was getting better."

"And I haven't been to see him since the holidays. I always meant to go."

"Had he any property?"

"I think his wife had a little—but something quite trifling."

"We shall have to go to see her, but they live so terribly far away."

"Far away from you, you mean. Everything's far away from your place."

"You see, he never can forgive my living on the other side of the river," said Peter Ivanovich, smiling at Shebek. Then, still talking of the distances between different parts of the city, they returned to the Court.

Besides considerations as to the possible transfers and promotions likely to result from Ivan Ilych's death, the mere fact of the death of a near acquaintance aroused, as usual, in all who heard of it the complacent feeling that, "it is he who is dead and not I."

Each one thought or felt, "Well, he's dead but I'm alive!" But the more intimate of Ivan Ilych's acquaintances, his so-called friends, could not help thinking also that they would now have to fulfil the very tiresome demands of propriety by attending the funeral service and paying a visit of condolence to the widow.

Fëdor Vasilievich and Peter Ivanovich had been his nearest acquaintances. Peter Ivanovich had studied law with Ivan Ilych and had considered himself to be under obligations to him.

Having told his wife at dinner-time of Ivan Ilych's death and of his conjecture that it might be possible to get her brother transferred to their circuit, Peter Ivanovich sacrificed his usual nap, put on his evening clothes, and drove to Ivan Ilych's house.

At the entrance stood a carriage and two cabs. Leaning against the wall in the hall downstairs near the cloak-stand was a coffin-lid covered with cloth of gold, ornamented with gold cord and tassels, that had been polished up with metal powder. Two ladies in black were taking off their fur cloaks. Peter Ivanovich recognized one of them as Ivan Ilych's sister, but the other was a stranger to him. His colleague Schwartz was just coming downstairs, but on seeing Peter Ivanovich enter he stopped and winked at him, as if to say: "Ivan Ilych has made a mess of things—not like you and me."

Schwartz's face with his Piccadilly whiskers and his slim figure in evening dress had as usual an air of elegant solemnity which contrasted with the playfulness of his character and had a special piquancy here, or so it seemed to Peter Ivanovich.

Peter Ivanovich allowed the ladies to procede him and slowly followed them upstairs. Schwartz did not come down but remained where he was, and Peter Ivanovich understood that he wanted to arrange where they should play bridge that evening. The ladies went upstairs to the widow's room, and Schwartz with seriously compressed lips but a playful look in his eyes, indicated by a twist of his eyebrows the room to the right where the body lay.

Peter Ivanovich, like everyone else on such occasions, entered feeling uncertain what he would have to do. All he knew was that at such times it is always safe to cross oneself. But he was not quite sure whether one should make obeisances while doing so. He therefore adopted a middle course. On entering the room he began crossing himself and made a slight movement resembling a bow. At the same time, as far as the motion of his head and arm allowed, he surveyed the room. Two young men—apparently nephews, one of whom was a high-school pupil—were leaving the room, crossing themselves as they did so. An old woman was standing motionless, and a lady with strangely arched eyebrows was saying something to her in a whisper. A vigorous, resolute Church Reader, in a frock-coat, was reading something in a loud voice with an expression that precluded any contradiction. The butler's assistant, Gerasim, stepping lightly in front of Peter Ivanovich, was strewing something on the floor. Noticing this, Peter Ivanovich was immediately aware of a faint odour of a decomposing body.

The last time he had called on Ivan Ilych, Peter Ivanovich had seen Gerasim in the study. Ivan Ilych had been particularly fond of him and he was performing the duty of a sick nurse.

Peter Ivanovich continued to make the sign of the cross, slightly inclining his head in an intermediate direction between the coffin, the Reader, and the icons on the table in a corner of the room. Afterwards, when it seemed to him that this movement of his arm in crossing himself had gone on too long, he stopped and began to look at the corpse.

The dead man lay, as dead men always lie, in a specially heavy way, his rigid limbs sunk in the soft cushions of the coffin, with the head forever bowed on the pillow. His yellow waxen brow with bald patches over his sunken temples was thrust up in the way peculiar to the dead, the protruding nose seeming to press on the upper lip. He was much changed and had grown even thinner since Peter Ivanovich had last seen him, but, as is always the case with the dead, his face was handsomer and above all more dignified than when he was alive. The expression on the face said that what was necessary had been accomplished, and accomplished rightly. Besides this there was in that expression a reproach and a warning to the living. This warning seemed to Peter Ivanovich out of place, or at least not applicable to him. He felt a certain discomfort and so he hurriedly crossed himself once more and turned and went out of the door—too hurriedly and too regardless of propriety, as he himself was aware.

Schwartz was waiting for him in the adjoining room with legs spread wide apart and both hands toying with his top-hat behind his back. The mere sight of that playful, well-groomed, and elegant figure refreshed Peter Ivanovich. He felt that Schwartz was above all these happenings and would not surrender to any depressing influences. His very look said that this incident of a church service for Ivan Ilych could not be a sufficient reason for infringing the order of the session—in other words, that it would certainly not prevent his unwrapping a new pack of cards and shuffling them that evening while a footman placed four fresh candles on the table: in fact, that there was no reason for supposing that this incident would hinder their spending the evening agreeably. Indeed he said this in a whisper as Peter Ivanovich passed him, proposing that they should meet for a game at Fëdor Vasilievich's. But apparently Peter Ivanovich was not destined to play bridge that evening. Praskovya Fëdorovna (a short, fat woman who despite all efforts to the contrary had continued to broaden steadily from her shoulders downwards and who had the same extraordinarily arched eyebrows as the lady who had been standing by the coffin), dressed all in black, her head covered with lace, came out of her own room with some other ladies, conducted them to the room where the dead body lay, and said: "The service will begin immediately. Please go in."

Schwartz, making an indefinite bow, stood still, evidently neither accepting nor declining this invitation. Praskovya Fëdorovna, recognizing Peter Ivanovich, sighed, went close up to him, took his hand, and said: "I know you were a true friend of Ivan Ilych . . ." and looked at him awaiting some suitable response. And Peter Ivanovich knew that, just as it had been the right thing to cross himself in that room, so what he had to do here was to press her hand, sigh, and say, "Believe me. . . ." So he did all this and as he did it felt that the desired result had been achieved: that both he and she were touched.

"Come with me. I want to speak to you before it begins," said the widow. "Give me your arm."

Peter Ivanovich gave her his arm and they went to the inner rooms, passing Schwartz, who winked at Peter Ivanovich compassionately.

"That does for our bridge! Don't object if we find another player. Perhaps you can cut in when you do escape," said his playful look.

Peter Ivanovich sighed still more deeply and despondently, and Praskovya Fëdorovna pressed his arm gratefully. When they reached the drawing-room, upholstered in pink cretonne and lighted by a dim lamp, they sat down at the table—she on a sofa and Peter Ivanovich on a low pouffe, the springs of which yielded spasmodically under his weight. Praskovya Fëdorovna had been on the point of warning him to take another seat, but felt that such a warning was out of keeping with her present condition and so changed her mind. As he sat down on the pouffe Peter Ivanovich recalled how Ivan Ilych had arranged this room and had consulted him regarding this pink cretonne with green leaves. The whole room was full of furniture and knick-knacks, and on her way to the sofa the lace of the widow's black shawl caught on the carved edge of the table. Peter Ivanovich rose to detach it, and the springs of the pouffe, relieved of his weight, rose also and gave him a push. The widow began detaching her shawl herself, and Peter Ivanovich again sat down, suppressing the rebellious springs of the pouffe under him. But the widow had not quite freed herself and Peter Ivanovich got up again, and again the pouffe rebelled and even creaked. When this was all over she took out a clean cambric handkerchief and began to weep. The episode with the shawl and the struggle with the pouffe had cooled Peter Ivanovich's emotions and he sat there with a sullen look on his face. This awkward situation was interrupted by Sokolov, Ivan Ilych's butler, who came to report that the plot in the cemetery that Praskovya Fëdorovna had chosen would cost two hundred rubles. She stopped weeping and, looking at Peter Ivanovich with the air of a victim, remarked in French that it was very hard for her. Peter Ivanovich made a silent gesture signifying his full conviction that it must indeed be so.

"Please smoke," she said in a magnanimous yet crushed voice, and turned to discuss with Sokolov the price of the plot for the grave.

Peter Ivanovich while lighting his cigarette heard her inquiring very circumstantially into the prices of different plots in the cemetery and finally decide which she would take. When that was done she gave instructions about engaging the choir. Sokolov then left the room.

"I look after everything myself," she told Peter Ivanovich, shifting the albums that lay on the table; and noticing that the table was endangered by his cigarette-ash, she immediately passed him an ash-tray, saying as she did so: "I consider it an affectation to say that my grief prevents my attending to practical affairs. On the contrary, if anything can—I won't say console me, but—distract me, it is seeing to everything concerning him." She again took out her handkerchief as if preparing to cry, but suddenly, as if mastering her feeling, she shook herself and began to speak calmly. "But there is something I want to talk to you about."

Peter Ivanovich bowed, keeping control of the springs of the pouffe, which immediately began quivering under him.

"He suffered terribly the last few days."

"Did he?" said Peter Ivanovich.

"Oh, terribly! He screamed unceasingly, not for minutes but for hours. For the last three days he screamed incessantly. It was unendurable. I cannot understand how I bore it; you could hear him three rooms off. Oh, what I have suffered!"

"Is it possible that he was conscious all that time?" asked Peter Ivanovich.

"Yes," she whispered. "To the last moment. He took leave of us a quarter of an hour before he died, and asked us to take Volodya away."

The thought of the sufferings of this man he had known so intimately, first as a merry little boy, then as a school-mate, and later as a grown-up colleague, suddenly struck Peter Ivanovich with horror, despite an unpleasant consciousness of his own and this woman's dissimulation. He again saw that brow, and that nose pressing down on the lip, and felt afraid for himself.

"Three days of frightful suffering and then death! Why, that might suddenly, at any time, happen to me, he thought, and for a moment felt terrified. But—he did not himself know how—the customary reflection at once occurred to him that this had happened to Ivan Ilych and not to him, and that it should not and could not happen to him, and that to think that it could would be yielding to depression which he ought not to do, as Schwartz's expression plainly showed. After which reflection Peter Ivanovich felt reassured, and began to ask with interest about the details of Ivan Ilych's death, as though death was an accident natural to Ivan Ilych but certainly not to himself.

After many details of the really dreadful physical sufferings Ivan Ilych had endured (which details he learnt only from the effect those sufferings had produced on Praskovya Fëdorovna's nerves) the widow apparently found it necessary to get to business.

"Oh, Peter Ivanovich, how hard it is! How terribly, terribly hard!" and she again began to weep.

Peter Ivanovich sighed and waited for her to finish blowing her nose. When she had done so he said, "Believe me . . ." and she again began talking and brought out what was evidently her chief concern with him—namely, to question him as to how she could obtain a grant of money from the government on the occasion of her husband's death. She made it appear that she was asking Peter Ivanovich's advice about her pension, but he soon saw that she already knew about that to the minutest detail, more even than he did himself. She knew how much could be got out of the government in consequence of her husband's death, but wanted to find out whether she could not possibly extract something more. Peter Ivanovich tried to think of some means of doing so, but after reflecting for a while and, out of propriety, condemning the government for its niggardliness, he said he thought that nothing more could be got. Then she sighed and evidently began to devise means of getting rid of her visitor. Noticing this, he put out his cigarette, rose, pressed her hand, and went out into the anteroom.

In the dining-room where the clock stood that Ivan Ilych had liked so much and had bought at an antique shop, Peter Ivanovich met a priest and a few acquaintances who had come to attend the service, and he recognized Ivan Ilych's daughter, a handsome young woman. She was in black and her slim figure appeared slimmer than ever. She had a gloomy, determined, almost angry expression, and bowed to Peter Ivanovich as though he were in some way to blame. Behind her, with the same offended look, stood a wealthy young man, an examining magistrate, whom Peter Ivanovich also knew and who was her fiancé, as he had heard. He bowed mournfully to them and was about to pass into the death-chamber, when from under the stairs appeared the figure of Ivan Ilych's schoolboy son, who was extremely like his father. He seemed a little Ivan Ilych, such as Peter Ivanovich remembered when they studied law together. His tear-stained eyes had in them the look that is seen in the eyes of boys of thirteen or fourteen who are not pure-minded. When he saw Peter Ivanovich he scowled morosely and shamefacedly. Peter Ivanovich nodded to him and entered the death-chamber. The service began: candles, groans, incense, tears, and sobs. Peter Ivanovich stood looking gloomily down at his feet. He did not look once at the dead man, did not yield to any depressing influence, and was one of the first to leave the room. There was no one in the anteroom, but Gerasim darted out of the dead man's room, rummaged with his strong hands among the fur coats to find Peter Ivanovich's, and helped him on with it.

"Well, friend Gerasim," said Peter Ivanovich, so as to say something. "It's a sad affair, isn't it?"

"It's God's will. We shall all come to it some day," said Gerasim, displaying his teeth—the even, white teeth of a healthy peasant—and, like a man in the thick of urgent work, he briskly opened the front door, called the coachman, helped Peter Ivanovich into the sledge, and sprang back to the porch as if in readiness for what he had to do next.

Peter Ivanovich found the fresh air particularly pleasant after the smell of incense, the dead body, and carbolic acid.

"Where to, sir?" asked the coachman.

"It's not too late even now. . . . I'll call round on Fëdor Vasilievich."

He accordingly drove there and found them just finishing the first rubber, so that it was quite convenient for him to cut in.

II

IVAN ILYCH'S life had been most simple and most ordinary and therefore most terrible.

He had been a member of the Court of Justice, and died at the age of forty-five. His father had been an official who after serving in various ministries and departments in Petersburg had made the sort of career which brings men to positions from which by reason of their long service they cannot be dismissed, though they are obviously unfit to hold any responsible position, and for whom therefore posts are specially created, which though fictitious carry salaries of from six to ten thousand rubles that are not fictitious, and in receipt of which they live on to a great age.

Such was the Privy Councillor and superfluous member of various superfluous institutions, Ilya Epimovich Golovin.

He had three sons, of whom Ivan Ilych was the second. The eldest son was following in his father's footsteps only in another department, and was already approaching that stage in the service at which a similar sinecure would be reached. The third son was a failure. He had ruined his prospects in a number of positions and was now serving in the railway department. His father and brothers, and still more their wives, not merely disliked meeting him, but avoided remembering his existence unless compelled to do so. His sister had married Baron Greff, a Petersburg official of her father's type. Ivan Ilych was *le phénix de la famille* as people said. He was neither as cold and formal as his elder brother nor as wild as the younger, but was a happy mean between them—an intelligent, polished, lively, and agreeable man. He had studied with his younger brother at the School of Law, but the latter had failed to complete the course and was expelled when he was in the fifth class. Ivan Ilych finished the course well. Even when he was at the School of Law he was just what he remained for the rest of his life: a capable, cheerful, good-natured, and sociable man, though strict in the fulfilment of what he considered to be his duty: and he considered his duty to be what was so considered by those in authority. Neither as a boy nor as a man was he a toady, but from early youth was by nature attracted to people of high station as a fly is drawn to the light, assimilating their ways and views of life and establishing friendly relations with them. All the enthusiasms of childhood and youth passed without leaving much trace on him; he succumbed to sensuality, to vanity, and latterly among the highest classes to liberalism, but always within limits which his instinct unfailingly indicated to him as correct.

At school he had done things which had formerly seemed to him very horrid and made him feel disgusted with himself when he did them; but when later on he saw that such actions were done by people of good position and that they did not regard them as wrong, he was able not exactly to regard them as right, but to forget about them entirely or not be at all troubled at remembering them.

Having graduated from the School of Law and qualified for the tenth rank of the civil service, and having received money from his father for his equipment, Ivan Ilych ordered himself clothes at Scharmer's, the fashionable tailor, hung a medallion inscribed *respice finem* on his watch-chain, took leave of his professor and the prince who was patron of the school, had a farewell dinner with his comrades at Donon's first-class restaurant, and with his new and fashionable portmanteau, linen, clothes, shaving and other toilet appliances, and a travelling rug, all purchased at the best shops, he set off for one of the provinces where, through his father's influence, he had been attached to the Governor as an official for special service.

In the province Ivan Ilych soon arranged as easy and agreeable a position for himself as he had had at the School of Law. He performed his official tasks, made his career, and at the same time amused himself pleasantly and decorously. Occasionally he paid official visits to country districts, where he behaved with dignity both to his superiors and inferiors, and performed the duties entrusted to him, which related chiefly to the sectarians, with an exactness and incorruptible honesty of which he could not but feel proud.

In official matters, despite his youth and taste for frivolous gaiety, he was exceedingly reserved, punctilious, and even severe; but in society he was often amusing and witty, and always good-natured, correct in his manner, and *bon enfant*, as the Governor and his wife—with whom he was like one of the family—used to say of him.

In the province he had an affair with a lady who made advances to the elegant young lawyer, and there was also a milliner; and there were carousals with aides-de-camp who visited the district, and after-supper visits to a certain outlying street of doubtful reputation; and there was too some obsequiousness to his chief and even to his chief's wife, but all this was done with such a tone of good breeding that no hard names could be applied to it. It all came under the heading of the French saying: "*Il faut que jeunesse se passe.*"[1] It was all done with clean hands, in clean linen, with French phrases, and above all among people

[1] Youth must have its fling.

of the best society and consequently with the approval of people of rank.

So Ivan Ilych served for five years and then came a change in his official life. The new and reformed judicial institutions were introduced, and new men were needed. Ivan Ilych became such a new man. He was offered the post of examining magistrate, and he accepted it though the post was in another province and obliged him to give up the connexions he had formed and to make new ones. His friends met to give him a send-off; they had a group-photograph taken and presented him with a silver cigarette-case, and he set off to his new post.

As examining magistrate Ivan Ilych was just as *comme il faut* and decorous a man, inspiring general respect and capable of separating his official duties from his private life, as he had been when acting as an official on special service. His duties now as examining magistrate were far more interesting and attractive than before. In his former position it had been pleasant to wear an undress uniform made by Scharmer, and to pass through the crowd of petitioners and officials who were timorously awaiting an audience with the Governor, and who envied him as with free and easy gait he went straight into his chief's private room to have a cup of tea and a cigarette with him. But not many people had been directly dependent on him—only police officials and the sectarians when he went on special missions—and he liked to treat them politely, almost as comrades, as if he were letting them feel that he who had the power to crush them was treating them in this simple, friendly way. There were then but few such people. But now, as an examining magistrate, Ivan Ilych felt that everyone without exception, even the most important and self-satisfied, was in his power, and that he need only write a few words on a sheet of paper with a certain heading, and this or that important, self-satisfied person would be brought before him in the role of an accused person or a witness, and if he did not choose to allow him to sit down, would have to stand before him and answer his questions. Ivan Ilych never abused his power; he tried on the contrary to soften its expression, but the consciousness of it and of the possibility of softening its effect, supplied the chief interest and attraction of his office. In his work itself, especially in his examinations, he very soon acquired a method of eliminating all considerations irrelevant to the legal aspect of the case, and reducing even the most complicated case to a form in which it would be presented on paper only in its externals, completely excluding his personal opinion of the matter, while above all observing every prescribed formality. The work was new and Ivan Ilych was one of the first men to apply the new Code of 1864.[2]

On taking up the post of examining magistrate in a new town, he made new acquaintances and connexions, placed himself on a new footing, and assumed a somewhat different tone. He took up an attitude of rather dignified aloofness towards the provincial authorities, but picked out the best circle of legal gentlemen and wealthy gentry living in the town and assumed a tone of slight dissatisfaction with the government, of moderate liberalism, and of enlightened citizenship. At the same time, without at all altering the elegance of his toilet, he ceased shaving his chin and allowed his beard to grow as it pleased.

Ivan Ilych settled down very pleasantly in this new town. The society there, which inclined towards opposition to the Governor, was friendly, his salary was larger, and he began to play *vint* [a form of bridge], which he found added not a little to the pleasure of life, for he had a capacity for cards, played good-humouredly, and calculated rapidly and astutely, so that he usually won.

After living there for two years he met his future wife, Praskovya Fëdorovna Mikhel, who was the most attractive, clever, and brilliant girl of the set in which he moved, and among other amusements and relaxations from his labours as examining magistrate, Ivan Ilych established light and playful relations with her.

While he had been an official on special service he had been accustomed to dance, but now as an examining magistrate it was exceptional for him to do so. If he danced now, he did it as if to show that though he served under the reformed order of things, and had reached the fifth official rank, yet when it came to dancing he could do it better than most people. So at the end of an evening he sometimes danced with Praskovya Fëdorovna, and it was chiefly during these dances that he captivated her. She fell in love with him. Ivan Ilych had at first no definite intention of marrying, but when the girl fell in love with him he said to himself: "Really, why shouldn't I marry?"

Praskovya Fëdorovna came of a good family, was not bad-looking, and had some little property. Ivan Ilych might have aspired to a more brilliant match, but even this was good. He had his salary, and she, he hoped, would have an equal income. She was well connected, and

[2] The emancipation of the serfs in 1861 was followed by a thorough all-round reform of judicial proceedings. A. M. [Translator's note.]

was a sweet, pretty, and thoroughly correct young woman. To say that Ivan Ilych married because he fell in love with Praskovya Fëdorovna and found that she sympathized with his views of life would be as incorrect as to say that he married because his social circle approved of the match. He was swayed by both these considerations: the marriage gave him personal satisfaction, and at the same time it was considered the right thing by the most highly placed of his associates.

So Ivan Ilych got married.

The preparations for marriage and the beginning of married life, with its conjugal caresses, the new furniture, new crockery, and new linen, were very pleasant until his wife became pregnant—so that Ivan Ilych had begun to think that marriage would not impair the easy, agreeable, gay, and always decorous character of his life, approved of by society and regarded by himself as natural, but would even improve it. But from the first months of his wife's pregnancy, something new, unpleasant, depressing, and unseemly, and from which there was no way of escape, unexpectedly showed itself.

His wife, without any reason—*de gaieté de cœur* as Ivan Ilych expressed it to himself—began to disturb the pleasure and propriety of their life. She began to be jealous without any cause, expected him to devote his whole attention to her, found fault with everything, and made coarse and ill-mannered scenes.

At first Ivan Ilych hoped to escape from the unpleasantness of this state of affairs by the same easy and decorous relation to life that had served him heretofore: he tried to ignore his wife's disagreeable moods, continued to live in his usual easy and pleasant way, invited friends to his house for a game of cards, and also tried going out to his club or spending his evenings with friends. But one day his wife began upbraiding him so vigorously, using such coarse words, and continued to abuse him every time he did not fulfil her demands, so resolutely and with such evident determination not to give way till he submitted—that is, till he stayed at home and was bored just as she was—that he became alarmed. He now realized that matrimony—at any rate with Praskovya Fëdorovna—was not always conducive to the pleasures and amenities of life, but on the contrary often infringed both comfort and propriety, and that he must therefore entrench himself against such infringement. And Ivan Ilych began to seek for means of doing so. His official duties were the one thing that imposed upon Praskovya Fëdorovna, and by means of his official work and the duties attached to it he began struggling with his wife to secure his own independence.

With the birth of their child, the attempts to feed it and the various failures in doing so, and with the real and imaginary illnesses of mother and child, in which Ivan Ilych's sympathy was demanded but about which he understood nothing, the need of securing for himself an existence outside his family life became still more imperative.

As his wife grew more irritable and exacting and Ivan Ilych transferred the centre of gravity of his life more and more to his official work, so did he grow to like his work better and become more ambitious than before.

Very soon, within a year of his wedding, Ivan Ilych had realized that marriage, though it may add some comforts to life, is in fact a very intricate and difficult affair towards which in order to perform one's duty, that is, to lead a decorous life approved of by society, one must adopt a definite attitude just as toward's one's official duties.

And Ivan Ilych evolved such an attitude towards married life. He only required of it those conveniences—dinner at home, housewife, and bed—which it could give him, and above all that propriety of external forms required by public opinion. For the rest he looked for light-hearted pleasure and propriety, and was very thankful when he found them, but if he met with antagonism and querulousness he at once retired into his separate fenced-off world of official duties, where he found satisfaction.

Ivan Ilych was esteemed a good official, and after three years was made Assistant Public Prosecutor. His new duties, their importance, the possibility of indicting and imprisoning anyone he chose, the publicity his speeches received, and the success he had in all these things, made his work still more attractive.

More children came. His wife became more and more querulous and ill-tempered, but the attitude Ivan Ilych had adopted towards his home life rendered him almost impervious to her grumbling.

After seven years' service in that town he was transferred to another province as Public Prosecutor. They moved, but were short of money and his wife did not like the place they moved to. Though the salary was higher the cost of living was greater, besides which two of their children died and family life became still more unpleasant for him.

Praskovya Fëdorovna blamed her husband for every inconvenience they encountered in their new home. Most of the conversations between husband and wife, especially as to the children's education, led to topics which recalled former disputes, and those disputes were apt to flare up again at any moment. There remained only those rare periods of amorousness which still came to them at times but did not last long. These were islets at which they anchored for a while and then again set out upon that ocean of veiled hostility which showed itself in their aloofness from one another. This aloofness might have grieved Ivan Ilych had he considered that it ought not to exist, but he now regarded the position as normal, and even made it the goal at which he aimed in family life. His aim was to free himself more and more from those unpleasantnesses and to give them a semblance of harmlessness and propriety. He attained this by spending less and less time with his family, and when obliged to be at home he tried to safeguard his position by the presence of outsiders. The chief thing however was that he had his official duties. The whole interest of his life now centered in the official world and that interest absorbed him. The consciousness of his power, being able to ruin anybody he wished to ruin, the importance, even the external dignity of his entry into court, or meetings with his subordinates, his success with superiors and inferiors, and above all his masterly handling of cases, of which he was conscious—all this gave him pleasure and filled his life, together with chats with his colleagues, dinners, and bridge. So that on the whole Ivan Ilych's life continued to flow as he considered it should do—pleasantly and properly.

So things continued for another seven years. His eldest daughter was already sixteen, another child had died, and only one son was left, a schoolboy and a subject of dissension. Ivan Ilych wanted to put him in the School Law, but to spite him Praskovya Fëdorovna entered him at the High School. The daughter had been educated at home and had turned out well: the boy did not learn badly either.

III

SO IVAN ILYCH lived for seventeen years after his marriage. He was already a Public Prosecutor of long standing, and had declined several proposed transfers while awaiting a more desirable post, when an unanticipated and unpleasant occurrence quite upset the peaceful course of his life. He was expecting to be offered the post of presiding judge in a University town, but Happe somehow came to the front and obtained the appointment instead. Ivan Ilych became irritable, reproached Happe, and quarrelled both with him and with his immediate superiors—who became colder to him and again passed him over when other appointments were made.

This was in 1880, the hardest year of Ivan Ilych's life. It was then that it became evident on the one hand that his salary was insufficient for them to live on, and on the other that he had been forgotten, and not only this, but that what was for him the greatest and most cruel injustice appeared to others a quite ordinary occurrence. Even his father did not consider it his duty to help him. Ivan Ilych felt himself abandoned by everyone, and that they regarded his position with a salary of 3,500 rubles as quite normal and even fortunate. He alone knew that with the consciousness of the injustices done him, with his wife's incessant nagging, and with the debts he had contracted by living beyond his means, his position was far from normal.

In order to save money that summer he obtained leave of absence and went with his wife to live in the country at her brother's place.

In the country, without his work, he experienced *ennui* for the first time in his life, and not only *ennui* but intolerable depression, and he decided that it was impossible to go on living like that, and that it was necessary to take energetic measures.

Having passed a sleepless night pacing up and down the veranda, he decided to go to Petersburg and bestir himself, in order to punish those who had failed to appreciate him and to get transferred to another ministry.

Next day, despite many protests from his wife and her brother, he started for Petersburg with the sole object of obtaining a post with a salary of five thousand rubles a year. He was no longer bent on any particular department, or tendency, or kind of activity. All he now wanted was an appointment to another post with a salary of five thousand rubles, either in the administration, in the banks, with the railways, in one of the Empress Marya's Institutions, or even in the customs—but it had to carry with it a salary of five thousand rubles and be in a ministry other than that in which they had failed to appreciate him.

And this quest of Ivan Ilych's was crowned with remarkable and unexpected success. At Kursk an acquaintance of his, F. I. Ilyin, got into the first-class carriage, sat down beside Ivan Ilych, and told him of a telegram just received by the Governor of Kursk announcing that a change was about to take place in the ministry: Peter Ivanovich was to be superseded by Ivan Semënovich.

The proposed change, apart from its significance for Russia, had a special significance for Ivan Ilych, because by bringing forward a new man, Peter Petrovich, and consequently his friend Zachar Ivanovich, it was highly favourable for Ivan Ilych, since Zachar Ivanovich was a friend and colleague of his.

In Moscow this news was confirmed, and on reaching Petersburg Ivan Ilych found Zachar Ivanovich and received a definite promise of an appointment in his former department of Justice.

A week later he telegraphed to his wife: "Zachar in Miller's place. I shall receive appointment on presentation of report."

Thanks to this change of personnel, Ivan Ilych had unexpectedly obtained an appointment in his former ministry which placed him two stages above his former colleagues besides giving him five thousand rubles salary and three thousand five hundred rubles for expenses connected with his removal. All his ill humour towards his former enemies and the whole department vanished, and Ivan Ilych was completely happy.

He returned to the country more cheerful and contented than he had been for a long time. Praskovya Fëdorovna also cheered up and a truce was arranged between them. Ivan Ilych told of how he had been fêted by everybody in Petersburg, how all those who had been his enemies were put to shame and now fawned on him, how envious they were of his appointment, and how much everybody in Petersburg had liked him.

Praskovya Fëdorovna listened to all this and appeared to believe it. She did not contradict anything, but only made plans for their life in the town to which they were going. Ivan Ilych saw with delight that these plans were his plans, that he and his wife agreed, and that, after a stumble, his life was regaining its due and natural character of pleasant lightheartedness and decorum.

Ivan Ilych had come back for a short time only, for he had to take up his new duties on the 10th of September. Moreover, he needed time to settle into the new place, to move all his belongings from the province, and to buy and order many additional things: in a word, to make such arrangements as he had resolved on, which were almost exactly what Praskovya Fëdorovna too had decided on.

Now that everything had happened so fortunately, and that he and his wife were at one in their aims and moreover saw so little of one another, they got on together better than they had done since the first years of marriage. Ivan Ilych had thought of taking his family away with him at once, but the insistence of his wife's brother and her sister-in-law, who had suddenly become particularly amiable and friendly to him and his family, induced him to depart alone.

So he departed, and the cheerful state of mind induced by his success and by the harmony between his wife and himself, the one intensifying the other, did not leave him. He found a delightful house, just the thing both he and his wife had dreamt of. Spacious, lofty reception rooms in the old style, a convenient and dignified study, rooms for his wife and daughter, a study for his son—it might have been specially built for them. Ivan Ilych himself superintended the arrangements, chose the wallpapers, supplemented the furniture (preferably with antiques which he considered particularly *comme il faut*), and supervised the upholstering. Everything progressed and progressed and approached the ideal he had set himself: even when things were only half completed they exceeded his expectations. He saw what a refined and elegant character, free from vulgarity, it would all have when it was ready. On falling asleep he pictured to himself how the reception-room would look. Looking at the yet unfinished drawing-room he could see the fireplace, the screen, the what-not, the little chairs dotted here and there, the dishes and plates on the walls, and the bronzes, as they would be when everything was in place. He was pleased by the thought of how his wife and daughter, who shared his taste in this matter, would be impressed by it. They were certainly not expecting as much. He had been particularly successful in finding, and buying cheaply, antiques which gave a particularly aristocratic character to the whole place. But in his letters he intentionally understated everything in order to be able to surprise them. All this so absorbed him that his new duties—though he liked his official work—interested him less than he had expected. Sometimes he even had moments of absent-mindedness during the Court Sessions, and would consider whether he should have straight or curved cornices for his curtains. He was so interested in it all that he often did things himself, rearranging the furniture, or rehanging the curtains. Once when mounting a step-ladder to show the upholsterer, who did not understand, how he wanted the hangings draped, he made a false step and slipped, but being a strong and agile man he clung on and only knocked his side against the knob of the window frame.

The bruised place was painful but the pain soon passed, and he felt particularly bright and well just then. He wrote: "I feel fifteen years younger." He thought he would have everything ready by September, but it dragged on till mid-October. But the result was charming not only in his eyes but to everyone who saw it.

In reality it was just what is usually seen in the houses of people of moderate means who want to appear rich, and therefore succeed only in resembling others like themselves: there were damasks, dark wood, plants, rugs, and dull and polished bronzes—all the things people of a certain class have in order to resemble other people of that class. His house was so like the others that it would never have been noticed, but to him it all seemed to be quite exceptional. He was very happy when he met his family at the station and brought them to the newly furnished house all lit up, where a footman in a white tie opened the door into the hall decorated with plants, and when they went on into the drawing-room, and the study uttering exclamations of delight. He conducted them everywhere, drank in their praises eagerly, and beamed with pleasure. At tea that evening, when Praskovya Fëdorovna among other things asked him about his fall, he laughed and showed them how he had gone flying and had frightened the upholsterer.

"It's a good thing I'm a bit of an athlete. Another man might have been killed, but I merely knocked myself, just here; it hurts when it's touched, but it's passing off already—it's only a bruise."

So they began living in their new home—in which, as always happens, when they got thoroughly settled in they found they were just one room short—and with the increased income, which as always was just a little (some five hundred rubles) too little, but it was all very nice.

Things went particularly well at first, before everything was finally arranged and while something had still to be done: this thing bought, that thing ordered, another thing moved, and something else adjusted. Though there were some disputes between husband and wife, they were both so well satisfied and had so much to do that it all passed off without any serious quarrels. When nothing was left to arrange it became rather dull and something seemed to be lacking, but they were then making acquaintances, forming habits, and life was growing fuller.

Ivan Ilych spent his mornings at the law courts and came home to dinner, and at first he was generally in a good humour, though he occasionally became irritable just on account of his house. (Every spot on the tablecloth or the upholstery, and every broken window-blind string, irritated him. He had devoted so much trouble to arranging it all that every disturbance of it distressed him.) But on the whole his life ran its course as he believed life should do: easily, pleasantly, and decorously.

He got up at nine, drank his coffee, read the paper, and then put on his undress uniform and went to the law courts. There the harness in which he worked had already been stretched to fit him and he donned it without a hitch: petitioners, inquiries at the chancery, the chancery itself, and the sittings public and administrative. In all this the thing was to exclude everything fresh and vital, which always disturbs the regular course of official business, and to admit only official relations with people, and then only on official grounds. A man would come, for instance, wanting some information. Ivan Ilych, as one in whose sphere the matter did not lie, would have nothing to do with him: but if the man had some business with him in his official capacity, something that could be expressed on officially stamped paper, he would do everything, positively everything he could within the limits of such relations, and in doing so would maintain the semblance of friendly human relations, that is, would observe the courtesies of life. As soon as the official relations ended, so did everything else. Ivan Ilych possessed this capacity to separate his real life from the official side of affairs and not mix the two, in the highest degree, and by long practice and natural aptitude had brought it to such a pitch that sometimes, in the manner of a virtuoso, he would even allow himself to let the human and official relations mingle. He let himself do this just because he felt that he could at any time he chose resume the strictly official attitude again and drop the human relation. And he did it all easily, pleasantly, correctly, and even artistically. In the intervals between the sessions he smoked, drank tea, chatted a little about politics, a little about general topics, a little about cards, but most of all about official appointments. Tired, but with the feelings of a virtuoso—one of the first violins who has played his part in an orchestra with precision—he would return home to find that his wife and daughter had been out paying calls, or had a visitor, and that his son had been to school, had done his homework with his tutor, and was duly learning what is taught at High Schools. Everything was as it should be. After dinner, if they had no visitors, Ivan Ilych sometimes read a book that was being much discussed at the time, and in the evening settled down to work, that is, read official papers, compared the depositions of witnesses, and noted paragraphs of the Code applying to them. This was neither dull nor amusing. It was dull when he might have been playing bridge, but if no bridge was available it was at any rate better than doing nothing or sitting with his wife. Ivan Ilych's chief pleasure was giving little dinners to which he invited men and women of good social position, and just as his drawing-room resembled all other drawing-rooms so did his enjoyable little parties resemble all other such parties.

Once they even gave a dance. Ivan Ilych enjoyed it and everything went off well, except that it led to a violent quarrel with his wife about the cakes and sweets. Praskovya Fëdorovna had made her own plans, but Ivan Ilych insisted on getting everything from an expensive confectioner and ordered too many cakes, and the quarrel occurred because some of those cakes were left over and the confectioner's bill came to forty-five rubles. It was a great and disagreeable quarrel. Praskovya Fëdorovna called him "a fool and an imbecile," and he clutched at his head and made angry allusions to divorce.

But the dance itself had been enjoyable. The best people were there, and Ivan Ilych had danced with Princess Trufonova, a sister of the distinguished founder of the Society "Bear my Burden."

The pleasures connected with his work were pleasures of ambition; his social pleasures were those of vanity; but Ivan Ilych's greatest pleasure was playing bridge. He acknowledged that whatever disagreeable incident happened in his life, the pleasure that beamed like a ray of light above everything else was to sit down to bridge with good players, not noisy partners, and of course to four-handed bridge (with five players it was annoying to have to stand out, though one pretended not to mind), to play a clever and serious game (when the cards allowed it), and then to have supper and drink a glass of wine. After a game of bridge, especially if he had won a little (to win a large sum was unpleasant), Ivan Ilych went to bed in specially good humour.

So they lived. They formed a circle of acquaintances among the best people and were visited by people of importance and by young folk. In their views as to their acqaintances, husband, wife, and daughter were entirely agreed, and tacitly and unanimously kept at arm's length and shook off the various shabby friends and relations who, with much show of affection, gushed into the drawing-room with its Japanese plates on the walls. Soon these shabby friends ceased to obtrude themselves and only the best people remained in the Golovins' set.

Young men made up to Lisa, and Petrishchev, an examining magistrate and Dmitri Ivanovich Petrishchev's son and sole heir, began to be so attentive to her that Ivan Ilych had already spoken to Praskovya Fëdorovna about it, and considered whether they should not arrange a party for them, or get up some private theatricals.

So they lived, and all went well, without change, and life flowed pleasantly.

IV

THEY WERE all in good health. It could not be called ill health if Ivan Ilych sometimes said that he had a queer taste in his mouth and felt some discomfort in his left side.

But this discomfort increased and, though not exactly painful, grew into a sense of pressure in his side accompanied by ill humour. And his irritability became worse and worse and began to mar the agreeable, easy, and correct life that had established itself in the Golovin family. Quarrels between husband and wife became more and more frequent, and soon the ease and amenity disappeared and even the decorum was barely maintained. Scenes again became frequent, and very few of those islets remained on which husband and wife could meet without an explosion. Praskovya Fëdorovna now had good reason to say that her husband's temper was trying. With characteristic exaggeration she said he had always had a dreadful temper, and that it had needed all her good nature to put up with it for twenty years. It was true that now the quarrels were started by him. His bursts of temper always came just before dinner, often just as he began to eat his soup. Sometimes he noticed that a plate or dish was chipped, or the food was not right, or his son put his elbow on the table, or his daughter's hair was not done as he liked it, and for all this he blamed Praskovya Fëdorovna. At first she retorted and said disagreeable things to him, but once or twice he fell into such a rage at the beginning of dinner that she realized it was due to some physical derangement brought on by taking food, and so she restrained herself and did not answer, but only hurried to get the dinner over. She regarded this self-restraint as highly praiseworthy. Having come to the conclusion that her husband had a dreadful temper and made her life miserable, she began to feel sorry for herself, and the more she pitied herself the more she hated her husband. She began to wish he would die; yet she did not want him to die because then his salary would cease. And this irritated her against him still more. She considered herself dreadfully unhappy just because not even his death could save her, and though she concealed her exasperation, that hidden exasperation of hers increased his irritation also.

After one scene in which Ivan Ilych had been particularly unfair and after which he had said in explanation that he certainly was irritable but that it was due to his not being well, she said that if he was ill it should be attended to, and insisted on his going to see a celebrated doctor.

He went. Everything took place as he had expected and as it always does. There was the usual waiting and the important air assumed by the doctor, with which he was so familiar (resembling that which he himself assumed in court), and the sounding and listening, and the questions which called for answers that were foregone conclusions and were evidently unnecessary, and the look of importance which implied that "if only you put yourself in our hands we will arrange everything—we know indubitably how it has to be done, always in the same way for everybody alike." It was all just as it was in the law courts. The doctor put on just the same air towards him as he himself put on towards an accused person.

The doctor said that so-and-so indicated that there was so-and-so inside the patient, but if the investigation of so-and-so did not confirm this, then he must assume that and that. If he assumed that and that, then . . . and so on. To Ivan Ilych only one question was important: was his case serious or not? But the doctor ignored that inappropriate question. From his point of view it was not the one under consideration, the real question was to decide between a floating kidney, chronic catarrh, or appendicitis. It was not a question of Ivan Ilych's life or death, but one between a floating kidney and appendicitis. And that question the doctor solved brilliantly, as it seemed to Ivan Ilych, in favour of the appendix, with the reservation that should an examination of the urine give fresh indications the matter would be reconsidered. All this was just what Ivan Ilych had himself brilliantly accomplished a thousand times in dealing with men on trial. The doctor summed up just as brilliantly, looking over his spectacles triumphantly and even gaily at the accused. From the doctor's summing up Ivan Ilych concluded that things were bad, but that for the doctor, and perhaps for everybody else, it was a matter of indifference, though for him it was bad. And this conclusion struck him painfully, arousing in him a great feeling of pity for himself and of bitterness towards the doctor's indifference to a matter of such importance.

He said nothing of this, but rose, placed the doctor's fee on the table, and remarked with a sigh: "We sick people probably often put inappropriate questions. But tell me, in general, is this complaint dangerous, or not? . . ."

The doctor looked at him sternly over his spectacles with one eye, as if to say: "Prisoner, if you will not keep to the questions put to you, I shall be obliged to have you removed from the court."

"I have already told you what I consider necessary and proper. The analysis may show something more." And the doctor bowed.

Ivan Ilych went out slowly, seated himself disconsolately in his sledge, and drove home. All the way home he was going over what the doctor had said, trying to translate those complicated, obscure, scientific phrases into plain language and find in them an answer to the question: "Is my condition bad? Is it very bad? Or is there as yet nothing much wrong?" And it seemed to him that the meaning of what the doctor had said was that it was very bad. Everything in the streets seemed depressing. The cabmen, the houses, the passers-by, and the shops, were dismal. His ache, this dull gnawing ache that never ceased for a moment, seemed to have acquired a new and more serious significance from the doctor's dubious remarks. Ivan Ilych now watched it with a new and oppressive feeling.

He reached home and began to tell his wife about it. She listened, but in the middle of his account his daughter came in with her hat on, ready to go out with her mother. She sat down reluctantly to listen to this tedious story, but could not stand it long, and her mother too did not hear him to the end.

"Well, I am very glad," she said. "Mind now to take your medicine regularly. Give me the prescription and I'll send Gerasim to the chemist's." And she went to get ready to go out.

While she was in the room Ivan Ilych had hardly taken time to breathe, but he sighed deeply when she left it.

"Well," he thought, "perhaps it isn't so bad after all."

He began taking his medicine and following the doctor's directions, which had been altered after the examination of the urine. But then it happened that there was a contradiction between the indications drawn from the examination of the urine and the symptoms that showed themselves. It turned out that what was happening differed from what the doctor had told him, and that he had either forgotten, or blundered, or hidden something from him. He could not, however, be blamed for that, and Ivan Ilych still obeyed his orders implicitly and at first derived some comfort from doing so.

From the time of his visit to the doctor, Ivan Ilych's chief occupation was the exact fulfilment of the doctor's instructions regarding hygiene and the taking of medicine, and the observation of his pain and his excretions. His chief interests came to be people's ailments and people's health. When sickness, deaths, or recoveries were mentioned in his presence, especially when the illness resembled his own, he listened with agitation which he tried to hide, asked questions, and applied what he heard to his own case.

The pain did not grow less, but Ivan Ilych made efforts to force himself to think that he was better. And he could do this so long as nothing agitated him. But as soon as he had any unpleasantness with his wife, any lack of success in his official work, or held bad cards at bridge, he was at once acutely sensible of his disease. He had formerly borne such mischances, hoping soon to adjust what was wrong, to master it and attain success, or make a grand slam. But now every mischance upset him and plunged him into despair. He would say to himself: "There now, just as I was beginning to get better and the medicine had begun to take effect, comes this accursed misfortune, or unpleasantness. . . ." And he was furious with the mishap, or with the people who were causing the unpleasantness and killing him, for he felt that this fury was killing him but could not restrain it. One would have thought that it should have been clear to him that this exasperation with circumstances and people aggravated his illness, and that he ought therefore to ignore unpleasant occurrences. But he drew the very opposite conclusion: he said that he needed peace, and he watched for everything that might disturb it and became irritable at the slightest infringement of it. His condition was rendered worse by the fact that he read medical books and consulted doctors. The progress of his disease was so gradual that he could deceive himself when comparing one day with another—the difference was so slight. But when he consulted the doctors it seemed to him that he was getting worse, and even very rapidly. Yet despite this he was continually consulting them.

That month he went to see another celebrity, who told him almost the same as the first had done but put his questions rather differently, and the interview with this celebrity only increased Ivan Ilych's doubts and fears. A friend of a friend of his, a very good doctor, diagnosed his illness again quite differently from the others, and though he predicted recovery, his questions and suppositions bewildered Ivan Ilych still more and increased his doubts. A homœopathist diagnosed the disease in yet another way, and prescribed medicine which Ivan Ilych took secretly for a week. But after a week, not feeling any improvement and having lost confidence both in the former doctor's treatment and in this one's, he became still more despondent. One day a lady acquaintance metioned a cure effected by a wonder-working icon. Ivan Ilych caught himself listening attentively and beginning to believe that it had occurred. This incident alarmed him. "Has my mind really weakened to such an extent?" he asked himself. "Nonsense! It's all rubbish. I mustn't give way to nervous fears but having chosen a doctor must keep strictly to his treatment. That is what I will do. Now it's all settled. I won't think about it, but will follow the treatment seriously till summer, and then we shall see. From now there must be no more of this wavering!" This was easy to say but impossible to carry out. The pain in his side oppressed him and seemed to grow worse and more incessant, while the taste in his mouth grew stranger and stranger. It seemed to him that his breath had a disgusting smell, and he was conscious of a loss of appetite and strength. There was no deceiving himself: something terrible, new, and more important than anything before in his life, was taking place within him of which he alone was aware. Those about him did not understand or would not understand it, but thought everything in the world was going on as usual. That tormented Ivan Ilych more than anything. He saw that his household, especially his wife and daughter who were in a perfect whirl of visiting, did not understand anything of it and were annoyed that he was so depressed and so exacting, as if he were to blame for it. Though they tried to disguise it he saw that he was an obstacle in their path, and that his wife had adopted a definite line in regard to his illness and kept to it regardless of anything he said or did. Her attitude was this: "You know," she would say to her friends, "Ivan Ilych can't do as other people do, and keep to the treatment prescribed for him. One day he'll take his drops and keep strictly to his diet and go to bed in good time, but the next day unless I watch him he'll suddenly forget his medicine, eat sturgeon—which is forbidden—and sit up playing cards till one o'clock in the morning."

"Oh, come, when was that?" Ivan Ilych would ask in vexation. "Only once at Peter Ivanovich's."

"And yesterday with Shebek."

"Well, even if I hadn't stayed up, this pain would have kept me awake."

"Be that as it may you'll never get well like that, but will always make us wretched."

Praskovya Fëdorovna's attitude to Ivan Ilych's illness, as she expressed it both to others and to him, was that it was his own fault and was another of the annoyances he caused her. Ivan Ilych felt that this opinion escaped her involuntarily—but that did not make it easier for him.

At the law courts too, Ivan Ilych noticed, or thought he noticed, a strange attitude towards himself. It sometimes seemed to him that people were watching him inquisitively as a man whose place might soon be vacant. Then again, his friends would suddenly begin to chaff him in a friendly way about his low spirits, as if the awful, horrible, and unheard-of thing that was going on within him, incessantly gnawing at him and irresistibly drawing him away, was a very agreeable subject for jests. Schwartz in particular irritated him by his jocularity, vivacity, and *savoir-faire*, which reminded him of what he himself had been ten years ago.

Friends came to make up a set and they sat down to cards. They dealt, bending the new cards to soften them, and he sorted the diamonds in his hand and found he had seven. His partner said "No trumps" and supported him with two diamonds. What more could be wished for? It ought to be jolly and lively. They would make a grand slam. But suddenly Ivan Ilych was conscious of that gnawing pain, that taste in his mouth, and it seemed ridiculous that in such circumstances he should be pleased to make a grand slam.

He looked at his partner Mikhail Mikhaylovich, who rapped the table with his strong hand and instead of snatching up the tricks pushed the cards courteously and indulgently towards Ivan Ilych that he might have the pleasure of gathering them up without the trouble of stretching out his hand for them. "Does he think I am too weak to stretch out my arm?" thought Ivan Ilych, and forgetting what he was doing he over-trumped his partner, missing the grand slam by three tricks. And what was most awful of all was that he saw how upset Mikhail Mikhaylovich was about it but did not himself care. And it was dreadful to realize why he did not care.

They all saw that he was suffering, and said: "We can stop if you are tired. Take a rest." Lie down? No, he was not at all tired, and he finished the rubber. All were gloomy and silent. Ivan Ilych felt that he had diffused this gloom over them and could not dispel it. They had supper and went away, and Ivan Ilych was left alone with the consciousness that his life was poisoned and was poisoning the lives of others, and that this poison did not weaken but penetrated more and more deeply into his whole being.

With this consciousness, and with physical pain besides the terror, he must go to bed, often to lie awake the greater part of the night. Next morning he had to get up again, dress, go to the law courts, speak, and write; or if he did not go out, spend at home those twenty-four hours a day each of which was a torture. And he had to live thus all alone on the brink of an abyss, with no one who understood or pitied him.

V

SO ONE month passed and then another. Just before the New Year his brother-in-law came to town and stayed at their house. Ivan Ilych was at the law courts and Praskovya Fëdorovna had gone shopping. When Ivan Ilych came home and entered his study he found his brother-in-law there—a healthy, florid man—unpacking his portmanteau himself. He raised his head on hearing Ivan Ilych's footsteps and looked up at him for a moment without a word. That stare told Ivan Ilych everything. His brother-in-law opened his mouth to utter an exclamation of surprise but checked himself, and that action confirmed it all.

"I have changed, eh?"

"Yes, there is a change."

And after that, try as he would to get his brother-in-law to return to the subjects of his looks, the latter would say nothing about it. Praskovya Fëdorovna came home and her brother went out to her. Ivan Ilych locked the door and began to examine himself in the glass, first full face, then in profile. He took up a portrait of himself taken with his wife, and compared it with what he saw in the glass. The change in him was immense. Then he bared his arms to the elbow, looked at them, drew the sleeves down again, sat down on an ottoman, and grew blacker than night.

"No, no, this won't do!" he said to himself, and jumped up, went to the table, took up some law papers, and began to read them, but could not continue. He unlocked the door and went into the reception-room. The door leading to the drawing-room was shut. He approached it on tiptoe and listened.

"No, you are exaggerating!" Praskovya Fëdorovna was saying.

"Exaggerating! Don't you see it? Why, he's a dead man! Look at his eyes—there's no light in them. But what is it that is wrong with him?"

"No one knows. Nikolaevich [that was another doctor] said something, but I don't know what. And Leshchetitsky [this was the celebrated specialist] said quite the contrary. . . ."

Ivan Ilych walked away, went to his own room, lay down, and began musing: "The kidney, a floating kidney." He recalled all the doctors had told him of how it detached itself and swayed about. And by an effort of imagination he tried to catch that kidney and arrest it and support it. So little was needed for this, it seemed to him. "No, I'll go to see Peter Ivanovich again." [That was the friend whose friend was a doctor.] He rang, ordered the carriage, and got ready to go.

"Where are you going, Jean?" asked his wife, with a specially sad and exceptionally kind look.

This exceptionally kind look irritated him. He looked morosely at her.

"I must go to see Peter Ivanovich."

He went to see Peter Ivanovich, and together they went to see his friend, the doctor. He was in, and Ivan Ilych had a long talk with him.

Reviewing the anatomical and physiological details of what in the doctor's opinion was going on inside him, he understood it all.

There was something, a small thing, in the vermiform appendix. It might all come right. Only stimulate the energy of one organ and check the activity of another, then absorption would take place and everything would come right. He got home rather late for dinner, ate his dinner, and conversed cheerfuly, but could not for a long time bring himself to go back to work in his room. At last, however, he went to his study and did what was necessary, but the consciousness that he had put something aside—an important, intimate matter which he would revert to when his work was done—never left him. When he had finished his work he remembered that this intimate matter was the thought of his vermiform appendix. But he did not give himself up to it, and went to the drawing-room for tea. There were callers there, including the examining magistrate who was a desirable match for his daughter, and they were conversing, playing the piano, and singing.

Ivan Ilych, as Praskovya Fëdorovna remarked, spent that evening more cheerfully than usual, but he never for a moment forgot that he had postponed the important matter of the appendix. At eleven o'clock he said good-night and went to his bedroom. Since his illness he had slept alone in a small room next to his study. He undressed and took up a novel by Zola, but instead of reading it he fell into thought, and in his imagination that desired improvement in the vermiform appendix occurred. There was the absorption and evacuation and the re-establishment of normal activity. "Yes, that's it!" he said to himself. "One need only assist nature, that's all." He remembered his medicine, rose, took it, and lay down on his back watching for the beneficent action of the medicine and for it to lessen the pain. "I need only take it regularly and avoid all injurious influences. I am already feeling better, much better." He began touching his side: it was not painful to the touch. "There, I really don't feel it. It's much better already." He put out the light and turned on his side. . . . "The appendix is getting better, absorption is occurring." Suddenly he felt the old, familiar, dull, gnawing pain, stubborn and serious. There was the same familiar loathsome taste in his mouth. His heart sank and he felt dazed. "My God! My God!" he muttered. "Again, again! and it will never cease." And suddenly the matter presented itself in a quite different aspect. "Vermiform appendix! Kidney!" he said to himself. "It's not a question of appendix or kidney, but of life and . . . death. Yes, life was there and now it is going, going and I cannot stop it. Yes. Why deceive myself? Isn't it obvious to everyone but me that I'm dying, and that it's only a question of weeks, days . . . it may happen this moment. There was light and now there is darkness. I was here and now I'm going there! Where?" A chill came over him, his breathing ceased, and he felt only the throbbing of his heart.

"When I am not, what will there be? There will be nothing. Then where shall I be when I am no more? Can this be dying? No, I don't want to!" He jumped up and tried to light the candle, felt for it with trembling hands, dropped candle and candlestick on the floor, and fell back on his pillow.

"What's the use? It makes no difference," he said to himself, staring with wide-open eyes into the darkness. "Death. Yes, death. And none of them know or wish to know it, and they have no pity for me. Now they are playing." (He heard through the door the distant sound of a song and its accompaniment.) "It's all the same to them, but they will die too! Fools! I first, and they later, but it will be the same for them. And now they are merry . . . the beasts!"

Anger choked him and he was agonizingly, unbearably miserable. "It is impossible that all men have been doomed to suffer this awful horror!" He raised himself.

"Something must be wrong. I must calm myself—must think it all over from the beginning." And he again began thinking. "Yes, the beginning of my illness: I knocked my side, but I was still quite well that day and the next. It hurt a little, then rather more. I saw the doctors, then followed despondency and anguish, more doctors, and I drew nearer to the abyss. My strength grew less and I kept coming nearer and nearer, and now I have wasted away and there is no light in my eyes. I think of the appendix—but this is death! I think of mending the appendix, and all the while here is death! Can it really be death?" Again terror seized him and he gasped for breath. He leant down and began feeling for the matches, pressing with his elbow on the stand beside the bed. It was in his way and hurt him, he grew furious with it, pressed on it still harder, and upset it. Breathless and in despair he fell on his back, expecting death to come immediately.

Meanwhile the visitors were leaving. Praskovya Fëdorovna was seeing them off. She heard something fall and came in.

"What has happened?"

"Nothing. I knocked it over accidentally."

She went out and returned with a candle. He lay there panting heavily, like a man who has run a thousand yards, and stared upwards at her with a fixed look.

"What is it, Jean?"

"No . . . o . . . thing. I upset it." ("Why speak of it? She won't understand," he thought.)

And in truth she did not understand. She picked up the stand, lit his candle, and hurried away to see another visitor off. When she came back he still lay on his back, looking upwards.

"What is it? Do you feel worse?"

"Yes."

She shook her head and sat down.

"Do you know, Jean, I think we must ask Leshchetitsky to come and see you here."

This meant calling in the famous specialist, regardless of expense. He smiled malignantly and said "No." She remained a little longer and then went up to him and kissed his forehead.

While she was kissing him he hated her from the bottom of his soul and with difficulty refrained from pushing her away.

"Good-night. Please God you'll sleep."

"Yes."

VI

IVAN ILYCH saw that he was dying, and he was in continual despair.

In the depth of his heart he knew he was dying, but not only was he not accustomed to the thought, he simply did not and could not grasp it.

The syllogism he had learnt from Kiezewetter's Logic: "Caius is a man, men are mortal, therefore Caius is mortal," had always seemed to him correct as applied to Caius, but certainly not as applied to himself. That Caius—man in the abstract—was mortal, was perfectly correct, but he was not Caius, not an abstract man, but a creature quite, quite separate from all others. He had been little Vanya, with a mamma and a papa, with Mitya and Volodya, with the toys, a coachman and a nurse, afterwards with Katenka and with all the joys, griefs, and delights of childhood, boyhood, and youth. What did Caius know of the smell of that striped leather ball Vanya had been so fond of? Had Caius kissed his mother's hand like that, and did the silk of her dress rustle so for Caius? Had he rioted like that at school when the pastry was bad? Had Caius been in love like that? Could Caius preside at a session as he did? "Caius really was mortal, and it was right for him to die; but for me, little Vanya, Ivan Ilych, with all my thoughts and emotions, it's altogether a different matter. It cannot be that I ought to die. That would be too terrible."

Such was his feeling.

"If I had to die like Caius I should have known it was so. An inner voice would have told me so, but there was nothing of the sort in me and I and all my friends felt that our case was quite different from that of Caius. And now here it is!" he said to himself. "It can't be. It's impossible! But here it is. How is this? How is one to understand it?"

He could not understand it, and tried to drive this false, incorrect, morbid thought away and to replace it by other proper and healthy thoughts. But that thought, and not the thought only but the reality iself, seemed to come and confront him.

And to replace that thought he called up a succession of others, hoping to find in them some support. He tried to get back into the former current of thoughts that had once screened the thought of death from him. But strange to say, all that had formerly shut off, hidden, and destroyed his consciousness of death, no longer had that effect. Ivan Ilych now spent most of his time in attempting to re-establish that old current. He would say to himself: "I will take up my duties again—after all I used to live by them." And banishing all doubts he would go to the law courts, enter into conversation with his colleagues, and sit carelessly as was his wont, scanning the crowd with a thoughtful look and leaning both his emaciated arms on the arms of his oak chair; beinding over as usual to a colleague and drawing his papers nearer he would interchange whispers with him, and then suddenly raising his eyes and sitting erect would pronounce certain words and open the proceedings. But suddenly in the midst of those proceedings the pain in his side, regardless of the stage the proceedings had reached, would begin its own gnawing work. Ivan Ilych would turn his attention to it and try to drive the thought of it away, but without success. *It* would come and stand before him and look at him, and he would be petrified and the light would die out of his eyes, and he would again begin asking himself whether *It* alone was true. And his colleagues and subordinates would see with surprise and distress that he, the brilliant and subtle judge, was becoming confused and making mistakes. He would shake himself, try to pull himself together, manage somehow to bring the sitting to a close, and return home with the sorrowful consciousness that his judicial labours could not as formerly hide from him what he wanted them to hide, and could not deliver him from *It*. And what was worst of all was that *It* drew his attention to itself not in order to make him take some action but only that he should look at *It*, look it straight in the face: look at it and, without doing anything, suffer inexpressibly.

And to save himself from this condition Ivan Ilych looked for consolations—new screens—and new screens were found and for a while seemed to save him, but then they immediately fell to pieces or rather became transparent, as if *It* penetrated them and nothing could veil *It*.

In these latter days he would go into the drawing-room he had arranged—that drawing-room where he had fallen and for the sake of which (how bitterly ridiculous it seemed) he had sacrificed his life—for he knew that his illness originated with that knock. He would enter and see that something had scratched the polished table. He would look for the cause of this and find that it was the bronze ornamentation of an album, that had got bent. He would take up the expensive album which he had lovingly arranged, and feel vexed with his daughter and her friends for their untidiness—for the album was torn here and there and some of the photographs turned upside down. He would put it carefully in order and bend the ornamentation back into position. Then it would occur to him to place all those things in another corner of the room, near the plants. He could call the footman, but his daughter or wife would come to help him. They would not agree, and his wife would contradict him, and he would dispute and grow angry. But that was all right, for then he did not think about *It*. *It* was invisible.

But then, when he was moving something himself, his wife would say: "Let the servants do it. You will hurt yourself again." And suddenly *It* would flash through the screen and he would see it. It was just a flash, and he hoped it would disappear, but he would involuntarily pay attention to his side. "It sits there as before, gnawing just the same!" And he could no longer forget *It*, but could distinctly see it looking at him from behind the flowers. "What is it all for?"

"It really is so! I lost my life over that curtain as I might have done when storming a fort. Is that possible? How terrible and how stupid. It can't be true! It can't, but it is."

He would go to his study, lie down, and again be alone with *It*: face to face with *It*. And nothing could be done with *It* except to look at it and shudder.

VII

HOW IT happened it is impossible to say because it came about step by step, unnoticed, but in the third month of Ivan Ilych's illness, his wife, his daughter, his son, his acquaintances, the doctors, the servants, and above all he himself, were aware that the whole interest he had for other people was whether he would soon vacate his place, and at last release the living from the discomfort caused by his presence and be himself released from his sufferings.

He slept less and less. He was given opium and hypodermic injections of morphine, but this did not relieve him. The dull depression he experienced in a somnolent condition at first gave him a little relief, but only as something new, afterwards it became as distressing as the pain itself or even more so.

Special foods were prepared for him by the doctors' orders, but all those foods became increasingly distasteful and disgusting to him.

For his excretions also special arrangements had to be made, and this was a torment to him every time—a torment from the uncleanliness, the unseemliness, and the smell, and from knowing that another person had to take part in it.

But just through this most unpleasant matter, Ivan Ilych obtained comfort. Gerasim, the butler's young assistant, always came in to carry the things out. Gerasim was a clean, fresh peasant lad, grown stout on town food and always cheerful and bright. At first the sight of him, in his clean Russian peasant costume, engaged on that disgusting task embarrassed Ivan Ilych.

Once when he got up from the commode too weak to draw up his trousers, he dropped into a soft armchair and looked with horror at his bare, enfeebled thighs with the muscles so sharply marked on them.

Gerasim with a firm light tread, his heavy boots emitting a pleasant smell of tar and fresh winter air, came in wearing a clean Hessian apron, the sleeves of his print shirt tucked up over his strong, bare young arms; and refraining from looking at his sick master out of consideration for his feelings, and restraining the joy of life that beamed from his face, he went up to the commode.

"Gerasim!" said Ivan Ilych in a weak voice.

Gerasim started, evidently afraid he might have committed some blunder, and with a rapid movement turned his fresh, kind, simple young face which just showed the first downy signs of a beard.

"Yes, sir?"

"That must be very unpleasant for you. You must forgive me. I am helpless."

"Oh, why, sir," and Gerasim's eyes beamed and he showed his glistening white teeth, "what's a little trouble? It's a case of illness with you sir."

And his deft strong hands did their accustomed task, and he went out of the room stepping lightly. Five minutes later he as lightly returned.

Ivan Ilych was still sitting in the same position in the armchair.

"Gerasim," he said when the latter had replaced the freshly-washed utensil. "Please come here and help me." Gerasim went up to him. "Lift me up. It is hard for me to get up, and I have sent Dmitri away."

Gerasim went up to him, grasped his master with his strong arms deftly but gently, in the same way that he stepped—lifted him, supported him with one hand, and with the other drew up his trousers and would have set him down again, but Ivan Ilych asked to be led to the sofa. Gerasim, without an effort and without apparent pressure, led him, almost lifting him, to the sofa and placed him on it.

"Thank you. How easily and well you do it all!"

Gerasim smiled again and turned to leave the room. But Ivan Ilych felt his presence such a comfort that he did not want to let him go.

"One thing more, please move up that chair. No, the other one—under my feet. It is easier for me when my feet are raised."

Gerasim brought the chair, set it down gently in place, and raised Ivan Ilych's legs on to it. It seemed to Ivan Ilych that he felt better while Gerasim was holding up his legs.

"It's better when my legs are higher," he said. "Place that cushion under them."

Gerasim did so. He again lifted the legs and placed them, and again Ivan Ilych felt better while Gerasim held his legs. When he set them down Ivan Ilych fancied he felt worse.

"Gerasim," he said. "Are you busy now?"

"Not at all, sir," said Gerasim, who had learnt from the townsfolk how to speak to gentlefolk.

"What have you still to do?"

"What have I to do? I've done everything except chopping the logs for tomorrow."

"Then hold my legs up a bit higher, can you?"

"Of course I can. Why not?" And Gerasim raised his master's legs higher and Ivan Ilych thought that in that position he did not feel any pain at all.

"And how about the logs?"

"Don't trouble about that, sir. There's plenty of time."

Ivan Ilych told Gerasim to sit down and hold his legs, and began to talk to him. And strange to say it seemed to him that he felt better while Gerasim held his legs up.

After that Ivan Ilych would sometimes call Gerasim and get him to hold his legs on his shoulders, and he liked talking to him. Gerasim did it all easily, willingly, simply, and with a good nature that touched Ivan Ilych. Health, strength, and vitality in other people were offensive to him, but Gerasim's strength and vitality did not mortify but soothed him.

What tormented Ivan Ilych most was the deception, the lie, which for some reason they all accepted, that he was not dying but was simply ill, and that he only need keep quiet and undergo a treatment and then something very good would result. He however knew that do what they would nothing would come of it, only still more agonizing suffering and death. This deception tortured him—their not wishing to admit what they all knew and what he knew, but wanting to lie to him concerning his terrible condition, and wishing and forcing him to participate in that lie. Those lies—lies enacted over him on the eve of his death and destined to degrade this awful, solemn act to the level of their visitings, their curtains, their sturgeon for dinner—were a terrible agony for Ivan Ilych. And strangely enough, many times when they were going through their antics over him he had been within a hairbreadth of calling out to them: "Stop lying! You know and I know that I am dying. Then at least stop lying about it!" But he had never had the spirit to do it. The awful, terrible act of his dying was, he could see, reduced by those about him to the level of a casual, unpleasant, and almost indecorous incident (as if someone entered a drawing-room diffusing an unpleasant odour) and this was done by that very decorum which he had served all his life long. He saw that no one felt for him, because no one even wished to grasp his position. Only Gerasim recognized it and pitied him. And so Ivan Ilych felt at ease only with him. He felt comforted when Gerasim supported his legs (sometimes all night long) and refused to go to bed, saying: "Don't you worry, Ivan Ilych. I'll get sleep enough later on," or when he suddenly became familiar and exclaimed: "If you weren't sick it would be another matter, but as it is, why should I grudge a little trouble?" Gerasim alone did not lie; everything showed that he alone understood the facts of the case and did not consider it necessary to disguise them, but simply felt sorry for his emaciated and enfeebled master. Once when Ivan Ilych was sending him away he even said straight out: "We shall all of us die, so why should I grudge a little trouble?" —expressing the fact that he did not think his work burdensome, because he was doing it for a dying man and hoped someone would do the same for him when his time came.

Apart from this lying, or because of it, what most tormented Ivan Ilych was that no one pitied him as he wished to be pitied. At certain moments after prolonged suffering he wished most of all (though he would have been ashamed to confess it) for someone to pity him as a sick child is pitied. He longed to be petted and comforted. He knew he was an important functionary, that he had a beard turning grey, and that therefore what he longed for was impossible, but still he longed for it. And in Gerasim's attitude towards him there was something akin to what he wished for, and so that attitude comforted him. Ivan Ilych wanted to weep, wanted to be petted and cried over, and then his colleague Shebek would come, and instead of weeping and being petted, Ivan Ilych would assume a serious, severe, and profound air, and by force of habit would express his opinion on a decision of the Court of Cassation and would stubbornly insist on that view. This falsity around him and within him did more than anything else to poison his last days.

VIII

IT WAS morning. He knew it was morning because Gerasim had gone, and Peter the footman had come and put out the candles, drawn back one of the curtains, and begun quietly to tidy up. Whether it was morning or evening, Friday or Sunday, made no difference, it was all just the same: the gnawing, unmitigated, agonizing pain, never ceasing for an instant, the consciousness of life inexorably waning but not yet extinguished, the approach of that ever dreaded and hateful Death which was the only reality, and always the same falsity. What were days, weeks, hours, in such a case?

"Will you have some tea, sir?"

"He wants things to be regular, and wishes the gentlefolk to drink tea in the morning," thought Ivan Ilych, and only said "No."

"Wouldn't you like to move onto the sofa, sir?"

"He wants to tidy up the room, and I'm in the way. I am uncleanliness and disorder," he thought, and said only:

"No, leave me alone."

The man went on bustling about. Ivan Ilych stretched out his hand. Peter came up, ready to help.

"What is it, sir?"

"My watch."

Peter took the watch which was close at hand and gave it to his master.

"Half-past eight. Are they up?"

"No, sir, except Vladimir Ivanovich" (the son) "who has gone to school. Praskovya Fëdorovna ordered me to wake her if you asked for her. Shall I do so?"

"No, there's no need to." "Perhaps I'd better have some tea," he thought, and added aloud: "Yes, bring me some tea."

Peter went to the door, but Ivan Ilych dreaded being left alone. "How can I keep him here? Oh yes, my medicine." "Peter, give me my medicine." "Why not? Perhaps it may still do me some good." He took a spoonful and swallowed it. "No, it won't help. It's all tomfoolery, all deception," he decided as soon as he became aware of the familiar, sickly, hopeless taste. "No, I can't believe in it any longer. But the pain, why this pain? If it would only cease just for a moment!" And he moaned. Peter turned towards him. "It's all right. Go and fetch me some tea."

Peter went out. Left alone Ivan Ilych groaned not so much with pain, terrible though that was, as from mental anguish. Always and for ever the same, always these endless days and nights. If only it would come quicker! If only *what* would come quicker? Death, darkness? . . . No, no! Anything rather than death!

When Peter returned with the tea on a tray, Ivan Ilych stared at him for a time in perplexity, not realizing who and what he was. Peter was disconcerted by that look and his embarrassment brought Ivan Ilych to himself.

"Oh, tea! All right, put it down. Only help me to wash and put on a clean shirt."

And Ivan Ilych began to wash. With pauses for rest, he washed his hands and then his face, cleaned his teeth, brushed his hair, and looked in the glass. He was terrified by what he saw, especially by the limp way in which his hair clung to his pallid forehead.

While his shirt was being changed he knew that he would be still more frightened at the sight of his body, so he avoided looking at it. Finally he was ready. He drew on a dressing-gown, wrapped himself in a plaid, and sat down in the armchair to take his tea. For a moment he felt refreshed, but soon as he began to drink the tea he was again aware of the same taste, and the pain also returned. He finished it with an effort, and then lay down stretching out his legs, and dismissed Peter.

Always the same. Now a spark of hope flashes up, then a sea of despair rages, and always pain; always pain, always despair, and always the same. When alone he had a dreadful and distressing desire to call someone, but he knew beforehand that with others present it would be still worse. "Another dose of morphine—to lose consciousness. I will tell him, the doctor, that he must think of something else. It's impossible, impossible, to go on like this."

An hour and another pass like that. But now there is a ring at the door bell. Perhaps it's the doctor? It is. He comes in fresh, hearty, plump, and cheerful, with that look on his face that seems to say: "There now, you're in a panic about something, but we'll arrange it all for you directly!" The doctor knows this expression is out of place here, but he has put it on once for all and can't take it off—like a man who has put on a frock-coat in the morning to pay a round of calls.

The doctor rubs his hands vigorously and reassuringly.

"Brr! How cold it is! There's such a sharp frost; just let me warm myself!" he says, as if it were only a matter of waiting till he was warm, and then he would put everything right.

"Well now, how are you?"

Ivan Ilych feels that the doctor would like to say: "Well, how are our affairs?" but that even he feels that this would not do, and says instead: "What sort of a night have you had?"

Ivan Ilych looks at him as much as to say: "Are you really never ashamed of lying?" But the doctor does not wish to understand this question, and Ivan Ilych says: "Just as terrible as ever. The pain never leaves me and never subsides. If only something . . ."

"Yes, you sick people are always like that. . . . There, now I think I am warm enough. Even Praskovya Fëdorovna, who is so particular, could find no fault with my temperature. Well, now I can say good-morning," and the doctor presses his patient's hand.

Then, dropping his former playfulness, he begins with a most serious face to examine the patient, feeling his pulse and taking his temperature, and then begins the sounding and auscultation.

Ivan Ilych knows quite well and definitely that all this is nonsense and pure deception, but when the doctor, getting down on his knee, leans over him, putting his ear first higher then lower, and performs various gymnastic movements over him with a significant expression on his face, Ivan Ilych submits to it all as he used to submit to the speeches of the lawyers, though he knew very well that they were all lying and why they were lying.

The doctor, kneeling on the sofa, is still sounding him when Praskovya Fëdorovna's silk dress rustles at the door and she is heard scolding Peter for not having let her know of the doctor's arrival.

She comes in, kisses her husband, and at once proceeds to prove that she has been up a long time already, and only owing to a misunderstanding failed to be there when the doctor arrived.

Ivan Ilych looks at her, scans her all over, sets against her the whiteness and plumpness and cleanness of her hands and neck, the gloss of her hair, and the sparkle of her vivacious eyes. He hates her with his whole soul. And the thrill of hatred he feels for her makes him suffer from her touch.

Her attitude towards him and his disease is still the same. Just as the doctor had adopted a certain relation to his patient which he could not abandon, so had she formed one towards him—that he was not doing something he ought to do and was himself to blame, and that she reproached him lovingly for this—and she could not now change that attitude.

"You see he doesn't listen to me and doesn't take his medicine at the proper time. And above all he lies in a position that is no doubt bad for him—with his legs up."

She described how he made Gerasim hold his legs up.

The doctor smiled with a contemptuous affability that said: "What's to be done? These sick people do have foolish fancies of that kind, but we must forgive them."

When the examination was over the doctor looked at his watch, and then Praskovya Fëdorovna announced to Ivan Ilych that it was of course as he pleased, but she had sent today for a celebrated specialist who would examine him and have a consultation with Michael Danilovich (their regular doctor).

"Please don't raise any objections. I am doing this for my own sake," she said ironically, letting it be felt that she was doing it all for his sake and only said this to leave him no right to refuse. He remained silent, knitting his brows. He felt that he was so surrounded and involved in a mesh of falsity that it was hard to unravel anything.

Everything she did for him was entirely for her own sake, and she told him she was doing for herself what she actually was doing for herself, as if that was so incredible that he must understand the opposite.

At half-past eleven the celebrated specialist arrived. Again the sounding began and the significant conversations in his presence and in another room, about the kidneys and the appendix, and the questions and answers, with such an air of importance that again, instead of the real question of life and death which now alone confronted him, the question arose of the kidney and appendix which were not behaving as they ought to and would now be attacked by Michael Danilovich and the specialist and forced to amend their ways.

The celebrated specialist took leave of him with a serious though not hopeless look, and in reply to the timid question Ivan Ilych, with eyes glistening with fear and hope, put to him as to whether there was a chance of recovery, said that he could not vouch for it but there was a possibility. The look of hope with which Ivan Ilych watched the doctor out was so pathetic that Praskovya Fëdorovna, seeing it, even wept as she left the room to hand the doctor his fee.

The gleam of hope kindled by the doctor's encouragement did not last long. The same room, the same pictures, curtains, wallpaper, medicine bottles, were all there, and the same aching suffering body, and Ivan Ilych began to moan. They gave him a subcutaneous injection and he sank into oblivion.

It was twilight when he came to. They brought him his dinner and he swallowed some beef tea with difficulty, and then everything was the same again and night was coming on.

After dinner, at seven o'clock, Praskovya Fëdorovna came into the room in evening dress, her full bosom pushed up by her corset, and with traces of powder on her face. She had reminded him in the morning that they were going to the theatre. Sarah Bernhardt was visiting the town and they had a box, which he had insisted on their taking. Now he had forgotten about it and her toilet offended him, but he concealed his vexation when he remembered that he had himself insisted on their securing a box and going because it would be an instructive and aesthetic pleasure for the children.

Praskovya Fëdorovna came in, self-satisfied but yet with a rather guilty air. She sat down and asked how he was, but, as he saw, only for the sake of asking and not in order to learn about it, knowing that there was nothing to learn—and then went on to what she really wanted to say: that she would not on any account have gone but that the box had been taken and Helen and their daughter were going, as well as Petrishchev (the examining magistrate, their daughter's fiancé), and that it was out of the question to let them go alone; but that she would have much preferred to sit with him for a while; and he must be sure to follow the doctor's orders while she was away.

"Oh, and Fëdor Petrovich" (the fiancé) "would like to come in. May he? And Lisa?"

"All right."

Their daughter came in in full evening dress, her fresh young flesh exposed (making a show of that very flesh which in his own case caused so much suffering), strong, healthy, evidently in love, and impatient with illness, suffering, and death, because they interfered with her happiness.

Fëdor Petrovich came in too, in evening dress, his hair curled *à la Capoul,* a tight stiff collar round his long sinewy neck, an enormous white shirt-front, and narrow black trousers tightly stretched over his strong thighs. He had one white glove tightly drawn on, and was holding his opera hat in his hand.

Following him the schoolboy crept in unnoticed, in a new uniform, poor little fellow, and wearing gloves. Terribly dark shadows showed under his eyes, the meaning of which Ivan Ilych knew well.

His son had always seemed pathetic to him, and now it was dreadful to see the boy's frightened look of pity. It seemed to Ivan Ilych that Vasya was the only one besides Gerasim who understood and pitied him.

They all sat down and again asked how he was. A silence followed. Lisa asked her mother about the opera-glasses, and there was an altercation between mother and daughter as to who had taken them and where they had been put. This occasioned some unpleasantness.

Fëdor Petrovich inquired of Ivan Ilych whether he had ever seen Sarah Bernhardt. Ivan Ilych did not at first catch the question, but then replied: "No, have you seen her before?"

"Yes, in *Adrienne Lecouvreur.*"

Praskovya Fëdorovna mentioned some rôles in which Sarah Bernhardt was particularly good. Her daughter disagreed. Conversation sprang up as to the elegance and realism of her acting—the sort of conversation that is always repeated and is always the same.

In the midst of the conversation Fëdor Petrovich glanced at Ivan Ilych and became silent. The others also looked at him and grew silent. Ivan Ilych was staring with glittering eyes straight before him, evidently indignant with them. This had to be rectified, but it was impossible to do so. The silence had to be broken, but for a time no one dared to break it and they all became afraid that the conventional deception would suddenly become obvious and the truth become plain to all. Lisa was the first to pluck up courage and break that silence, but by trying to hide what everybody was feeling, she betrayed it.

"Well, if we are going it's time to start," she said, looking at her watch, a present from her father, and with a faint and significant smile at Fëdor Petrovich relating to something known only to them. She got up with a rustle of her dress.

They all rose, said good-night, and went away.

When they had gone it seemed to Ivan Ilych that he felt better; the falsity had gone with them. But the pain remained—that same pain and that same fear that made everything monotonously alike, nothing harder and nothing easier. Everything was worse.

Again minute followed minute and hour followed hour. Everything remained the same and there was no cessation. And the inevitable end of it all became more and more terrible.

"Yes, send Gerasim here," he replied to a question Peter asked.

IX

HIS WIFE returned late at night. She came in on tiptoe, but he heard her, opened his eyes, and made haste to close them again. She wished to send Gerasim away and to sit with him herself, but he opened his eyes and said: "No, go away."

"Are you in great pain?"

"Always the same."

"Take some opium."

He agreed and took some. She went away.

Till about three in the morning he was in a state of stupefied misery. It seemed to him that he and his pain were being thrust into a narrow, deep black sack, but though they were pushed further and further in they could not be pushed to the bottom. And this, terrible enough in itself, was accompanied by suffering. He was frightened yet wanted to fall through the sack, he struggled but yet co-operated. And suddenly he broke through, fell, and regained consciousness. Gerasim was sitting at the foot of the bed dozing quietly and patiently, while he himself lay with his emaciated stockinged legs resting on Gerasim's shoulders; the same shaded candle was there and the same unceasing pain.

"Go away, Gerasim," he whispered.

"It's all right, sir. I'll stay a while."

"No. Go away."

He removed his legs from Gerasim's shoulders, turned sideways onto his arm, and felt sorry for himself. He only waited till Gerasim had gone into the next room and then restrained himself no longer but wept like a child. He wept on account of his helplessness, his terrible loneliness, the cruelty of man, the cruelty of God, and the absence of God.

"Why hast Thou done all this? Why hast Thou brought me here? Why, why dost Thou torment me so terribly?"

He did not expect an answer and yet wept because there was no answer and could be none. The pain again grew more acute, but he did not stir and did not call. He said to himself: "Go on! Strike me! But what is it for? What have I done to Thee? What is it for?"

Then he grew quiet and not only ceased weeping but even held his breath and became all attention. It was as though he were listening not to an audible voice but to the voice of his soul, to the current of thoughts arising within him.

"What is it you want?" was the first clear conception capable of expression in words, that he heard.

"What do you want? What do you want?" he repeated to himself.

"What do I want? To live and not to suffer," he answered.

And again he listened with such concentrated attention that even his pain did not distract him.

"To live? How?" asked his inner voice.

"Why, to live as I used to—well and pleasantly."

"As you lived before, well and pleasantly?" the voice repeated.

And in imagination he began to recall the best moments of his pleasant life. But strange to say none of those best moments of his pleasant life now seemed at all what they had then seemed—none of them except the first recollections of childhood. There, in childhood, there had been something really pleasant with which it would be possible to live if it could return. But the child who had experienced that happiness existed no longer, it was like a reminiscence of somebody else.

As soon as the period began which had produced the present Ivan Ilych, all that had then seemed joys now melted before his sight and turned into something trivial and often nasty.

And the further he departed from childhood and the nearer he came to the present the more worthless and doubtful were the joys. This began with the School of Law. A little that was really good was still found there—there was lightheartedness, friendship, and hope. But in the upper classes there had already been fewer of such good moments. Then during the first years of his official career, when he was in the service of the Governor, some pleasant moments again occurred; they were the memories of love for a woman. Then all became confused and there was still less of what was good; later on again there was still less that was good, and the further he went the less there was. His marriage, a mere accident, then the disenchantment that followed it, his wife's bad breath and the sensuality and hypocrisy: then that deadly official life and those preoccupations about money, a year of it, and two, and ten, and twenty, and always the same thing. And the longer it lasted the more deadly it became. "It is as if I had been going downhill while I imagined I was going up. And that is really what it was. I was going up in public opinion, but to the same extent life was ebbing away from me. And now it is all done and there is only death."

"Then what does it mean? Why? It can't be that life is so senseless and horrible. But if it really has been so horrible and senseless, why must I die and die in agony? There is something wrong!"

"Maybe I did not live as I ought to have done," it suddenly occurred to him. "But how could that be, when I did everything properly?" he replied, and immediately dismissed from his mind this, the sole solution of all the riddles of life and death, as something quite impossible.

"Then what do you want now? To live? Live how? Live as you lived in the law courts when the usher proclaimed 'The judge is coming!' The judge is coming, the judge!" he repeated to himself. "Here he is, the judge. But I am not guilty!" he exclaimed angrily. "What is it for?" And he ceased crying, but turning his face to the wall continued to ponder on the same question: Why, and for what purpose, is there all this horror? But however much he pondered he found no answer. And whenever the thought occurred to him, as it often did, that it all resulted from his not having lived as he ought to have done, he at once recalled the correctness of his whole life and dismissed so strange an idea.

X

ANOTHER fortnight passed. Ivan Ilych now no longer left his sofa. He would not lie in bed but lay on the sofa, facing the wall nearly all the time. He suffered ever the same unceasing agonies and in his loneliness pondered always on the same insoluble question: "What is this? Can it be that it is Death?" And the inner voice answered: "Yes, it is Death."

"Why these sufferings?" And the voice answered, "For no reason—they just are so." Beyond and besides this there was nothing.

From the very beginning of his illness, ever since he had first been to see the doctor, Ivan Ilych's life had been divided between two contrary and alternating moods: now it was despair and the expectation of this uncomprehended and terrible death, and now hope and an intently interested observation of the functioning of his organs. Now before his eyes there was only a kidney or an intestine that temporarily evaded its duty, and now only that incomprehensible and dreadful death from which it was impossible to escape.

These two states of mind had alternated from the very beginning of his illness, but the further it progressed the more doubtful and fantastic became the conception of the kidney, and the more real the sense of impending death.

He had but to call to mind what he had been three months before and what he was now, to call to mind with what regularity he had been going downhill, for every possibility of hope to be shattered.

Latterly during that loneliness in which he found himself as he lay facing the back of the sofa, a loneliness in the midst of a populous town and surrounded by numerous acquaintances and relations but that yet could not have been more complete anywhere—either at the bottom of the sea or under the earth—during that terrible loneliness Ivan Ilych had lived only in memories of the past. Pictures of his past rose before him one after another. They always began with what was nearest in time and then went back to what was most remote—to his childhood—and rested there. If he thought of the stewed prunes that had been offered him that day, his mind went back to the raw shrivelled French plums of his childhood, their peculiar flavour and the flow of saliva when he sucked their stones, and along with the memory of that taste came a whole series of memories of those days: his nurse, his brother, and their toys. "No, I mustn't think of that. . . . It is too painful," Ivan Ilych said to himself, and brought himself back to the present—to the button on the back of the sofa and the creases in its morocco. "Morocco is expensive, but it does not wear well: there had been a quarrel about it. It was a different kind of quarrel and a different kind of morocco that time when we tore father's portfolio and were punished, and mamma brought us some tarts. . . ." And again his thoughts dwelt on his childhood, and again it was painful and he tried to banish them and fix his mind on something else.

Then again together with that chain of memories another series passed through his mind—of how his illness had progressed and grown worse. There also the further back he looked the more life there had been. There had been more of what was good in life and more of life itself. The two merged together. "Just as the pain went on getting worse and worse, so my life grew worse and worse," he thought. "There is one bright spot there at the back, at the beginning of life, and afterwards all becomes blacker and blacker and proceeds more and more rapidly—in inverse ratio to the square of the distance from death," thought Ivan Ilych. And the example of a stone falling downwards with increasing velocity entered his mind. Life, a series of increasing sufferings, flies further and further towards its end—the most terrible suffering. "I am flying. . . ." He shuddered, shifted himself, and tried to resist, but was already aware that resistance was impossible, and again, with eyes weary of gazing but unable to cease seeing what was before them, he stared at the back of the sofa and waited—awaiting that dreadful fall and shock and destruction.

"Resistance is impossible!" he said to himself. "If I could only understand what it is all for! But that too is impossible. An explanation would be possible if it could be said that I have not lived as I ought to. But it is impossible to say that," and he remembered all the legality, correctitude, and propriety of his life. "That at any rate can certainly not be admitted," he thought, and his lips smiled ironically as if someone could see that smile and be taken in by it. "There is no explanation! Agony, death . . . What for?"

XI

ANOTHER two weeks went by in this way and during that fortnight an event occurred that Ivan Ilych and his wife had desired. Petrishchev formally proposed. It happened in the evening. The next day Praskovya Fëdorovna came into her husband's room considering how best to inform him of it, but that very night there had been a fresh change for the worse in his condition. She found him still lying on the sofa but in a different position. He lay on his back, groaning and staring fixedly straight in front of him.

She began to remind him of his medicines, but he turned his eyes towards her with such a look that she did not finish what she was saying; so great an animosity, to her in particular, did that look express.

"For Christ's sake let me die in peace!" he said.

She would have gone away, but just then their daughter came in and went up to say good morning. He looked at her as he had done at his wife, and in reply to her inquiry about his health said dryly that he would soon free them all of himself. They were both silent and after sitting with him for a while went away.

"Is it our fault?" Lisa said to her mother. "It's as if we were to blame! I am sorry for papa, but why should we be tortured?"

The doctor came at his usual time. Ivan Ilych answered "Yes" and "No," never taking his angry eyes from him, and at last said: "You know you can do nothing for me, so leave me alone."

"We can ease your sufferings."

"You can't even do that. Let me be."

The doctor went into the drawing-room and told Praskovya Fëdorovna that the case was very serious and that the only resource left was opium to allay her husband's sufferings, which must be terrible.

It was true, as the doctor said, that Ivan Ilych's physical sufferings were terrible, but worse than the physical sufferings were his mental sufferings, which were his chief torture.

His mental sufferings were due to the fact that that night, as he looked at Gerasim's sleepy, good-natured face with its prominent cheek-bones, the question suddenly occurred to him: "What if my whole life has really been wrong?"

It occurred to him that what had appeared perfectly impossible before, namely that he had not spent his life as he should have done, might after all be true. It occurred to him that his scarcely perceptible attempts to struggle against what was considered good by the most highly placed people, those scarcely noticeable impulses which he had immediately suppressed, might have been the real thing, and all the rest false. And his professional duties and the whole arrangement of his life and of his family, and all his social and official interests, might all have been false. He tried to defend all those things to himself and suddenly felt the weakness of what he was defending. There was nothing to defend.

"But if that is so," he said to himself, "and I am leaving this life with the consciousness that I have lost all that was given me and it is impossible to rectify it—what then?"

He lay on his back and began to pass his life in review in quite a new way. In the morning when he saw first his footman, then his wife, then his daughter, and then the doctor, their every word and movement confirmed to him the awful truth that had been revealed to him during the night. In them he saw himself—all that for which he had lived—and saw clearly that it was not real at all, but a terrible and huge deception which had hidden both life and death. This consciousness intensified his physical suffering tenfold. He groaned and tossed about, and pulled at his clothing which choked and stifled him. And he hated them on that account.

He was given a large dose of opium and became unconscious, but at noon his sufferings began again. He drove everybody away and tossed from side to side.

His wife came to him and said:

"Jean, my dear, do this for me. It can't do any harm and often helps. Healthy people often do it."

He opened his eyes wide.

"What? Take communion? Why? It's unnecessary! However . . ."

She began to cry.

"Yes, do, my dear. I'll send for our priest. He is such a nice man."

"All right. Very well," he muttered.

When the priest came and heard his confession, Ivan Ilych was softened and seemed to feel a relief from his doubts and consequently from his sufferings, and for a moment there came a ray of hope. He again began to think of the vermiform appendix and the possibility of correcting it. He received the sacrament with tears in his eyes.

When they laid him down again afterwards he felt a moment's ease, and the hope that he might live awoke in him again. He began to think of the operation that had been suggested to him. "To live! I want to live!" he said to himself.

His wife came in to congratulate him after his communion, and when uttering the usual conventional words she added:

"You feel better, don't you?"

Without looking at her he said "Yes."

Her dress, her figure, the expression of her face, the tone of her voice, all revealed the same thing. "This is wrong, it is not as it should be. All you have lived for and still live for is falsehood and deception, hiding life and death from you." And as soon as he admitted that thought, his hatred and his agonizing physical suffering again sprang up, and with that suffering a consciousness of the unavoidable, approaching end. And to this was added a new sensation of grinding shooting pain and a feeling of suffocation.

The expression of his face when he uttered that "yes" was dreadful. Having uttered it, he looked her straight in the eyes, turned on his face with a rapidity extraordinary in his weak state and shouted:

"Go away! Go away and leave me alone!"

XII

FROM THAT moment the screaming began that continued for three days, and was so terrible that one could not hear it through two closed doors without horror. At the moment he answered his wife he realized that he was lost, that there was no return, that the end had come, the very end, and his doubts were still unsolved and remained doubts.

"Oh! Oh! Oh!" he cried in various intonations. He had begun by screaming "I won't!" and continued screaming on the letter O.

For three whole days, during which time did not exist for him, he struggled in that black sack into which he was being thrust by an invisible, resistless force. He struggled as a man condemned to death struggles in the hands of the executioner, knowing that he cannot save himself. And every moment he felt that despite all his efforts he was drawing nearer and nearer to what terrified him. He felt that his agony was due to his being thrust into that black hole and still more to his not being able to get right into it. He was hindered from getting into it by his conviction that his life had been a good one. That very justification of his life held him fast and prevented his moving forward, and it caused him most torment of all.

Suddenly some force struck him in the chest and side, making it still harder to breathe, and he fell through the hole and there at the bottom was a light. What had happened to him was like the sensation one sometimes experiences in a railway carriage when one thinks one is going backwards while one is really going forwards and suddenly becomes aware of the real direction.

"Yes, it was all not the right thing," he said to himself, "but that's no matter. It can be done. But what *is* the right thing?" he asked himself, and suddenly grew quiet.

This occurred at the end of the third day, two hours before his death. Just then his schoolboy son had crept softly in and gone up to the bedside. The dying man was still screaming desperately and waving his arms. His hand fell on the boy's head, and the boy caught it, pressed it to his lips, and began to cry.

At that very moment Ivan Ilych fell through and caught sight of the light, and it was revealed to him that though his life had not been what it should have been, this could still be rectified. He asked himself, "What *is* the right thing?" and grew still, listening. Then he felt that someone was kissing his hand. He opened his eyes, looked at his son, and felt sorry for him. His wife came up to him and he glanced at her. She was gazing at him open-mouthed, with undried tears on her nose and cheek and a despairing look on her face. He felt sorry for her too.

"Yes, I am making them, wretched," he thought. "They are sorry, but it will be better for them when I die." He wished to say this but had not the strength to utter it. "Besides, why speak? I must act," he thought. With a look at his wife he indicated his son and said: "Take him away . . . sorry for him . . . sorry for you too. . . ." He tried to add, "Forgive me," but said "forgo" and waved his hand, knowing that He whose understanding mattered would understand.

And suddenly it grew clear to him that what had been oppressing him and would not leave him was all dropping away at once from two sides, from ten sides, and from all sides. He was sorry for them, he must act so as not to hurt them: release them and free himself from these sufferings. "How good and how simple!" he thought. "And the pain?" he asked himself. "What has become of it? Where are you, pain?"

He turned his attention to it.

"Yes, here it is. Well, what of it? Let the pain be."

"And death . . . where is it?

He sought his former accustomed fear of death and did not find it. "Where is it? What death?" There was no fear because there was no death.

In place of death there was light.

"So that's what it is!" he suddenly exclaimed aloud. "What joy!"

To him all this happened in a single instant, and the meaning of that instant did not change. For those present his agony continued for another two hours. Something rattled in his throat, his emaciated body twitched, then the gasping and rattle became less and less frequent.

"It is finished!" said someone near him.

He heard these words and repeated them in his soul.

"Death is finished," he said to himself. "It is no more!"

He drew in a breath, stopped in the midst of a sigh, stretched out, and died.

HERMAN MELVILLE

Herman Melville was one of the few literary giants in America during the nineteenth century, but in his own time he remained little known and little appreciated. As late as 1919, there was no widespread interest in the work of one of America's most gifted writers. Five decades of critical evaluation have been required to achieve for Melville's fiction the high place it now enjoys.

Although admiration of Moby-Dick *as a masterpiece is now quite general, Melville's other novels and short novels still invite strong critical disagreement. For example, Professor Newton Arvin, one of Melville's biographers, considers* Benito Cereno *to be "unduly celebrated, surely. For neither the conception nor the actual composition and texture of 'Benito' are of anything like the brilliance that has been repeatedly attributed to them. The story is an artistic miscarriage, with moments of undeniable power. . . . A greater portentousness of moral meaning is constantly suggested than is ever actually present. Of moral meaning, indeed, there is singularly little."*

In Benito Cereno *Melville was retelling an old story he had read in* Narrative of Voyages and Travels in the Northern and Southern Hemispheres, *written by one Amasa Delano and published in 1817, though of course the novelist took liberties with the original where he wished. In defense of what Melville did achieve, Professor Warner Berthoff, another Melville student, believes* Benito *must fairly be judged as a "paradigm of the inward life of ordinary consciousness, with all its mysterious shifts, penetrations, and side-steppings, in a world in which this ambiguity of appearances is the baffling norm."*

The individual will like to make his own judgment about the success of the tale; Benito *does sharpen the critical acumen. There is a great fascination in the progressive steps by which Melville alternately strains our credulity and then quickly satisfies it. The disturbing differences between appearance and reality are pointedly dramatized. When the unusual turning point is reached, the observer, taken by surprise, is still momentarily at a loss for explanation. Not until the strange events are recapitulated, at a different pace and from another point of view, does one know fully the meaning of what has been experienced. Finally, the reader may ponder, even in this day of psychiatric shibboleths, the compassionate understanding in the thesis: "To such degree may malign machinations and deceptions impose. So far may even the best man err, in judging the conduct of one with the recesses of whose condition he is not acquainted."*

Benito Cereno

IN THE year 1799, Captain Amasa Delano, of Duxbury, in Massachusetts, commanding a large sealer and general trader, lay at anchor with a valuable cargo, in the harbour of Santa Maria—a small, desert, uninhabited island toward the southern extremity of the long coast of Chile. There he had touched for water.

On the second day, not long after dawn, while lying in his berth, his mate came below, informing him that a strange sail was coming into the bay. Ships were then not so plenty in those waters as now. He rose, dressed, and went on deck.

The morning was one peculiar to that coast. Everything was mute and calm; everything grey. The sea, though undulated into long roods of swells, seemed fixed, and was sleeked at the surface like waved lead that has cooled and set in the smelter's mould. The sky seemed a grey surtout. Flights of troubled grey fowl, kith and kin with flights of troubled grey vapours among which they were mixed, skimmed low and fitfully over the waters, as swallows over meadows before storms. Shadows present, foreshadowing deeper shadows to come.

To Captain Delano's surprise, the stranger, viewed through the glass, showed no colours; though to do so upon entering a haven, however uninhabited in its shores, where but a single other ship might be lying, was the custom among peaceful seamen of all nations. Considering the lawlessness and loneliness of the spot, and the sort of stories, at that day, associated with those seas, Captain Delano's surprise might have deepened into some uneasiness had he not been a person of a singularly undistrustful good nature, not liable, except on extraordinary and repeated incentives, and hardly then, to indulge in personal alarms, any way involving the imputation of malign evil in man. Whether, in view of what humanity is capable, such a trait implies, along with a benevolent heart, more than ordinary quickness and accuracy of intellectual perception, may be left to the wise to determine.

But whatever misgivings might have obtruded on first seeing the stranger, would almost, in any seaman's mind, have been dissipated by observing, that the ship, in navigating into the harbour, was drawing too near the land; a sunken reef making out off her bow. This seemed to prove her a stranger, indeed, not only to the sealer, but the island; consequently, she could be no wonted freebooter on that ocean. With no small interest, Captain Delano continued to watch her—a proceeding not much facilitated by the vapours partly mantling the hull, through which the far matin light from her cabin streamed equivocally enough; much like the sun—by this time hemisphered on the rim of the horizon, and, apparently, in company with the strange ship entering the harbour—which, wimpled by the same low, creeping clouds, showed not unlike a Lima intriguante's one sinister eye peering across the Plaza from the Indian loop-hole of her dusk *saya-y-manto.*

It might have been but a deception of the vapours, but, the longer the stranger was watched the more singular appeared her manœuvres. Ere long it seemed hard to decide whether she meant to come in or no—what she wanted, or what she was about. The wind, which had breezed up a little during the night, was not extremely light and baffling, which the more increased the apparent uncertainty of her movements.

Surmising, at last, that it might be a ship in distress, Captain Delano ordered his whale-boat to be dropped, and, much to the wary opposition of his mate, prepared to board her, and, at the least, pilot her in. On the night previous, a fishing party of the seamen had gone a long distance to some detached rocks out of sight from the sealer, and, an hour or two before daybreak, had returned, having met with no small success. Presuming that the stranger might have been long off soundings, the good captain put several baskets of the fish, for presents, into his boat, and so pulled away. From her continuing too near the sunken reef, deeming her in danger, calling to his men, he made all haste to apprise those on board of their situation. But, some time ere the boat came up, the wind, light though it was, having shifted, had headed the vessel off, as well as partly broken the vapours from about her.

Upon gaining a less remote view, the ship, when made signally visible on the verge of the leaden-hued swells, with the shreds of fog here and there raggedly furring her, appeared like a whitewashed monastery after a thunder-storm, seen perched upon some dun cliff among the Pyrenees. But it was no purely fanciful resemblance which now, for a moment, almost led Captain Delano to think that nothing less than a ship-load of monks was before him. Peering over the bulwarks were what really seemed, in the hazy distance, throngs of dark cows; while, fitfully revealed through the open port-holes, other dark moving figures were dimly descried, as of Black Friars pacing the cloisters.

Upon a still nigher approach, this appearance was modified, and the true character of the vessel was plain—a Spanish merchantman of the first class, carrying negro slaves, amongst other valuable freight, from one colonial port to another. A very large, and, in its time, a very fine vessel, such as in those days were at intervals encountered along that main; sometimes superseded Acapulco treasure-ships, or retired frigates of the Spanish king's navy, which, like superannuated Italian palaces, still, under a decline of masters, preserved signs of former state.

As the whale-boat drew more and more nigh, the cause of the peculiar pipe-clayed aspect of the stranger was seen in the slovenly neglect pervading her. The spars, ropes, and great part of the bulwarks, looked woolly, from long unacquaintance with the scraper, tar, and the brush. Her keel seemed laid, her ribs put together, and she launched, from Ezekiel's Valley of Dry Bones.

In the present business in which she was engaged, the ship's general model and rig appeared to have undergone no material change from their original warlike and Froissart pattern. However, no guns were seen.

The tops were large, and were railed about with what had once been octagonal net-work, all now in sad disrepair. These tops hung overhead like three ruinous aviaries, in one of which was seen perched, on a ratlin, a white noddy, a strange fowl, so called from its lethargic, somnambulistic character, being frequently caught by hand at sea. Battered and mouldy, the castellated forecastle seemed some ancient turret, long ago taken by assault, and then left to decay. Toward the stern, two high-raised quarter-galleries—the balustrades here and there covered with dry, tindery sea-moss—opening out from the unoccupied state-cabin, whose dead-lights, for all the mild weather, were hermetically closed and calked—these tenantless balconies hung over the sea as if it were the grand Venetian canal. But the principal relic of faded grandeur was the ample oval of the shield-like stern-piece, intricately carved with the arms of Castile and León, medallioned about by groups of mythological or symbolical devices; uppermost and central of which was a dark satyr in a mask, holding his foot on the prostrate neck of a writhing figure, likewise masked.

Whether the ship had a figure-head, or only a plain beak, was not quite certain, owing to canvas wrapped about that part, either to protect it while undergoing a refurbishing, or else decently to hide its decay. Rudely painted or chalked, as in a sailor freak, along the forward side of a sort of pedestal below the canvas was the sentence, "*Sequid vuestro jefe*," (follow your leader); while upon the tarnished headboards, near by, appeared, in stately capitals, once gilt, the ship's name, "SAN DOMINICK," each letter streakingly corroded with tricklings of copper-spike rust; while, like mourning weeds, dark festoons of sea-grass slimily swept to and fro over the name, with every hearse-like roll of the hull.

As, at last, the boat was hooked from the bow along toward the gangway amidship, its keel, while yet some inches separated from the hull, harshly grated as on a sunken coral reef. It proved a huge bunch of conglobated barnacles adhering below the water to the side like a wen—a token of baffling airs and long calms passed somewhere in those seas.

Climbing the side, the visitor was at once surrounded by a clamorous throng of whites and blacks, but the latter outnumbering the former more than could have been expected, negro transportation-ship as the stranger in port was. But, in one language, and as with one voice, all poured out a common tale of suffering; in which the negresses, of whom there were not a few, exceeded the others in their dolorous vehemence. The survey, together with the fever, had swept off a great part of their number, more especially the Spaniards. Off Cape Horn they had narrowly escaped shipwreck; then, for days together, they had lain tranced without wind; their provisions were low; their water next to none; their lips that moment were baked.

While Captain Delano was thus made the mark of all eager tongues, his one eager glance took in all faces, with every other object about him.

Always upon first boarding a large and populous ship at sea, especially a foreign one, with a nondescript crew such as Lascars or Manila men, the impression varies in a peculiar way from that produced by first entering a strange house with strange inmates in a strange land. Both house and ship—the one by its walls and blinds, the other by its high bulwarks like ramparts—hoard from view their interiors till the last moment; but in the case of the ship there is this addition: that the living spectacle it contains, upon its sudden and complete disclosure, has, in contrast with the blank ocean which zones it, something of the effect of enchantment. The ship seems unreal; these strange costumes, gestures, and faces, but a shadowy tableau just emerged from the deep, which directly must receive back what it gave.

Perhaps it was some such influence, as above is attempted to be described, which, in Captain Delano's mind, heightened whatever, upon a staid scrutiny, might have seemed unusual; especially the conspicuous figures of four elderly grizzled negroes, their heads like black, doddered willow tops, who, in venerable contrast to the tumult below them, were couched, sphynx-like, one on the starboard cat-head, another on the larboard, and the remaining pair face to face on the opposite bulwarks above the main-chains. They each had bits of unstranded old junk in their hands, and, with a sort of stoical self-content, were picking the junk into oakum, a small heap of which lay by their sides. They accompanied the task with a continuous, low, monotonous chant; droning and drooling away like so many grey-headed bag-pipers playing a funeral march.

The quarter-deck rose into an ample elevated poop, upon the forward verge of which, lifted, like the oakum-pickers, some eight feet above the general throng, sat along in a row, separated by regular spaces, the cross-legged figures of six other blacks; each with a rusty hatchet in his hand, which, with a bit of brick and a rag, he was engaged like a scullion in scouring; while between each two was a small stack of hatchets, their rusted edges turned forward awaiting a like operation. Though occasionally the four oakum-pickers would briefly address some person or persons in the crowd below, yet the six hatchet-polishers neither spoke to others, nor breathed a whisper among themselves, but sat intent upon their task, except at intervals, when, with the peculiar love in negroes of uniting industry with pastime, two and two they sideways clashed their hatchets together, like cymbals, with a barbarous din. All six, unlike the generality, had the raw aspect of unsophisticated Africans.

But that first comprehensive glance which took in those ten figures, with scores less conspicuous, rested but an instant upon them, as, impatient of the hubbub of voices, the visitor turned in quest of whomsoever it might be that commanded the ship.

But as if not unwilling to let nature make known her own case among his suffering charge, or else in despair of restraining it for the time, the Spanish captain, a gentlemanly, reserved-looking, and rather young man to a stranger's eye, dressed with singular richness, but bearing plain traces of recent sleepless cares and disquietudes, stood passively by, leaning against the main-mast, at one moment casting a dreary, spiritless look upon his excited people, at the next an unhappy glance toward his visitor. By his side stood a black of small stature, in whose rude face, as occasionally, like a shepherd's dog, he mutely turned it up into the Spaniard's, sorrow and affection were equally blended.

Struggling through the throng, the American advanced to the Spaniard, assuring him of his sympathies, and offering to render whatever assistance might be in his power. To which the Spaniard returned for the present but grave and ceremonious acknowledgments, his national formality ducked by the saturnine mood of ill-health.

But losing no time in mere compliments, Captain Delano, returning to the gangway, had his baskets of fish brought up; and as the wind still continued light, so that some hours at least must elapse ere the ship could be brought to the anchorage, he bade his men return to the sealer, and fetch back as much water as the whale-boat could carry, with whatever soft bread the steward might have, all the remaining pumpkins on board, with a box of sugar, and a dozen of his private bottles of cider.

Not many minutes after the boat's pushing off, to the vexation of all, the wind entirely died away, and the tide turning, began drifting back the ship helplessly seaward. But trusting this would not long last, Captain Delano sought, with good hopes, to cheer up the strangers, feeling no small satisfaction that, with persons in their condition, he could—thanks to his frequent voyages along the Spanish Main—converse with some freedom in their native tongue.

While left alone with them, he was not long in observing some things tending to heighten his first impressions; but surprise was lost in pity, both for the Spaniards and blacks, alike evidently reduced from scarcity of water and provisions; while long-continued suffering seemed to have brought out the less good-natured qualities of the negroes, besides, at the same time, impairing the Spaniard's authority over them. But, under the circumstances, precisely this condition of things was to have been anticipated. In armies, navies, cities, or families, in nature herself, nothing more relaxes good order than misery. Still, Captain Delano was not without the idea, that had Benito Cereno been a man of greater energy, misrule would hardly have come to the present pass. But the debility, constitutional or induced by hardships, bodily and mental, of the Spanish captain, was too obvious to be overlooked. A prey to settled dejection, as if long mocked with hope he would not now indulge it, even when it had ceased to be a mock, the prospect of that day, or evening at furthest, lying at anchor, with plenty of water for his people, and a brother captain to counsel and befriend, seemed in no perceptible degree to encourage him. His mind appeared unstrung, if not still more seriously affected. Shut up in these oaken walls, chained to one dull round of command, whose unconditionality cloyed him, like some hypochondriac abbot he moved slowly about, at times suddenly pausing, starting, or staring, biting his lip, biting his finger-nail, flushing, paling, twitching his beard, with other symptoms of an absent or moody mind. This distempered spirit was lodged, as before hinted, in as distempered a frame. He was rather tall, but seemed never to have been robust, and now with nervous suffering was almost worn to a skeleton. A tendency to some pulmonary complaint appeared to have been lately confirmed. His voice was like that of one with lungs half gone—hoarsely suppressed, a husky whisper. No wonder that, as in this state he tottered about, his private servant apprehensively followed him. Sometimes the negro gave his master his arm, or took his handkerchief out of his pocket for him; performing these and similar offices with that affectionate zeal which transmutes into something filial or fraternal acts in themselves but menial; and which has gained for the negro the repute of making the most pleasing body-servant in the world; one, too, whom a master need be on no stiffly superior terms with, but may treat with familiar trust; less a servant than a devoted companion.

Marking the noisy indocility of the blacks in general, as well as what seemed the sullen inefficiency of the whites, it was not without humane satisfaction that Captain Delano witnessed the steady good conduct of Babo.

But the good conduct of Babo, hardly more than the ill-behaviour of others, seemed to withdraw the half-lunatic Don Benito from his cloudy languor. Not that such precisely was the impression made by the Spaniard on the mind of his visitor. The Spaniard's individual unrest was, for the present, but noted as a conspicuous feature in the ship's general affliction. Still, Captain Delano was not a little concerned at what he could not help taking for the time to be Don Benito's unfriendly indifference towards himself. The Spaniard's manner, too, conveyed a sort of sour and gloomy disdain, which he seemed at no pains to disguise. But this the American in charity ascribed to the harassing effects of sickness, since, in former instances, he had noted that there are peculiar natures on whom prolonged physical suffering seems to cancel every social instinct of kindness; as if, forced to black bread themselves, they deemed it but equity that each person coming nigh them should, indirectly, by some slight or affront, be made to partake of their fare.

But ere long Captain Delano bethought him that, indulgent as he was at the first, in judging the Spaniard, he might not, after all, have exercised charity enough. At bottom it was Don Benito's reserve which displeased him; but the same reserve was shown towards all but his faithful personal attendant. Even the formal reports which, according to sea-usage, were, at stated times, made to him by some petty underling, either a white, mulatto or black, he hardly had patience enough to listen to, without betraying contemptuous aversion. His manner upon such occasions was, in its degree, not unlike that which might be supposed to have been his imperial countryman's, Charles V, just previous to the anchoritish retirement of that monarch from the throne.

This splenetic disrelish of his place was evinced in almost every function pertaining to it. Proud as he was moody, he condescended to no personal mandate. Whatever special orders were necessary, their delivery was delegated to his body-servant, who in turn transferred them to their ultimate destination, through runners, alert Spanish boys or slave boys, like pages or pilot-fish within easy call continually hovering round Don Benito. So that to have beheld this undemonstrative invalid gliding about, apathetic and mute, no landsman could have dreamed that in him was lodged a dictatorship beyond which, while at sea, there was no earthly appeal.

Thus, the Spaniard, regarded in his reserve, seemed the involuntary victim of mental disorder. But, in fact, his reserve might, in some degree, have proceeded from design. If so, then here was evinced the unhealthy climax of that icy though conscientious policy, more or less adopted by all commanders of large ships, which, except in signal emergencies, obliterates alike the manifestation of sway with every trace of sociality; transforming the man into a block, or rather into a loaded cannon, which, until there is call for thunder, has nothing to say.

Viewing him in this light, it seemed but a natural token of the perverse habit induced by a long course of such hard self-restraint, that, notwithstanding the present condition of his ship, the Spaniard should still persist in a demeanour, which, however harmless, or, it may be, appropriate, in a well-appointed vessel, such as the *San Dominick* might have been at the outset of the voyage, was anything but judicious now. But the Spaniard, perhaps, thought that it was with captains as with gods: reserve, under all events, must still be their cue. But probably this appearance of slumbering dominion might have been but an attempted disguise to conscious imbecility—not deep policy, but shallow device. But be all this as it might, whether Don Benito's manner was designed or not, the more Captain Delano noted its pervading reserve, the less he felt uneasiness at any particular manifestation of that reserve towards himself.

Neither were his thoughts taken up by the captain alone. Wonted to the quiet orderliness of the sealer's comfortable family of a crew, the noisy confusion of the *San Dominick's* suffering host repeatedly challenged his eye. Some prominent breaches, not only of discipline but of decency, were observed. These Captain Delano could not but ascribe, in the main, to the absence of those subordinate deck-officers to whom, along with higher duties, is intrusted what may be styled the police department of a populous ship. True, the old oakum-pickers appeared at times to act the part of monitorial constables to their countrymen, the blacks; but though occasionally succeeding in allaying trifling outbreaks now and then between man and man, they could do little or nothing toward establishing general quiet. The *San Dominick* was in the condition of a transatlantic emigrant ship, among whose multitude of living freight are some individuals, doubtless, as little troublesome as crates and bales; but the friendly remonstrances of such with their ruder companions are of not so much avail as the unfriendly arm of the mate. What the *San Dominick* wanted was, what the emigrant ship has, stern superior officers. But on these decks not so much as a fourth-mate was to be seen.

The visitor's curiosity was roused to learn the particulars of those mishaps which had brought about such absenteeism, with its consequences; because, though deriving some inkling of the voyage from the wails which at the first moment had greeted him, yet of the details no clear understanding had been had. The best account would, doubtless, be given by the captain. Yet at first the visitor was loth to ask it, unwilling to provoke some distant rebuff. But plucking up courage, he at last accosted Don Benito, renewing the expression of his benevolent interest, adding, that did he (Captain Delano) but know the particulars of the ship's misfortunes, he would, perhaps, be better able in the end to relieve them. Would Don Benito favour him with the whole story?

Don Benito faltered; then, like some somnambulist suddenly interfered with, vacantly stared at his visitor, and ended by looking down on the deck. He maintained this posture so long, that Captain Delano, almost equally disconcerted, and involuntarily almost as rude, turned suddenly from him, walking forward to accost one of the Spanish seamen for the desired information. But he had hardly gone five paces, when, with a sort of eagerness, Don Benito invited him back, regretting his momentary absence of mind, and professing readiness to gratify him.

While most part of the story was being given, the two captains stood on the after part of the main-deck, a privileged spot, no one being near but the servant.

"It is now a hundred and ninety days," began the Spaniard, in his husky whisper, "that this ship, well officered and well manned, with several cabin passengers—some fifty Spaniards in all—sailed from Buenos Ayres bound to Lima, with a general cargo, hardware, Paraguay tea and the like—and," pointing forward, "that parcel of negroes, now not more than a hundred and fifty, as you see, but then numbering over three hundred souls. Off Cape Horn we had heavy gales. In one moment, by night, three of my best officers, with fifteen sailors, were lost, with the main-yard; the spar snapping under them in the slings, as they sought, with heavers, to beat down the icy sail. To lighten the hull, the heavier sacks of maté were thrown into the sea, with most of the water-pipes lashed on deck at the time. And this last necessity it was, combined with the prolonged detentions afterwards experienced, which eventually brought about our chief causes of suffering. When——"

Here there was a sudden fainting attack of his cough, brought on, no doubt, by his mental distress. His servant sustained him, and drawing a cordial from his pocket placed it to his lips. He a little revived. But unwilling to leave him unsupported while yet imperfectly restored, the black with one arm still encircled his master, at the same time keeping his eye fixed on his face, as if to watch for the first sign of complete restoration, or relapse, as the event might prove.

The Spaniard proceeded, but brokenly and obscurely, as one in a dream.

—"Oh, my God! rather than pass through what I have, with joy I would have hailed the most terrible gales; but——"

His cough returned and with increased violence; this subsiding, with reddened lips and closed eyes he fell heavily against his supporter.

"His mind wanders. He was thinking of the plague that followed the gales," plaintively sighed the servant; "my poor, poor master!" wringing one hand, and with the other wiping the mouth. "But be patient, Señor," again turning to Captain Delano, "these fits do not last long; master will soon be himself."

Don Benito reviving, went on; but as this portion of the story was very brokenly delivered, the substance only will here be set down.

It appeared that after the ship had been many days tossed in storms off the Cape, the scurvy broke out, carrying off numbers of the whites and blacks. When at last they had worked round into the Pacific, their spars and sails were so damaged, and so inadequately handled by the surviving mariners, most of whom were become invalids, that, unable to lay her northerly course by the wind which was powerful, the unmanageable ship, for successive days and nights, was blown northwestward, where the breeze suddenly deserted her, in unknown waters, to sultry calms. The absence of the water-pipes now proved as fatal to life as before their presence had menaced it. Induced, or at least aggravated, by the more than scanty allowance of water, a malignant fever followed the scurvy; with the excessive heat of the lengthened calm, making such short work of it as to sweep away, as by billows, whole families of the Africans, and a yet larger number, proportionably, of the Spaniards, including, by a luckless fatality, every remaining officer on board. Consequently, in the smart west winds eventually following the calm, the already rent sails, having to be simply dropped, not furled, at need, had been gradually reduced to the beggars' rags they were now. To procure substitutes for his lost sailors, as well as supplies of water and sails, the captain, at the earliest opportunity, had made for Valdivia, the southernmost civilized port of Chile and South America; but upon nearing the coast the thick weather had prevented him from so much as sighting that harbour. Since which period, almost without a crew, and almost without canvas and almost without water, and, at intervals, giving its added dead to the sea, the *San Dominick* had been battledored about by contrary winds, inveigled by currents, or grown weedy in calms. Like a man lost in woods, more than once she had doubled upon her own track.

"But throughout these calamities," huskily continued Don Benito, painfully turning in the half embrace of his servant, "I have to thank those negroes you see, who, though to your inexperienced eyes appearing unruly, have, indeed, conducted themselves with less of restlessness than even their owner could have thought possible under such circumstances."

Here he again fell faintly back. Again his mind wandered; but he rallied, and less obscurely proceeded.

"Yes, their owner was quite right in assuring me that no fetters would be needed with his blacks; so that while, as is wont in this transportation, those negroes have always remained upon deck—not thrust below, as in the Guineamen—they have, also, from the beginning, been freely permitted to range within given bounds at their pleasure."

Once more the faintness returned—his mind roved—but, recovering, he resumed:

"But it is Babo here to whom, under God, I owe not only my own preservation, but likewise to him, chiefly, the merit is due, of pacifying his more ignorant brethren, when at intervals tempted to murmurings."

"Ah, master," sighed the black, bowing his face, "don't speak of me; Babo is nothing; what Babo has done was but duty."

"Faithful fellow!" cried Captain Delano. "Don Benito, I envy you such a friend; slave I cannot call him."

As master and man stood before him, the black upholding the white, Captain Delano could not but bethink him of the beauty of that relationship which could present such a spectacle of fidelity on the one hand and confidence on the other. The scene was heightened by the contrast in dress, denoting their relative positions. The Spaniard wore a loose Chile jacket of dark velvet, white small-clothes and stockings, with silver buckles at the knee and instep; a high-crowned sombrero, of fine grass; a slender sword, silver mounted, hung from a knot in his sash—the last being an almost invariable adjunct, more for utility than ornament, of a South American gentleman's dress to this hour. Excepting when his occasional nervous contortions brought about disarray, there was a certain precision in his attire curiously at variance with the unsightly disorder around; especially in the belittered Ghetto, forward of the main-mast, wholly occupied by the blacks.

The servant wore nothing but wide trousers, apparently, from their coarseness and patches, made out of some old topsail; they were clean, and confined at the waist by a bit of unstranded rope, which, with his composed, deprecatory air at times, made him look something like a begging friar of St. Francis.

However unsuitable for the time and place, at least in the blunt-thinking American's eyes, and however strangely surviving in the midst of all his afflictions, the toilette of Don Benito might not, in fashion at least, have gone beyond the style of the day among South Americans of his class. Though on the present voyage sailing from Buenos Ayres, he had avowed himself a native and resident of Chile, whose inhabitants had not so generally adopted the plain coat and once plebeian pantaloons; but, with a becoming modification, adhered to their provincial costume, picturesque as any in the world. Still, relatively to the pale history of the voyage, and his pale face, there seemed something so incongruous in the Spaniard's apparel, as almost to suggest the image of an invalid courtier tottering about London streets in the time of the plague.

The portion of the narrative which, perhaps, most excited interest, as well as some surprise, considering the latitudes in question, was the long calms spoken of, and more particularly the ship's so long drifting about. Without communicating the opinion, of course, the American could not but impute at least part of the detentions both to clumsy seamanship and faulty navigation. Eyeing Don Benito's small, yellow hands, he easily inferred that the young captain had not got into command at the hawse-hole, but the cabin-window; and if so, why wonder at incompetence, in youth, sickness, and gentility united?

But drowning criticism in compassion, after a fresh repetition of his sympathies, Captain Delano, having heard out his story, not only engaged, as in the first place, to see Don Benito and his people supplied in their immediate bodily needs, but, also, now further promised to assist him in procuring a large permanent supply of water, as well as some sails and rigging, and, though it would involve no small embarrassment to himself, yet he would spare three of his best seamen for temporary deck-officers; so that without delay the ship might proceed to Concepcion, there fully to refit for Lima, her destined port.

Such generosity was not without its effect, even upon the invalid. His face lighted up; eager and hectic, he met the honest glance of his visitor. With gratitude he seemed overcome.

"This excitement is bad for master," whispered the servant, taking his arm, and with soothing words gently drawing him aside.

When Don Benito returned, the American was pained to observe that his hopefulness, like the sudden kindling in his cheek, was but febrile and transient.

Ere long, with a joyless mien, looking up towards the poop, the host invited his guest to accompany him there, for the benefit of what little breath of wind might be stirring.

As during the telling of the story, Captain Delano had once or twice started at the occasional cymballing of the hatchet-polishers, wondering why such an interruption should be allowed, especially in that part of the ship, and in the ears of an invalid; and moreover, as the hatchets had anything but an attractive look, and the handlers of them still less so, it was, therefore, to tell the truth, not without some lurking reluctance, or even shrinking, it may be, that Captain Delano, with apparent complaisance, acquiesced in his host's invitation. The more so, since, with an untimely caprice of punctilio, rendered distressing by his cadaverous aspect, Don Benito, with Castilian bows, solemnly insisted upon his guest's preceding him up the ladder leading to the elevation; where, one on each side of the last step, sat for armorial supporters and sentries two of the ominous file. Gingerly enough stepped good Captain Delano between them, and in the instant of leaving them behind, like one running the gauntlet, he felt an apprehensive twitch in the calves of his legs.

But when, facing about, he saw the whole file, like so many organ-grinders, still stupidly intent on their work, unmindful of everything beside, he could not but smile at his late fidgety panic.

Presently, while standing with his host, looking forward upon the decks below, he was struck by one of those instances of insubordination previously alluded to. Three black boys, with two Spanish boys, were sitting together on the hatches, scraping a rude wooden platter, in which some scanty mess had recently been cooked. Suddenly, one of the black boys, enraged at a word dropped by one of his white companions, seized a knife, and, though called to forbear by one of the oakum-pickers, struck the lad over the head, inflicting a gash from which blood flowed.

In amazement, Captain Delano inquired what this meant. To which the pale Don Benito dully muttered, that it was merely the sport of the lad.

"Pretty serious sport, truly," rejoined Captain Delano. "Had such a thing happened on board the *Bachelor's Delight*, instant punishment would have followed."

At these words the Spaniard turned upon the American one of his sudden, staring, half-lunatic looks; then, relapsing into his torpor, answered, "Doubtless, doubtless, Señor."

Is it, thought Captain Delano, that this hapless man is one of those paper captains I've known, who by policy wink at what by power they cannot put down? I know no sadder sight than a commander who has little of command but the name.

"I should think, Don Benito," he now said, glancing towards the oakum-picker who had sought to interfere with the boys, "that you would find it advantageous to keep all your blacks employed, especially the younger ones, no matter at what useless task, and no matter what happens to the ship. Why, even with my little band, I find such a course indispensable. I once kept a crew on my quarter-deck thrumming mats for my cabin, when, for three days, I had given up my ship —mats, men, and all—for a speedy loss, owing to the violence of a gale, in which we could do nothing but helplessly drive before it."

"Doubtless, doubtless," muttered Don Benito.

"But," continued Captain Delano, again glancing upon the oakum-pickers and then at the hatchet-polishers, near by, "I see you keep some, at least, of your host employed."

"Yes," was again the vacant response.

"Those old men there, shaking their pows from their pulpits," continued Captain Delano, pointing to the oakum-pickers, "seem to act the part of old dominies to the rest, little heeded as their admonitions are at times. Is this voluntary on their part, Don Benito, or have you appointed them shepherds to your flock of black sheep?"

"What posts they fill, I appointed them," rejoined the Spaniard, in an acrid tone, as if resenting some supposed satiric reflection.

"And these others, these Ashantee conjurors here," continued Captain Delano, rather uneasily eyeing the brandished steel of the hatchet-polishers, where, in spots, it had been brought to a shine, "this seems a curious business they are at, Don Benito?"

"In the gales we met," answered the Spaniard, "what of our general cargo was not thrown overboard was much damaged by the brine. Since coming into calm weather, I have had several cases of knives and hatchets daily brought up for overhauling and cleaning."

"A prudent idea, Don Benito. You are part owner of ship and cargo, I presume; but none of the slaves, perhaps?"

"I am owner of all you see," impatiently returned Don Benito, "except the main company of blacks, who belonged to my late friend, Alexandro Aranda."

As he mentioned this name, his air was heart-broken; his knees shook; his servant supported him.

Thinking he divined the cause of such unusual emotion, to confirm his surmise, Captain Delano, after a pause, said: "And may I ask, Don Benito, whether—since a while ago you spoke of some cabin passengers —the friend, whose loss so afflicts you, at the outset of the voyage accompanied his blacks?"

"Yes."

"But died of the fever?"

"Died of the fever. Oh, could I but——" Again quivering, the Spaniard paused.

"Pardon me," said Captain Delano, lowly, "but I think that, by a sympathetic experience, I conjecture, Don Benito, what it is that gives the keener edge to your grief. It was once my hard fortune to lose, at sea, a dear friend, my own brother then supercargo. Assured of the welfare of his spirit, its departure I could have borne like a man; but that honest eye, that honest hand—both of which had so often met mine—and that warm heart; all, all—like scraps to the dogs—to throw all to the sharks! It was then I vowed never to have for fellow-voyager a man I loved, unless, unbeknown to him, I had provided every requisite, in case of a fatality, for embalming his mortal part for interment on shore. Were your friend's remains now on board this ship, Don Benito, not thus strangely would the mention of his name affect you."

"On board this ship?" echoed the Spaniard. Then, with horrified gestures, as directed against some spectre, he unconsciously fell into the ready arms of his attendant, who, with a silent appeal toward Captain Delano, seemed beseeching him not again to broach a theme so unspeakably distressing to his master.

This poor fellow now, thought the pained American, is the victim of that sad superstition which associates goblins with the deserted body of man, as ghosts with an abandoned house. How unlike are we made! What to me, in like case, would have been a solemn satisfaction, the bare suggestion, even, terrifies the Spaniard into this trance. Poor Alexandro Aranda! what would you say could you here see your friend —who, on former voyages, when you, for months, were left behind, has, I dare say, often longed, and longed, for one peep at you—now transported with terror at the least thought of having you anyway nigh him.

At this moment, with a dreary grave-yard toll, betokening a flaw, the ship's forecastle bell, smote by one of the grizzled oakum-pickers, proclaimed ten o'clock, through the leaden calm; when Captain Delano's attention was caught by the moving figure of a gigantic black, emerging from the general crowd below, and slowly advancing towards the elevated poop. An iron collar was about his neck, from which depended a chain, thrice wound round his body; the terminating links padlocked together at a broad band of iron, his girdle.

"How like a mute Atufal moves," murmured the servant.

The black mounted the steps of the poop, and, like a brave prisoner, brought up to receive sentence, stood in unquailing muteness before Don Benito, now recovered from his attack.

At the first glimpse of his approach, Don Benito had started, a resentful shadow swept over his face; and, as with the sudden memory of bootless rage, his white lips glued together.

This is some mulish mutineer, thought Captain Delano, surveying, not without a mixture of admiration, the colossal form of the negro.

"See, he waits your question, master," said the servant.

Thus reminded, Don Benito, nervously averting his glance, as if shunning, by anticipation, some rebellious response, in a disconcerted voice, thus spoke:—

"Atufal, will you ask my pardon, now?"

The black was silent.

"Again, master," murmured the servant, with bitter upbraiding eyeing his countryman. "Again, master; he will bend to master yet."

"Answer," said Don Benito, still averting his glance, "say but the one word, *pardon*, and your chains shall be off."

Upon this, the black, slowly raising both arms, let them lifelessly fall, his links clanking, his head bowed; as much as to say, "No, I am content."

"Go," said Don Benito, with inkept and unknown emotion.

Deliberately as he had come, the black obeyed.

"Excuse me, Don Benito," said Captain Delano, "but this scene surprises me; what means it, pray?"

"It means that the negro alone, of all the band, has given me peculiar cause of offence. I have put him in chains; I——"

Here he paused; his hand to his head, as if there were a swimming there, or a sudden bewilderment of memory had come over him; but meeting his servant's kindly glance seemed reassured, and proceeded:—

"I could not scourge such a form. But I told him he must ask my pardon. As yet he has not. At my command, every two hours he stands before me."

"And how long has this been?"

"Some sixty days."

"And obedient in all else? And respectful?"

"Yes."

"Upon my conscience, then," exclaimed Captain Delano, impulsively, "he has a royal spirit in him, this fellow."

"He may have some right to it," bitterly returned Don Benito, "he says he was king in his own land."

"Yes," said the servant, entering a word, "those slits in Atufal's ears once held wedges of gold; but poor Babo here, in his own land, was only a poor slave; a black man's slave was Babo, who now is the white's."

Somewhat annoyed by these conversational familiarities, Captain Delano turned curiously upon the attendant, then glanced inquiringly at his master; but, as if long wonted to these little informalities, neither master nor man seemed to understand him.

"What, pray, was Atufal's offence, Don Benito?" asked Captain Delano; "if it was not something very serious, take a fool's advice, and, in view of his general docility, as well as in some natural respect for his spirit, remit him his penalty."

"No, no, master never will do that," here murmured the servant to himself, "proud Atufal must first ask master's pardon. The slave there carries the padlock, but master here carries the key."

His attention thus directed, Captain Delano now noticed for the first time that, suspended by a slender silken cord, from Don Benito's neck, hung a key. At once, from the servant's muttered syllables, divining the key's purpose, he smiled and said:—"So, Don Benito—padlock and key—significant symbols, truly."

Biting his lip, Don Benito faltered.

Though the remark of Captain Delano, a man of such native simplicity as to be incapable of satire or irony, had been dropped in playful allusion to the Spaniard's singularly evidenced lordship over the black; yet the hypochondriac seemed some way to have taken it as a malicious reflection upon his confessed inability thus far to break down, at least, on a verbal summons, the entrenched will of the slave. Deploring this supposed misconception, yet despairing of correcting it, Captain Delano shifted the subject; but finding his companion more than ever withdrawn, as if still sourly digesting the lees of the presumed affront above-mentioned, by and by Captain Delano likewise became less talkative, oppressed, against his own will, by what seemed the secret vindictiveness of the morbidly sensitive Spaniard. But the good sailor, himself of a quite contrary disposition, refrained, on his part, alike from the appearance as from the feeling of resentment, and if silent, was only so from contagion.

Presently the Spaniard, assisted by his servant, somewhat discourteously crossed over from his guest; a procedure which, sensibly enough, might have been allowed to pass from idle caprice of ill-humour, had not master and man, lingering round the corner of the elevated skylight, begun whispering together in low voices. This was unpleasing. And more: the moody air of the Spaniard, which at times had not been without a sort of valetudinarian stateliness, now seemed anything but dignified; while the menial familiarity of the servant lost its original charm of simple-hearted attachment.

In his embarrassment, the visitor turned his face to the other side of the ship. By so doing, his glance accidentally fell on a young Spanish sailor, a coil of rope in his hand, just stepped from the deck to the first round of the mizzen-rigging. Perhaps the man would not have been particularly noticed, were it not that, during his ascent to one of the yards, he, with a sort of covert intentness, kept his eye fixed on Captain Delano, from whom, presently, it passed, as if by a natural sequence, to the two whisperers.

His own attention thus redirected to that quarter, Captain Delano gave a slight start. From something in Don Benito's manner just then, it seemed as if the visitor had, at least partly, been the subject of the withdrawn consultation going on—a conjecture as little agreeable to the guest as it was little flattering to the host.

The singular alternations of courtesy and ill-breeding in the Spanish captain were unaccountable, except on one of two suppositions—innocent lunacy, or wicked imposture.

But the first idea, though it might naturally have occurred to an indifferent observer, and, in some respect, had not hitherto been wholly a stranger to Captain Delano's mind, yet, now that, in an incipient way, he began to regard the stranger's conduct something in the light of an intentional affront, of course the idea of lunacy was virtually vacated. But if not a lunatic, what then? Under the circumstances, would a gentleman, nay, any honest boor, act the part now acted by his host? The man was an imposter. Some low-born adventurer, masquerading as an oceanic grandee; yet so ignorant of the first requisites of mere gentlemanhood as to be betrayed into the present remarkable indecorum. The strange ceremoniousness, too, at other times evinced, seemed not uncharacteristic of one playing a part above his real level. Benito Cereno—Don Benito Cereno—a sounding name. One, too, at that period, not unknown, in the surname, to supercargoes and sea captains trading along the Spanish Main, as belonging to one of the most enterprising and extensive mercantile families in all those provinces; several members of it having titles; a sort of Castilian Rothschild, with a noble brother, or cousin, in every great trading town of South America. The alleged Don Benito was in early manhood, about twenty-nine or thirty. To assume a sort of roving cadetship in the maritime affairs of such a house, what more likely scheme for a young knave of talent and spirit? But the Spaniard was a pale invalid. Never mind. For even to the degree of simulating mortal disease, the craft of some tricksters had been known to attain. To think that, under the aspect of infantile weakness, the most savage energies might be couched—those velvets of the Spaniard but the silky paw to his fangs.

From no train of thought did these fancies come; not from within, but from without; suddenly, too, and in one throng, like hoar frost; yet as soon to vanish as the mild sun of Captain Delano's good nature regained its meridian.

Glancing over once more towards his host—whose sideface, revealed above the skylight, was now turned towards him—he was struck by the profile, whose clearness of cut was refined by the thinness, incident to ill-health, as well as ennobled about the chin by the beard. Away with suspicion. He was a true off-shoot of a true hidalgo Cereno.

Relieved by these and other better thoughts, the visitor, lightly humming a tune, now began indifferently pacing the poop, so as not to betray to Don Benito that he had at all mistrusted incivility, much less duplicity; for such mistrust would yet be proved illusory, and by the event; though, for the present, the circumstance which had provoked that distrust remained unexplained. But when that little mystery should have been cleared up, Captain Delano thought he might extremely regret it, did he allow Don Benito to become aware that he had indulged in ungenerous surmises. In short, to the Spaniard's black-letter text, it was best, for a while, to leave open margin.

Presently, his pale face twitching and overcast, the Spaniard, still supported by his attendant, moved over towards his guest, when, with even more than his usual embarrassment, and a strange sort of intriguing intonation in his husky whisper, the following conversation began:—

"Señor, may I ask how long you have lain at this isle?"

"Oh, but a day or two, Don Benito."

"And from what port are you last?"

"Canton."

"And there, Señor, you exchanged your sealskins for teas and silks, I think you said?"

"Yes. Silks mostly."

"And the balance you took in specie, perhaps?"

Captain Delano, fidgeting a little, answered—

"Yes; some silver; not a very great deal, though."

"Ah—well. May I ask how many men have you, Señor?"

Captain Delano slightly started, but answered—

"About five-and-twenty, all told."

"And at present, Señor, all on board, I suppose?"

"All on board, Don Benito," replied the Captain, now with satisfaction.

"And will be to-night, Señor?"

At this last question, following so many pertinacious ones, for the soul of him Captain Delano could not but look very earnestly at the questioner, who, instead of meeting the glance, with every token of craven discomposure dropped his eyes to the deck; presenting an unworthy contrast to his servant, who, just then, was kneeling at his feet, adjusting a loose shoe-buckle; his disengaged face meantime, with humble curiosity, turned openly up into his master's downcast one.

The Spaniard, still with a guilty shuffle, repeated his question:

"And—and will be to-night, Señor?"

"Yes, for aught I know," returned Captain Delano—"but nay," rallying himself into fearless truth, "some of them talked of going off on another fishing party about midnight."

"Your ships generally go—go more or less armed, I believe, Señor?"

"Oh, a six-pounder or two, in case of emergency," was the intrepidly indifferent reply, "with a small stock of muskets, sealing-spears, and cutlasses, you know."

As he thus responded, Captain Delano again glanced at Don Benito, but the latter's eyes were averted; while abruptly and awkwardly shifting the subject, he made some peevish allusion to the calm, and then, without apology, once more, with his attendant, withdrew to the opposite bulwarks, where the whispering was resumed.

At this moment, and ere Captain Delano could cast a cool thought upon what had just passed, the young Spanish sailor, before mentioned, was seen descending from the rigging. In act of stooping over to spring inboard to the deck, his voluminous, unconfined frock, or shirt, of coarse woolen, much spotted with tar, opened out far down the chest, revealing a soiled undergarment of what seemed the finest linen, edged, about the neck, with a narrow blue ribbon, sadly faded and worn. At this moment the young sailor's eye was again fixed on the whisperers, and Captain Delano thought he observed a lurking significance in it, as if silent signs, of some Freemason sort, had that instant been interchanged.

This once more impelled his own glance in the direction of Don Benito, and, as before, he could not but infer that himself formed the subject of the conference. He paused. The sound of the hatchet-polishing fell on his ears. He cast another swift side-look at the two. They had the air of conspirators. In connection with the late questionings, and the incident of the young sailor, these things now begat such return of involuntary suspicion, that the singular guilelessness of the American could not endure it. Plucking up a gay and humorous expression, he crossed over to the two rapidly, saying:—"Ha, Don Benito, your black here seems high in your trust; a sort of privy-counsellor, in fact."

Upon this, the servant looked up with a good-natured grin, but the master started as from a venomous bite. It was a moment or two before the Spaniard sufficiently recovered himself to reply; which he did, at last, with cold constraint:—"Yes, Señor, I have trust in Babo."

Here Babo, changing his previous grin of mere animal humour into an intelligent smile, not ungratefully eyed his master.

Finding that the Spaniard now stood silent and reserved, as if involuntarily, or purposely giving hint that his guest's proximity was inconvenient just then, Captain Delano, unwilling to appear uncivil even to incivility itself, made some trivial remark and moved off; again and again turning over in his mind the mysterious demeanor of Don Benito Cereno.

He had descended from the poop, and, wrapped in thought, was passing near a dark hatchway, leading down into the steerage, when, perceiving motion there, he looked to see what moved. The same instant there was a sparkle in the shadowy hatchway, and he saw one of the Spanish sailors, prowling there, hurriedly placing his hand in the bosom of his frock, as if hiding something. Before the man could have been certain who it was that was passing, he slunk below out of sight. But enough was seen of him to make it sure that he was the same young sailor before noticed in the rigging.

What was that which so sparkled? thought Captain Delano. It was no lamp—no match—no live coal. Could it have been a jewel? But how come sailors with jewels?—or with silk-trimmed under-shirts either? Has he been robbing the trunks of the dead cabin-passengers? But if so, he would hardly wear one of the stolen articles on board ship here. Ah, ah—if, now, that was, indeed, a secret sign I saw passing between this suspicious fellow and his captain awhile since; if I could only be certain that, in my uneasiness, my senses did not deceive me, then——

Here, passing from one suspicious thing to another, his mind revolved the strange questions put to him concerning his ship.

By a curious coincidence, as each point was recalled, the black wizards of Ashantee would strike up with their hatchets, as in ominous comment on the white stranger's thoughts. Pressed by such enigmas and portents, it would have been almost against nature, had not, even into the least distrustful heart, some ugly misgivings obtruded.

Observing the ship, now helplessly fallen into a current, with enchanted sails, drifting with increased rapidity seaward; and noting that, from a lately intercepted projection of the land, the sealer was hidden, the stout mariner began to quake at thoughts which he barely durst confess to himself. Above all, he began to feel a ghostly dread of Don Benito. And yet, when he roused himself, dilated his chest, felt himself strong on his legs, and coolly considered it—what did all these phantoms amount to?

Had the Spaniard any sinister scheme, it must have reference not so much to him (Captain Delano) as to his ship (the *Bachelor's Delight*). Hence the present drifting away of the one ship from the other, instead of favouring any such possible scheme, was, for the time, at least, opposed to it. Clearly any suspicion, combining such contradictions, must need be delusive. Besides, was it not absurd to think of a vessel in distress—a vessel by sickness almost dismanned of her crew—a vessel whose inmates were parched for water—was it not a thousand times absurd that such a craft should, at present, be of a piratical character; or her commander, either for himself or those under him, cherish any desire but for speedy relief and refreshment? But then, might not general distress, and thirst in particular, be affected? And might not that same undiminished Spanish crew, alleged to have perished off to a remnant, be at that very moment lurking in the hold? On heart-broken pretence of entreating a cup of cold water, fiends in human form had got into lonely dwellings, nor retired until a dark deed had been done. And among the Malay pirates, it was no unusual thing to lure ships after them into their treacherous harbours, or entice boarders from a declared enemy at sea, by the spectacle of thinly manned or vacant decks, beneath which prowled a hundred spears with yellow arms ready to upthrust them through the mats. Not that Captain Delano had entirely credited such things. He had heard of them—and now, as stories, they recurred. The present destination of the ship was the anchorage. There she would be near his own vessel. Upon gaining that vicinity, might not the *San Dominick*, like a slumbering volcano, suddenly let loose energies now hid?

He recalled the Spaniard's manner while telling his story. There was a gloomy hesitancy and subterfuge about it. It was just the manner of one making up his tale for evil purposes, as he goes. But if that story was not true, what was the truth? That the ship had unlawfully come into the Spaniard's possession? But in many of its details, especially in reference to the more calamitous parts, such as the fatalities among the seamen, the consequent prolonged beating about, the past sufferings from obstinate calms, and still continued suffering from thirst;

in all these points, as well as others, Don Benito's story had corroborated not only the wailing ejaculations of the indiscriminate multitude, white and black, but likewise—what seemed impossible to be counterfeit—by the very expression and play of every human feature, which Captain Delano saw. If Don Benito's story was, throughout, an invention, then every soul on board, down to the youngest negress, was his carefully drilled recruit in the plot: an incredible inference. And yet, if there was ground for mistrusting his veracity, that inference was a legitimate one.

But those questions of the Spaniard. There, indeed, one might pause. Did they not seem put with much the same object with which the burglar or assassin, by day-time, reconnoitres the walls of a house? But, with ill purposes, to solicit such information openly of the chief person endangered, and so, in effect, setting him on his guard; how unlikely a procedure was that? Absurd, then, to suppose that those questions had been prompted by evil designs. Thus, the same conduct, which, in this instance, had raised the alarm, served to dispel it. In short, scarce any suspicion or uneasiness, however apparently reasonable at the time, which was not now, with equally apparent reason, dismissed.

At last he began to laugh at his former forebodings; and laugh at the strange ship for, in its aspect, someway siding with them, as it were; and laugh, to, at the odd-looking blacks, particularly those old scissors-grinders, the Ashantees; and those bed-ridden old knitting women, the oakum-pickers; and almost at the dark Spaniard himself, the central hobgoblin of all.

For the rest, whatever in a serious way seemed enigmatical, was now good-naturedly explained away by the thought that, for the most part, the poor invalid scarcely knew what he was about; either sulking in black vapours, or putting idle questions without sense of object. Evidently, for the present, the man was not fit to be intrusted with the ship. On some benevolent plea withdrawing the command from him, Captain Delano would yet have to send her to Conception, in charge of his second mate, a worthy person and good navigator—a plan not more convenient for the *San Dominick* than for Don Benito; for, relieved from all anxiety, keeping wholly to his cabin, the sick man, under the good nursing of his servant, would, probably, by the end of the passage, be in a measure restored to health, and with that he should also be restored to authority.

Such were the American's thoughts. They were tranquilizing. There was a difference between the idea of Don Benito's darkly pre-ordaining Captain Delano's fate, and Captain Delano's lightly arranging Don Benito's. Nevertheless, it was not without something of relief that the good seaman presently perceived his whale-boat in the distance. Its absence had been prolonged by unexpected detention at the sealer's side, as well as its returning trip lengthened by the continual recession of the goal.

The advancing speck was observed by the blacks. Their shouts attracted the attention of Don Benito, who, with a return of courtesy, approaching Captain Delano, expressed satisfaction at the coming of some supplies, slight and temporary as they must necessarily prove.

Captain Delano responded; but while doing so, his attention was drawn to something passing on the deck below: among the crowd climbing the landward bulwarks, anxiously watching the coming boat, two blacks, to all appearances accidentally incommoded by one of the sailors, violently pushed him aside, which the sailor someway resenting, they dashed him to the deck, despite the earnest cries of the oakum-pickers.

"Don Benito," said Captain Delano quickly, "do you see what is going on there? Look!"

But, seized by his cough, the Spaniard staggered, with both hands to his face, on the point of falling. Captain Delano would have supported him, but the servant was more alert, who, with one hand sustaining his master, with the other applied the cordial. Don Benito restored, the black withdrew his support, slipping aside a little, but dutifully remaining within call of a whisper. Such discretion was here evinced as quite wiped away, in the visitor's eyes, any blemish of impropriety which might have attached to the attendant, from the indecorous conferences before mentioned; showing, too, that if the servant were to blame, it might be more the master's fault than his own, since, when left to himself, he could conduct thus well.

His glance called away from the spectacle of disorder to the more pleasing one before him, Captain Delano could not avoid again congratulating his host upon possessing such a servant, who, though perhaps a little too forward now and then, must upon the whole be invaluable to one in the invalid's situation.

"Tell me, Don Benito," he added, with a smile—"I should like to have your man here, myself—what will you take for him? Would fifty doubloons be any object?"

"Master wouldn't part with Babo for a thousand doubloons," murmured the black, overhearing the offer, and taking it in earnest, and, with the strange vanity of a faithful slave, appreciated by his master, scorning to hear so paltry a valuation put upon him by a stranger. But Don Benito, apparently hardly yet completely restored, and again interrupted by his cough, made but some broken reply.

Soon his physical distress became so great, affecting his mind, too, apparently, that, as if to screen the sad spectacle, the servant gently conducted his master below.

Left to himself, the American, to while away the time till his boat should arrive, would have pleasantly accosted some one of the few Spanish seamen he saw; but recalling something that Don Benito had said touching their ill conduct, he refrained; as a shipmaster indisposed to countenance cowardice or unfaithfulness in seamen.

While, with these thoughts, standing with eye directed forward towards that handful of sailors, suddenly he thought that one or two of them returned the glance and with a sort of meaning. He rubbed his eyes, and looked again; but again seemed to see the same thing. Under a new form, but more obscure than any previous one, the old suspicions recurred, but, in the absence of Don Benito, with less of panic than before. Despite the bad account given of the sailors, Captain Delano resolved forthwith to accost one of them. Descending the poop, he made his way through the blacks, his movement drawing a queer cry from the oakum-pickers, prompted by whom, the negroes, twitching each other aside, divided before him; but, as if curious to see what was the object of this deliberate visit to their Ghetto, closing in behind, in tolerable order, followed the white stranger up. His progress thus proclaimed as by mounted kings-at-arms, and escorted as by a Kaffir guard of honour, Captain Delano, assuming a good-humoured, off-handed air, continued to advance; now and then saying a blithe word to the negroes, and his eye curiously surveying the white faces, here and there sparsely mixed in with the blacks, like stray white pawns venturously involved in the ranks of the chess-men opposed.

While thinking which of them to select for his purpose, he chanced to observe a sailor seated on the deck engaged in tarring the strap of a large block, a circle of blacks squatted round him inquisitively eyeing the process.

The mean employment of the man was in contrast with something superior in his figure. His hand, black with continually thrusting it into the tar-pot held for him by a negro, seemed not naturally allied to his face, a face which would have been a very fine one but for its haggardness. Whether this haggardness had aught to do with criminality, could not be determined; since, as intense heat and cold, though unlike, produce like sensations, so innocence and guilt, when, through casual association with mental pain, stamping any visible impress, use one seal—a hacked one.

Not again that this reflection occurred to Captain Delano at the time, charitable man as he was. Rather another idea. Because observing so singular a haggardness combined with a dark eye, averted as in trouble and shame, and then again recalling Don Benito's confessed ill opinion of his crew, insensibly he was operated upon by certain general notions which, while disconnecting pain and abashment from virtue, invariably link them with vice.

If, indeed, there be any wickedness on board this ship, thought Captain Delano, be sure that man there has fouled his hand in it, even as now he fouls it in the pitch. I don't like to accost him. I will speak to this other, this old Jack here on the windlass.

He advanced to an old Barcelona tar, in ragged red breeches and dirty night-cap, cheeks trenched and bronzed, whiskers dense as thorn hedges. Seated between two sleepy-looking Africans, this mariner, like his younger shipmate, was employed upon some rigging—splicing a cable—the sleepy-looking blacks performing the inferior function of holding the outer parts of the ropes for him.

Upon Captain Delano's approach, the man at once hung his head below its previous level; the one necessary for business. It appeared as if he desired to be thought absorbed, with more than common fidelity, in his task. Being addressed, he glanced up, but with what seemed a furtive, diffident air, which sat strangely enough on his weather-beaten visage, much as if a grizzly bear, instead of growling and biting, should simper and cast sheep's eyes. He was asked several questions concerning the voyage—questions purposely referring to several particulars in Don Benito's narrative, not previously corroborated by those impulsive cries greeting the visitor on the first coming on board. The questions were briefly answered, confirming all that remained to be confirmed of the story. The negroes about the windlass joined in with the old sailor; but, as they became talkative, he by degrees became mute, and at length quite glum, seemed morosely unwilling to answer more questions, and yet, all the while, this ursine air was somehow mixed with his sheeepish one.

Despairing of getting into unembarrassed talk with such a centaur, Captain Delano, after glancing round for a more promising countenance, but seeing none, spoke pleasantly to the blacks to make way for him; and so, amid various grins and grimaces, returned to the poop, feeling a little strange at first, he could hardly tell why, but upon the whole with regained confidence in Benito Cereno.

How plainly, thought he, did that old whiskerando yonder betray a consciousness of ill desert. No doubt, when he saw me coming, he dreaded lest I, appraised by his captain of the crew's general misbehaviour, came with sharp words for him, and so down with his head. And yet—and yet, now that I think of it, that very fellow, if I err not, was one of those who seemed so earnestly eyeing me here awhile since. Ah, these currents spin one's head round almost as much as they do the ship. Ha, there now's a pleasant sort of sunny sight; quite sociable, too.

His attention had been drawn back to a slumbering negress, partly disclosed through the lacework of some rigging, lying, with youthful limbs carelessly disposed, under the lee of the bulwarks, like a doe in the shade of a woodland rock. Sprawling at her lapped breasts was her wide-awake fawn, stark naked, its black little body half lifted from the deck, crosswise with its dam's; its hands, like two paws, clambering upon her; its mouth and nose ineffectually rooting to get at the mark; and meantime giving a vexatious half-grunt, blending with the composed snore of the negress.

The uncommon vigour of the child at length roused the mother. She started up, at a distance facing Captain Delano. But as if not at all concerned at the attitude in which she had been caught, delightedly she caught the child up, with maternal transports, covering it with kisses.

There's naked nature, now; pure tenderness and love, thought Captain Delano, well pleased.

This incident prompted him to remark the other negresses more particularly than before. He was gratified with their manners: like most uncivilized women, they seemed at once tender of heart and tough of constitution; equally ready to die for their infants or fight for them. Unsophisticated as leopardesses; loving as doves. Ah! thought

Captain Delano, these, perhaps, are some of the very women whom Ledyard saw in Africa, and gave such a noble account of.

These natural sights somehow insensibly deepened his confidence and ease. At last he looked to see how his boat was getting on; but it was still pretty remote. He turned to see if Don Benito had returned; but he had not.

To change the scene, as well as to please himself with a leisurely observation of the coming boat, stepping over into the mizzen-chains, he clambered his way into the starboard quarter-gallery—one of those abandoned Venetian-looking water-balconies previously mentioned—retreats cut off from the deck. As his foot pressed the half-damp, half-dry sea-mosses matting the place, and a chance phantom cats-paw—an islet of breeze, unheralded, unfollowed—as this ghostly cats-paw came fanning his cheek; as his glance fell upon the row of small, round dead-lights—all closed like coppered eyes of the coffined—and the state-cabin door, once connecting with the gallery, even as the dead-lights had once looked out upon it, but now calked fast like a sarcophagus lid; and to a purple-black, tarred-over panel, threshold, and post; and he bethought him of the time, when that state-cabin and this state-balcony had heard the voices of the Spanish king's officers, and the forms of the Lima viceroy's daughters had perhaps leaned where he stood—as these and other images flitted through his mind, as the cats-paw through the calm, gradually he felt rising a dreamy inquietude, like that of one who alone on the prairie feels unrest from the repose of the noon.

He leaned against the carved balustrade, again looking off toward his boat; but found his eye falling upon the ribbon grass, trailing along the ship's water-line, straight as a border of green box; and parterres of sea-weed, broad ovals and crescents, floating nigh and far, with what seemed long formal alleys between, crossing the terraces of swells, and sweeping round as if leading to the grottoes below. And overhanging all was the balustrade by his arm, which, partly stained with pitch and partly embossed with moss, seemed the charred ruin of some summer-house in a grand garden long running to waste.

Trying to break one charm, he was but becharmed anew. Though upon the wide sea, he seemed in some far inland country; prisoner in some deserted château, left to stare at empty grounds, and peer out at vague roads, where never wagon or wayfarer passed.

But these enchantments were a little disenchanted as his eye fell on the corroded main-chains. Of an ancient style, massy and rusty in link, shackle and bolt, they seemed even more fit for the ship's present business than the one for which she had been built.

Presently he thought something moved nigh the chains. He rubbed his eyes, and looked hard. Groves of rigging were about the chains; and there, peering from behind a great stay, like an Indian from behind a hemlock, a Spanish sailor, a marlingspike in his hand, was seen, who made what seemed an imperfect gesture towards the balcony, but immediately, as if alarmed by some advancing step along the deck within, vanished into the recesses of the hempen forest, like a poacher.

What meant this? Something the man had sought to communicate, unbeknown to any one, even to his captain. Did the secret involve aught unfavourable to his captain? Were those previous misgivings of Captain Delano's about to be verified? Or, in his haunted mood at the moment, had some random, unintentional motion of the man, while busy with the stay, as if repairing it, been mistaken for a significant beckoning?

Not unbewildered, again he gazed off for his boat. But it was temporarily hidden by a rocky spur of the isle. As with some eagerness he bent forward, watching for the first shooting view of its beak, the balustrade gave way before him like charcoal. Had he not clutched an outreaching rope he would have fallen into the sea. The crash, though feeble, and the fall, though hollow, of the rotten fragments, must have been overheard. He glanced up. With sober curiosity peering down upon him was one of the old oakum-pickers, slipped from his perch to an outside boom; while below the old negro, and, invisible to him, reconnoitring from a port-hole like a fox from the mouth of its den, crouched the Spanish sailor again. From something suddenly suggested by the man's air, the mad idea now darted into Captain Delano's mind, that Don Benito's plea of indisposition, in withdrawing below, was but a pretence: that he was engaged there maturing his plot, of which the sailor, by some means gaining an inkling, had a mind to warn the stranger against; incited, it may be, by gratitude for a kind word on first boarding the ship. Was it from foreseeing some possible interference like this, that Don Benito had, beforehand, given such a bad character of his sailors, while praising the negroes; though, indeed, the former seemed as docile as the latter the contrary? The whites, too, by nature, were the shrewder race. A man with some evil design, would he not be likely to speak well of that stupidity which was blind to his depravity, and malign that intelligence from which it might not be hidden? Not unlikely, perhaps. But if the whites had dark secrets concerning Don Benito, could then Don Benito be any way in complicity with the blacks? But they were too stupid. Besides, who ever heard of a white so far a renegade as to apostatize from his very species almost, by leaguing in against it with negroes? These difficulties recalled former ones. Lost in their mazes, Captain Delano, who had now regained the deck, was uneasily advancing along it, when he observed a new face; an aged sailor seated cross-legged near the main hatchway. His skin was shrunk up with wrinkles like a pelican's empty pouch; his hair frosted; his countenance grave and composed. His hands were full of ropes, which he was working into a large knot. Some blacks were about him obligingly dipping the strands for him, here and there, as the exigencies of the operation demanded.

Captain Delano crossed over to him, and stood in silence surveying the knot; his mind, by a not uncongenial transition, passing from its own entanglements to those of the hemp. For intricacy, such a knot he had never seen in an American ship, nor indeed any other. The old man looked like an Egyptian priest, making Gordian knots for the temple of Ammon. The knot seemed a combination of double-bowline-knot, treble-crown-knot, back-handed-well-not, knot-in-and-out-knot, and jamming knot.

At last, puzzled to comprehend the meaning of such a knot, Captain Delano addressed the knotter:—

"What are you knotting there, my man?"

"The knot," was the brief reply, without looking up.

"So it seems; but what is it for?"

"For some one else to undo," muttered back the old man, plying his fingers harder than ever, the knot being now nearly completed.

While Captain Delano stood watching him, suddenly the old man threw the knot towards him, saying in broken English—the first heard in the ship—something to this effect: "Undo it, cut it, quick." It was said lowly, but with such condensation of rapidity, that the long, slow words in Spanish, which had preceded and followed, almost operated as covers to the brief English between.

For a moment, knot in hand, and knot in head, Captain Delano stood mute; while, without further heeding him, the old man was now intent upon other ropes. Presently there was a slight stir behind Captain Delano. Turning, he saw the chained negro, Atufal, standing quietly there. The next moment the old sailor rose, muttering, and, followed by his subordinate negroes, removed to the forward part of the ship, where in the crowd he disappearred.

An elderly negro, in a clout like an infant's, and with a pepper-and-salt head, and a kind of attorney air, now approached Captain Delano. In tolerable Spanish, and with a good-natured, knowing wink, he informed him that the old knotter was simple-witted, but harmless; often playing his odd tricks. The negro concluded by begging the knot, for of course the stranger would not care to be troubled with it. Unconsciously, it was handed to him. With a sort of congé, the negro received it, and, turning his back, ferreted into it like a detective custom-house officer after smuggled laces. Soon, with some African word, equivalent to pshaw, he tossed the knot overboard.

All this is very queer now, thought Captain Delano, with a qualmish sort of emotion; but, as one feeling incipient sea-sickness, he strove, by ignoring the symptoms, to get rid of the malady. Once more he looked off for his boat. To his delight, it was now again in view, leaving the rocky spur astern.

The sensation here experienced, after at first relieving his uneasiness, with unforeseen efficacy soon began to remove it. The less distant sight of that well-known boat—showing it, not as before, half blended with the haze, but with outline defined, so that its individuality, like a man's, was manifest; that boat, *Rover* by name, which, though now in strange seas, had often pressed the beach of Captain Delano's home, and, brought to its threshold for repairs, had familiarly lain there, as a Newfoundland dog; the sight of that household boat evoked a thousand trustful associations, which, contrasted with previous suspicions, filled him not only with lightsome confidence, but somehow with half-humorous self-reproaches at his former lack of it.

"What, I, Amasa Delano—Jack of the Beach, as they called me when a lad—I, Amasa; the same that, duck-satchel in hand, used to paddle along the water-side to the school-house made from the old hulk—I, little Jack of the Beach, that used to go berrying with cousin Nat and the rest; I to be murdered here at the ends of the earth, on board a haunted pirate-ship by a horrible Spaniard? Too nonsensical to think of! Who would murder Amasa Delano? His conscience is clean. There is someone above. Fie, fie, Jack of the Beach! you are a child indeed; a child of the second childhood, old boy; you are beginning to dote and drool, I'm afraid."

Light of heart and foot, he stepped aft, and there was met by Don Benito's servant, who, with a pleasing expression, responsive to his own present feelings, informed him that his master had recovered from the effects of his coughing fit, and had just ordered him to go present his compliments to his good guest, Don Amasa, and say that he (Don Benito) would soon have the happiness to rejoin him.

There now, do you mark that? again thought Captain Delano, walking the poop. What a donkey I was. This kind gentleman who here sends me his kind compliments, he, but ten minutes ago, dark-lantern in hand, was dodging round some old grindstone in the hold, sharpening a hatchet for me, I thought. Well, well; these long calms have a morbid effect on the mind, I've often heard, though I never believed it before. Ha! glancing towards the boat; there's *Rover*; good dog; a white bone in her mouth. A pretty big bone though, seems to me.—What? Yes, she has fallen afoul of the bubbling tide-rip there. It sets her the other way, too, for the time. Patience.

It was now about noon, though, from the greyness of everything, it seemed to be getting towards dusk.

The calm was confirmed. In the far distance away from the influence of land, the leaden ocean seemed laid out and leaded up, its course finished, soul gone, defunct. But the current from landward, where the ship was, increased; silently sweeping her further and further towards the tranced waters beyond.

Still, from his knowledge of those latitudes, cherishing hopes of a breeze, and a fair and fresh one, at any moment, Captain Delano, despite present prospects, buoyantly counted upon bringing the *San Dominick* safely to anchor ere night. The distance swept over was nothing; since, with a good wind, ten minutes' sailing would retrace more than sixty minutes' drifting. Meantime, one moment turning to mark *Rover* fighting the tide-rip, and the next to see Don Benito approaching, he continued walking the poop.

Gradually he felt a vexation arising from the delay of his boat; this soon merged into uneasiness; and at last—his eye falling continually, as from a stage-box into the pit, upon the strange crowd before and below him, and, by and by, recognizing there the face—now composed to indifference—of the Spanish sailor who had seemed to beckon from the main-chains—something of his old trepidations returned.

Ah, thought he—gravely enough—this is like the ague: because it went off, it follows not that it won't come back.

Though ashamed of the relapse, he could not altogether subdue it; and so, exerting his good nature to the utmost, insensibly he came to a compromise.

Yes, this is a strange craft; a strange history, too, and strange folks on board. But—nothing more.

By way of keeping his mind out of mischief till the boat should arrive, he tried to occupy it with turning over and over, in a purely speculative sort of way, some lesser peculiarities of the captain and crew. Among others, four curious points recurred:

First, the affair of the Spanish lad assailed with a knife by the slave boy; an act winked at by Don Benito. Second, the tyranny in Don Benito's treatment of Atufal, the black; as if a child should lead a bull of the Nile by the ring in his nose. Third, the trampling of the sailor by the two negroes; a piece of insolence passed over without so much as a reprimand. Fourth, the cringing submission to their master, of all the ship's underlings, mostly blacks; as if by the least inadvertence they feared to draw down his despotic displeasure.

Coupling these points, they seemed somewhat contradictory. But what then, though Captain Delano, glancing towards his now nearing boat—what then? Why, Don Benito is a very capricious commander. But he is not the first of the sort I have seen; though it's true he rather exceeds any other. But as a nation—continued he in his reveries—these Spaniards are all an odd set; the very word Spaniard has a curious, conspirator, Guy-Fawkish twang to it. And yet, I dare say, Spaniards in the main are as good folks as any in Duxbury, Massachusetts. Ah, good! At last *Rover* has come.

As, with its welcome freight, the boat touched the side, the oakum-pickers, with venerable gestures, sought to restrain the blacks, who, at the sight of three gurried water casks in its bottom, and a pile of wilted pumpkins in its bow, hung over the bulwarks in disorderly raptures.

Don Benito, with his servant, now appeared; his coming, perhaps, hastened by hearing the noise. Of him Captain Delano sought permission to serve out the water, so that all might share alike, and none injure themselves by unfair excess. But sensible, and, on Don Benito's account, kind as this offer was, it was received with what seemed impatience; as if aware that he lacked energy as a commander, Don Benito, with the true jealousy of weakness, resented as an affront any interference. So, at least, Captain Delano inferred.

In another moment the casks were being hoisted in, when some of the eager negroes accidently jostled Captain Delano, where he stood by the gangway; so that, unmindful of Don Benito, yielding to the impulse of the moment, with good-natured authority he bade the blacks stand back; to enforce his words making use of a half-mirthful, half-menacing gesture. Instantly the blacks paused, just where they were, each negro and negress suspended in his or her posture, exactly as the word had found them—for a few seconds continuing so—while, as between the responsive posts of a telegraph, an unknown syllable ran from man to man among the perched oakum-pickers. While the visitor's attention was fixed by this scene, suddenly the hatchet-polishers half rose, and a rapid cry came from Don Benito.

Thinking that at the signal of the Spaniard he was about to be massacred, Captain Delano would have sprung for his boat, but paused, as the oakum-pickers, dropping down into the crowd with earnest exclamations, forced every white and every negro back, at the same moment, with gestures friendly and familiar, almost jocose, bidding him, in substance, not be a fool. Simultaneously the hatchet-polishers resumed their seats, quietly as so many tailers, and at once, as if nothing had happened, the work of hoisting in the casks was resumed, whites and blacks singing at the tackle.

Captain Delano glanced towards Don Benito. As he saw his meagre form in the act of recovering itself from reclining in the servant's arms, into which the agitated invalid had fallen, he could not but marvel at the panic by which himself had been surprised, on the darting supposition that such a commander, who, upon a legitimate occasion, so trivial, too, as it now appeared, could lose all self-command, was, with energetic iniquity, going to bring about his murder.

The casks being on deck, Captain Delano was handed a number of jars and cups by one of the steward's aids, who, in the name of his captain, entreated him to do as he had proposed—dole out the water. He complied, with republican impartiality as to this republican element, which always seeks one level, serving the oldest white no better than the youngest black; excepting, indeed, poor Don Benito, whose condition, if not rank, demanded an extra allowance. To him, in the first place, Captain Delano presented a fair pitcher of the fluid; but, thirsting as he was for it, the Spaniard quaffed not a drop until after several grave bows and salutes. A reciprocation of courtesies which the sight-loving Africans hailed with clapping of hands.

Two of the less wilted pumpkins being reserved for the cabin table, the residue were minced up on the spot for the general regalement. But the soft bread, sugar, and bottled cider, Captain Delano would have given the whites alone, and in chief Don Benito; but the latter objected; which disinterestedness not a little pleased the American; and so mouthfuls all around were given alike to whites and blacks; excepting one bottle of cider, which Babo insisted upon setting aside for his master.

Here it may be observed that as, on the first visit of the boat, the American had not permitted his men to board the ship, neither did he now; being unwilling to add to the confusion of the decks.

Not uninfluenced by the peculiar good-humor at present prevailing, and for the time oblivious of any but benevolent thoughts, Captain Delano, who, from recent indications, counted upon a breeze within an hour or two at furthest, dispatched the boat back to the sealer, with orders for all the hands that could be spared immediately to set about rafting casks to the watering-place and filling them. Likewise he bade word be carried to his chief officer, that if, against present expectation, the ship was not brought to anchor by sunset, he need be under no concern; for as there was to be a full moon that night, he (Captain Delano) would remain on board ready to play the pilot, come the wind soon or late.

As the two captains stood together, observing the departing boat—the servant, as it happened, having just spied a spot on his master's velvet sleeve, and silently engaged rubbing it out—the American expressed his regrets that the *San Dominick* had no boats; none, at least, but the unseaworthy old hulk of the long-boat, which, warped as a camel's skeleton in the desert, and almost as bleached, lay pot-wise inverted amid-ships, one side a little tipped, furnishing a subterraneous sort of den for family groups of the blacks, mostly women and small children; who, squatting on old mats below, or perched above in the dark dome, on the elevated seats, were descried, some distance within, like a social circle of bats, sheltering in some friendly cave; at intervals, ebon flights of naked boys and girls, three or four years old, darting in and out of the den's mouth.

"Had you three or four boats now, Don Benito," said Captain Delano, "I think that, by tugging at the oars, your negroes here might help along matters some. Did you sail from port without boats, Don Benito?"

"They were stove in the gales, Señor."

"That was bad. Many men, too, you lost then. Boats and men. Those must have been hard gales, Don Benito."

"Past all speech," cringed the Spaniard.

"Tell me, Don Benito," continued his companion with increased interest, "tell me, were these gales immediately off the pitch of Cape Horn?"

"Cape Horn?—who spoke of Cape Horn?"

"Yourself did, when giving me an account of your voyage," answered Captain Delano, with almost equal astonishment at this eating of his own words, even as he ever seemed eating his own heart, on the part of the Spaniard. "You yourself, Don Benito, spoke of Cape Horn," he emphatically repeated.

The Spaniard turned, in a sort of stooping posture, pausing an instant, as one about to make a plunging exchange of elements, as from air to water.

At this moment a messenger-boy, a white, hurried by, in the regular performance of his function carrying the last expired half-hour forward to the forecastle, from the cabin time-piece, to have it struck at the ship's large bell.

"Master," said the servant, discontinuing his work on the coat sleeve, and addressing the rapt Spaniard with a sort of timid apprehensiveness, as one charged with a duty, the discharge of which, it was foreseen, would prove irksome to the very person who had imposed it, and for whose benefit it was intended, "master told me never mind where he was, or how engaged, always to remind him, to a minute, when shaving-time comes. Miguel has gone to strike the half-hour afternoon. It is *now*, master. Will master go into the cuddy?"

"Ah—yes," answered the Spaniard, starting, as from dreams into realities; then turning upon Captain Delano, he said that ere long he would resume the conversation.

"Then if master means to talk more to Don Amasa," said the servant, "why not let Don Amasa sit by master in the cuddy, and master can talk, and Don Amasa can listen, while Babo here lathers and strops."

"Yes," said Captain Delano, not unpleased with this sociable plan, "yes, Don Benito, unless you had rather not, I will go with you."

"Be it so , Señor."

As the three passed aft, the American could not but think it another strange instance of his host's capriciousness, this being shaved with such uncommon punctuality in the middle of the day. But he deemed it more than likely that the servant's anxious fidelity had something to do with the matter; inasmuch as the timely interruption served to rally his master from the mood which had evidently been coming upon him.

The place called the cuddy was a light deck-cabin formed by the poop, a sort of attic to the large cabin below. Part of it had formerly been the quarters of the officers; but since their death all the partitionings had been thrown down, and the whole interior converted into one spacious and airy marine hall; for absence of fine furniture and picturesque disarray of odd appurtenances, somewhat answering to the wide, cluttered hall of some eccentric bachelor-squire in the country, who hangs his shooting-jacket and tobacco-pouch on deer antlers, and keeps his fishing-rod, tongs, and walking-stick in the same corner.

The similitude was heightened, if not originally suggested, by glimpses of the surrounding sea; since, in one aspect, the country and the ocean seem cousins-german.

The floor of the cuddy was matted. Overhead, four or five old muskets were stuck into horizontal holes along the beams. On one side was a claw-footed old table lashed to the deck; a thumbed missal on it, and over it a small, meagre crucifix attached to the bulk-head. Under the table lay a dented cutlass or two, with a hacked harpoon, among some melancholy old rigging, like a heap of poor friars' girdles. There were also two long, sharp, ribbed settees of Malacca cane, black with age, and uncomfortable to look at as inquisitors' racks, with a large, misshapen arm-chair, which, furnished with a rude barber's crotch at the back, working with a screw, seemed some grotesque engine of torment. A flag-locker was in one corner, open, exposing various coloured bunting, some rolled up, others half unrolled, still others tumbled. Opposite was a cumbrous washstand, of black mahogany, all of one block, with a pedestal, like a font, and over it a railed shelf, containing combs, brushes, and other implements of the toilet. A torn hammock of stained grass swung near; the sheets tossed, and the pillow wrinkled up like a brow, as if whoever slept here slept but illy, with alternate visitations of sad thoughts and bad dreams.

The further extremity of the cuddy, overhanging the ship's stern, was pierced with three openings, windows or port-holes, according as men or cannon might peer, socially or unsocially, out of them. At present neither men nor cannon were seen, though huge ring-bolts and other rusty iron fixtures of the woodwork hinted of twenty-four-pounders.

Glancing towards the hammock as he entered, Captain Delano said, "You sleep here, Don Benito?"

"Yes, Señor, since we got into mild weather."

"This seems a sort of dormitory, sitting-room, sail-loft, chapel, armoury, and private closet altogether, Don Benito," added Captain Delano, looking round.

"Yes, Señor; events have not been favourable to much order in my arrangements."

Here the servant, napkin on arm, made a motion as if waiting his master's good pleasure. Don Benito signified his readiness, when, seating him in the Malacca arm-chair, and for the guest's convenience drawing opposite one of the settees, the servant commenced operations by throwing back his master's collar and loosening his cravat.

There is something in the negro which, in a peculiar way, fits him for avocations about one's person. Most negroes are natural valets and hair-dressers; taking to the comb and brush congenially as to the castanets, and flourishing them apparently with almost equal satisfaction. There is, too, a smooth tact about them in this employment, with a marvellous, noiseless, gliding briskness, not ungraceful in its way, singularly pleasing to behold, and still more so to be the manipulated subject of. And above all is the great gift of good-humour. Not the mere grin or laugh is here meant. Those were unsuitable. But a certain easy cheerfulness, harmonious in every glance and gesture; as though God had set the whole negro to some pleasant tune.

When to this is added the docility arising from the unaspiring contentment of a limited mind, and that susceptibility of blind attachment sometimes inhering in indisputable inferiors, one readily perceives why those hypochondriacs, Johnson and Byron—it may be, something like the hypochondriac Benito Cereno—took to their hearts, almost to the exclusion of the entire white race, their serving men, the negroes, Barber and Fletcher. But if there be that in the negro which exempts him from the inflicted sourness of the morbid or cynical mind, how, in his most prepossessing aspects, must he appear to a benevolent one? When at ease with respect to exterior things, Captain Delano's nature was not only benign, but familiarly and humorously so. At home, he had often taken rare satisfaction in sitting in his door, watching some free man of colour at his work or play. If on a voyage he chanced to have a black sailor, invariably he was on chatty and half-gamesome terms with him. In fact, like most men of a good, blithe heart, Captain Delano took to negroes, not philanthropically, but genially, just as other men to Newfoundland dogs.

Hitherto, the circumstances in which he found the *San Dominick* had repressed the tendency. But in the cuddy, relieved from his former uneasiness, and, for various reasons, more sociably inclined than at any previous period of the day, and seeing the coloured servant, napkin on arm, so debonair about his master, in a business so familiar as that of shaving, too, all his old weakness for negroes returned.

Among other things, he was amused with an odd instance of the African love of bright colours and fine shows, in the black's informally taking from the flag-locker a great piece of bunting of all hues, and lavishly tucking it under his master's chin for an apron.

The mode of shaving among the Spaniards is a little different from what it is with other nations. They have a basin, specifically called a barber's basin, which on one side is scooped out, so as accurately to receive the chin, against which it is closely held in lathering; which is done, not with a brush, but with soap dipped in the water of the basin and rubbed on the face.

In the present instance salt-water was used for lack of better; and the parts lathered were only the upper lip, and low down under the throat, all the rest being cultivated beard.

The preliminaries being somewhat novel to Captain Delano, he sat curiously eyeing them, so that no conversation took place, nor, for the present, did Don Benito appear disposed to renew any.

Setting down his basin, the negro searched among the razors, as for the sharpest, and having found it, gave it an additional edge by expertly stropping it on the firm, smooth, oily skin of his open palm; he then made a gesture as if to begin, but midway stood suspended for an instant, one hand elevating the razor, the other professionally dabbling among the bubbling suds on the Spaniard's lank neck. Not unaffected by the close sight of the gleaming steel, Don Benito nervously shuddered; his usual ghastliness was heightened by the lather, which lather, again, was intensified in its hue by the contrasting sootiness of the negro's body. Altogether the scene was somewhat peculiar, at least to Captain Delano, nor, as he saw the two thus postured, could he resist the vagary, that in the black he saw a headsman, and in the white a man at the block. But this was one of those antic conceits, appearing and vanishing in a breath, from which, perhaps, the best regulated mind is not always free.

Meantime the agitation of the Spaniard had a little loosened the bunting from around him, so that one broad fold swept curtain-like over the chair-arm to the floor, revealing, amid a profusion of armorial bars and ground-colours—black, blue, and yellow—a closed castle in a blood-red field diagonal with a lion rampant in a white.

"The castle and the lion," exclaimed Captain Delano—"why, Don Benito, this is the flag of Spain you use here. It's well it's only I, and not the King, that sees this," he added, with a smile, "but,"—turning towards the black—"it's all one, I suppose, so the colours be gay," which playful remark did not fail somewhat to tickle the negro.

"Now, master," he said, readjusting the flag, and pressing the head gently further back into the crotch of the chair; "now, master," and the steel glanced nigh that throat.

Again Don Benito faintly shuddered.

"You must not shake so, master. See, Don Amasa, master always shakes when I shave him. And yet master knows I never yet have drawn blood, though it's true, if master will shake so, I may some of these times. Now, master," he continued. "And now, Don Amasa, please go on with your talk about the gale, and all that; master can hear, and, between times, master can answer."

"Ah yes, these gales," said Captain Delano; "but the more I think of your voyage, Don Benito, the more I wonder, not at the gales, terrible as they must have been, but at the disastrous interval following them. For here, by your account, have you been these two months and more getting from Cape Horn to Santa Maria, a distance which I myself, with a good wind, have sailed in a few days. True, you had calms, and long ones, but to be becalmed for two months, that is, at least, unusual. Why, Don Benito, had almost any other gentleman told me such a story, I should have been half disposed to a little incredulity."

Here an involuntary expression came over the Spaniard, similiar to that just before on the deck, and whether it was the start he gave, or a sudden gawky roll of the hull in the calm, or a momentary unsteadiness of the servant's hand, however it was, just then the razor drew blood, spots of which stained the creamy lather under the throat: immediately the black barber drew back his steel, and, remaining in his professional attitude, back to Captain Delano, and face to Don Benito, held up the trickling razor, saying, with a sort of half-humorous sorrow, "See, master—you shook so—here's Babo's first blood."

No sword drawn before James the First of England, no assassination in that timid King's presence, could have produced a more terrified aspect than was now presented by Don Benito.

Poor fellow, thought Captain Delano, so nervous he can't even bear the sight of barber's blood; and this unstrung, sick man, is it credible that I should have imagined he meant to spill all my blood, who can't endure the sight of one little drop of his own? Surely, Amasa Delano, you have been beside yourself this day. Tell it not when you get home, sappy Amasa. Well, he looks like a murderer, doesn't he? More like as if himself were to be done for. Well, well, this day's experience shall be a good lesson.

Meantime, while these things were running through the honest seaman's mind, the servant had taken the napkin from under his arm, and to Don Benito had said—"But answer Don Amasa, please, master, while I wipe this ugly stuff off the razor, and strop it again."

As he said the words, his face was turned half round, so as to be alike visible to the Spaniard and the American, and seemed, by its expression, to hint, that he was desirous, by getting his master to go on with the conversation, considerately to withdraw his attention from the recent annoying accident. As if glad to snatch the offered relief, Don Benito resumed, rehearsing to Captain Delano, that not only were the calms of unusual duration, but the ship had fallen in with obstinate currents; and other things he added, some of which were but repetitions of former statements, to explain how it came to pass that the passage from Cape Horn to Santa Maria had been so exceedingly long; now and then mingling with his words, incidental praises, less qualified than before, to the blacks, for their general good conduct. These particulars were not given consecutively, the servant, at convenient times, using his razor, and so, between the intervals of shaving, the story and panegyric went on with more than usual huskiness.

To Captain Delano's imagination, now again not wholly at rest, there was something so hollow in the Spaniard's manner, with apparently some reciprocal hollowness in the servant's dusky comment of silence, that the idea flashed across him, that possibly master and man, for some unknown purpose, were acting out, both in word and deed, nay, to the very tremor of Don Benito's limbs, some juggling play before him. Neither did the suspicion of collusion lack apparent support, from the fact of those whispered conferences before mentioned. But then, what could be the object of enacting this play of the barber before him? At last, regarding the notion as a whimsy, insensibly suggested, perhaps, by the theatrical aspect of Don Benito in his harlequin ensign, Captain Delano speedily banished it.

The shaving over, the servant bestirred himself with a small bottle of scented waters, pouring a few drops on the head, and then diligently rubbing; the vehemence of the exercise causing the muscles of his face to twitch rather strangely.

His next operation was with comb, scissors, and brush; going round and round, smoothing a curl here, clipping an unruly whisker-hair there, giving a graceful sweep to the temple-lock, with other impromptu touches evincing the hand of a master; while, like any resigned gentleman in barber's hands, Don Benito bore all, much less uneasily, at least, than he had done the razoring; indeed, he sat so pale and rigid now, that the negro seemed a Nubian sculptor finishing off a white statue-head.

All being over at last, the standard of Spain removed, tumbled up, and tossed back into the flag-locker, the negro's warm breath blowing away any stray hair which might have lodged down his master's neck; collar and cravat readjusted; a speck of lint whisked off the velvet lapel; all this being done, backing off a little space, and pausing with an expression of subdued self-complacency, the servant for a moment surveyed his master, as, in toilet at least, the creature of his own tasteful hands.

Captain Delano playfully complimented him upon his achievement; at the same time congratulating Don Benito.

But neither sweet waters, nor shampooing, nor fidelity, nor sociality, delighted the Spaniard. Seeing him relapsing into forbidding gloom, and still remaining seated, Captain Delano, thinking that his presence was undesired just then, withdrew, on pretence of seeing whether, as he had prophesied, any signs of a breeze were visible.

Walking forward to the main-mast, he stood awhile thinking over the scene, and not without some undefined misgivings, when he heard a noise near the cuddy, and turning, saw the negro, his hand to his cheek. Advancing, Captain Delano perceived that the cheek was bleeding. He was about to ask the cause, when the negro's wailing soliloquy enlightened him.

"Ah, when will master get better from his sickness; only the sour heart that sour sickness breeds made him serve Babo so; cutting Babo with the razor, because, only by accident, Babo had given master one little scratch; and for the first time in so many a day, too. Ah, ah, ah," holding his hand to his face.

It is possible, thought Captain Delano; was it to wreak in private his Spanish spite against this poor friend of his, that Don Benito, by his sullen manner, impelled me to withdraw? Ah, this slavery breeds ugly passions in man.—Poor fellow!

He was about to speak in sympathy to the negro, but with a timid reluctance he now re-entered the cuddy.

Presently master and man came forth; Don Benito leaning on his servant as if nothing had happened.

But a sort of love-quarrel, after all, thought Captain Delano.

He accosted Don Benito, and they slowly walked together. They had gone but a few paces, when the steward—a tall, rajah-looking mulatto, orientally set off with a pagoda turban formed by three or four Madras handkerchiefs wound about his head, tier on tier—approaching with a salaam, announced lunch in the cabin.

On their way thither, the two captains were preceded by the mulatto, who, turning round as he advanced, with continual smiles and bows, ushered them on, a display of elegance which quite completed the insignificance of the small bare-headed Babo, who, as if not unconscious of inferiority, eyed askance the graceful steward. But in part, Captain Delano imputed his jealous watchfulness to that peculiar feeling which the full-blooded African entertains from the adulterated one. As for the steward, his manner, if not bespeaking much dignity of self-respect, yet evidenced his extreme desire to please; which is doubly meritorious, as at once Christian and Chesterfieldian.

Captain Delano observed with interest that while the complexion of the mulatto was hybrid, his physiognomy was European—classically so.

"Don Benito," whispered he, "I am glad to see this usher-of-the-golden-rod of yours; the sight refutes an ugly remark once made to me by a Barbados planter; that when a mulatto has a regular European face, look out for him; he is a devil. But see, your steward here has features more regular than King George's of England; and yet there he nods, and bows, and smiles; a king, indeed—the king of kind hearts and polite fellows. What a pleasant voice he has, too."

"He has, Señor."

"But tell me, has he not, so far as you have known him, always proved a good, worthy fellow?" said Captain Delano, pausing, while with a final genuflexion the steward disappeared into the cabin; "come, for the reason just mentioned, I am curious to know."

"Francesco is a good man," sort of sluggishly responded Don Benito, like a phlegmatic appreciator, who would neither find fault nor flatter.

"Ah, I thought so. For it were strange, indeed, and not very creditable to us white-skins, if a little of our blood mixed with the African's, should, far from improving the latter's quality, have the sad effect of pouring vitriolic acid into black broth; improving the hue, perhaps, but not the wholesomeness."

"Doubtless, doubtless, Señor, but,"—glancing at Babo—"not to speak of negroes, your planter's remark I have heard applied to the Spanish and Indian intermixtures in our provinces. But I know nothing about the matter," he listlessly added.

And here they entered the cabin.

The lunch was a frugal one. Some of Captain Delano's fresh fish and pumpkins, biscuit and salt beef, the reserved bottle of cider, and the *San Dominick's* last bottle of Canary.

As they entered, Francesco, with two or three coloured aids, was hovering over the table giving the last adjustments. Upon perceiving their master they withdrew, Francesco making a smiling congé, and the Spaniard, without condescending to notice it, fastidiously remarking to his companion that he relished not superfluous attendance.

Without companions, host and guest sat down, like a childless married couple, at opposite ends of the table, Don Benito waving Captain Delano to his place, and, weak as he was, insisting upon that gentleman being seated before himself.

The negro placed a rug under Don Benito's feet, and a cushion behind his back, and then stood behind, not his master's chair, but Captain Delano's. At first, this a little surprised the latter. But it was soon evident that, in taking his position, the black was still true to his master; since by facing him he could the more readily anticipate his slightest want.

"This is an uncommonly intelligent fellow of yours, Don Benito," whispered Captain Delano across the table.

"You say true, Señor."

During the repast, the guest again reverted to parts of Don Benito's story, begging further particulars here and there. He inquired how it was that the scurvy and fever should have committed such wholesale havoc upon the whites, while destroying less than half the blacks. As if this question reproduced the whole scenes of plague before the Spaniard's eyes, miserably reminding him of his solitude in a cabin where before he had had so many friends and officers round him, his hand shook, his face became hueless, broken words escaped; but directly the sane memory of the past seemed replaced by insane terrors of the present. With starting eyes he stared before him at vacancy. For nothing was to be seen but the hand of his servant pushing the Canary over towards him. At length a few sips served partially to restore him. He made random reference to the different constitution of races, enabling one to offer more resistance to certain maladies than another. The thought was new to his companion.

Presently Captain Delano, intending to say something to his host concerning the pecuniary part of the business he had undertaken for him, especially—since he was strictly accountable to his owners—with reference to the new suite of sails, and other things of that sort; and naturally preferring to conduct such affairs in private, was desirous that the servant should withdraw; imagining that Don Benito for a few minutes could dispense with his attendance. He, however, waited awhile; thinking that, as the conversation proceeded, Don Benito, without being prompted, would perceive the propriety of the step.

But it was otherwise. At last catching his host's eye, Captain Delano, with a slight backward gesture of his thumb, whispered, "Don Benito, pardon me, but there is an interference with the full expression of what I have to say to you."

Upon this the Spaniard changed countenance; which was imputed to his resenting the hint, as in some way a reflection upon his servant. After a moment's pause, he assured his guest that the black's remaining with them could be of no disservice; because since losing his officers he had made Babo (whose original office, it now appeared, had been captain of the slaves) not only his constant attendant and companion, but in all things his confidant.

After this, nothing more could be said; though, indeed, Captain Delano could hardly avoid some little tinge of irritation upon being left ungratified in so inconsiderable a wish, by one, too, for whom he intended such solid services. But it is only his querulousness, thought he; and so filling his glass he proceeded to business.

The price of the sails and other matters were fixed upon. But while this was being done, the American observed that, though his original offer of assistance had been hailed with hectic animation, yet now, when it was reduced to a business transaction, indifference and apathy were betrayed. Don Benito, in fact, appeared to submit to hearing the details more out of regard to common propriety, than from any impression that weighty benefit to himself and his voyage was involved.

Soon, his manner became still more reserved. The effort was vain to seek to draw him into social talk. Gnawed by his splenetic mood, he sat twitching his beard, while to little purpose the hand of his servant, mute as that on the wall, slowly pushed over the Canary.

Lunch being over, they sat down on the cushioned transom; the servant placing a pillow behind his master. The long continuance of the calm had now affected the atmosphere. Don Benito sighed heavily, as if for breath.

"Why not adjourn to the cuddy?" said Captain Delano. "There is more air there." But the host sat silent and motionless.

Meantime his servant knelt before him, with a large fan of feathers. And Francesco coming in on tiptoes, handed the negro a little cup of aromatic waters, with which at intervals he chafed his master's brow; smoothing the hair along the temples as a nurse does a child's. He spoke no word. He only rested his eye on his master's, as if, amid all Don Benito's distress, a little to refresh his spirit by the silent sight of fidelity.

Presently the ship's bell sounded two o'clock; and through the cabin windows a slight rippling of the sea was discerned; and from the desired direction.

"There," exclaimed Captain Delano, "I told you so, Don Benito, look!"

He had risen to his feet, speaking in a very animated tone, with a view the more to rouse his companion. But though the crimson curtain of the stern-window near him that moment fluttered against his pale cheek, Don Benito seemed to have even less welcome for the breeze than the calm.

Poor fellow, thought Captain Delano, bitter experience has taught him that one ripple does not make a wind, any more than one swallow a summer. But he is mistaken for once. I will get his ship in for him, and prove it.

Briefly alluding to his weak condition, he urged his host to remain quietly where he was, since he (Captain Delano) would with pleasure take upon himself the responsibility of making the best use of the wind.

Upon gaining the deck, Captain Delano started at the unexpected figure of Atufal, monumentally fixed at the threshold, like one of those sculptured porters of black marble guarding the porches of Egyptian tombs.

But this time the start was, perhaps, purely physical. Atufal's presence, singularly attesting docility even in sullenness, was contrasted with that of the hatchet-polishers, who in patience evinced their industry; while both spectacles showed, that lax as Don Benito's general authority might be, still, whenever he chose to exert it, no man so savage or colossal but must, more or less, bow.

Snatching a trumpet which hung from the bulwarks, with a free step Captain Delano advanced to the forward edge of the poop, issuing his orders in his best Spanish. The few sailors and many negroes, all equally pleased, obediently set about heading the ship towards the harbour.

While giving some directions about setting a lower stu'n-sail, suddenly Captain Delano heard a voice faithfully repeating his orders. Turning, he saw Babo, now for the time acting, under the pilot, his original part of captain of the slaves. This assistance proved valuable. Tattered sails and warped yards were soon brought into some trim. And no brace or halyard was pulled but to the blithe songs of the inspirited negroes.

Good fellows, thought Captain Delano, a little training would make fine sailors of them. Why see, the very women pull and sing too. These must be some of those Ashantee negresses that make such capital soldiers, I've heard. But who's at the helm? I must have a good hand there.

He went to see.

The *San Dominick* steered with a cumbrous tiller, with large horizontal pulleys attached. At each pulley-end stood a subordinate black, and between them, at the tiller-head, the responsible post, a Spanish seaman, whose countenance evinced his due share in the general hopefulness and confidence at the coming of the breeze.

He proved the same man who had behaved with so shame-faced an air on the windlass.

"Ah,—it is you, my man," exclaimed Captain Delano—"well, no more sheep's eyes now;—look straight forward and keep the ship so. Good hand, I trust? And want to get into the harbour, don't you?"

The man assented with an inward chuckle, grasping the tiller-head firmly. Upon this, unperceived by the American, the two blacks eyed the sailor intently.

Finding all right at the helm, the pilot went forward to the forecastle, to see how matters stood there.

The ship now had way enough to breast the current. With the approach of evening, the breeze would be sure to freshen.

Having done all that was needed for the present, Captain Delano, giving his last orders to the sailors, turned aft to report affairs to Don Benito in the cabin; perhaps additionally incited to rejoin him by the hope of snatching a moment's private chat while the servant was engaged upon deck.

From opposite sides, there were, beneath the poop, two approaches to the cabin; one further forward than the other, and consequently communicating with a longer passage. Marking the servant still above, Captain Delano, taking the nighest entrance—the one last named, and at whose porch Atufal still stood—hurried on his way, till, arrived at the cabin threshold, he paused an instant, a little to recover from his eagerness. Then, with the words of his intended business upon his lips, he entered. As he advanced toward the seated Spaniard, he heard another footstep, keeping time with his. From the opposite door, a salver in hand, the servant was likewise advancing.

"Confound the faithful fellow," thought Captain Delano; "what a vexatious coincidence."

Possibly, the vexation might have been something different, were it not for the brisk confidence inspired by the breeze. But even as it was, he felt a slight twinge, from a sudden indefinite association in his mind of Babo with Atufal.

"Don Benito," said he, "I give you joy; the breeze will hold, and will increase. By the way, your tall man and time-piece, Atufal, stands without. By your order, of course?"

Don Benito recoiled, as if at some bland satirical touch, delivered with such adroit garnish of apparent good breeding as to present no handle for retort.

He is like one flayed alive, thought Captain Delano; where may one touch him without causing a shrink?

The servant moved before his master, adjusting a cushion; recalled to civility, the Spaniard stiffly replied: "You are right. The slave appears where you saw him, according to my command; which is, that if at the given hour I am below, he must take his stand and abide my coming."

"Ah now, pardon me, but that is treating the poor fellow like an ex-king indeed. Ah, Don Benito," smiling, "for all the licence you permit in some things, I fear lest, at bottom, you are a bitter hard master."

Again Don Benito shrank; and this time, as the good sailor thought, from a genuine twinge of his conscience.

Again conversation became constrained. In vain Captain Delano called attention to the now perceptible motion of the keel gently cleaving the sea; with lack-lustre eye, Don Benito returned words few and reserved.

By and by, the wind having steadily risen, and still blowing right into the harbour, bore the *San Dominick* swiftly on. Rounding a point of land, the sealer at distance came into open view.

Meantime Captain Delano had again repaired to the deck, remaining there some time. Having at last altered the ship's course, so as to give the reef a wide berth, he returned for a few moments below.

I will cheer up my poor friend, this time, thought he.

"Better and better, Don Benito," he cried as he blithely re-entered: "there will soon be an end to your cares, at least for a while. For when, after a long, sad voyage, you know, the anchor drops into the haven, all its vast weight seems lifted from the captain's heart. We are getting on famously, Don Benito. My ship is in sight. Look through this side-light here; there she is; all a-taunt-o! The *Bachelor's Delight*, my good friend. Ah, how this wind braces one up. Come, you must take a cup of coffee with me this evening. My old steward will give you as fine a cup as ever any sultan tasted. What say you, Don Benito, will you?"

At first, the Spaniard glanced feverishly up, casting a longing look towards the sealer, while with mute concern his servant gazed into his face. Suddenly the old ague of coldness returned, and dropping back to his cushions he was silent.

"You do not answer. Come, all day you have been my host; would you have hospitality all on one side?"

"I cannot go," was the response.

"What? It will not fatigue you. The ships will lie together as near as they can, without swinging foul. It will be little more than stepping from deck to deck; which is but as from room to room. Come, come, you must not refuse me."

"I cannot go," decisively and repulsively repeated Don Benito.

Renouncing all but the last appearance of courtesy, with a sort of cadaverous sullenness, and biting his thin nails to the quick he glanced, almost glared, at his guest, as if impatient that a stranger's presence should interfere with the full indulgence of his morbid hour. Meantime the sound of the parted waters came more and more gurglingly and merrily in at the windows; as reproaching him for his dark spleen; as telling him that, sulk as he might, and go mad with it, nature cared not a jot; since, whose fault was it, pray?

But the foul mood was now at its depth, as the fair wind at its height.

There was something in the man so far beyond any mere unsociality or sourness previously evinced, that even the forbearing good-nature of his guest could no longer endure it. Wholly at a loss to account for such demeanour, and deeming sickness with eccentricity, however, extreme, no adequate excuse, well satisfied, too, that nothing in his own conduct could justify it, Captain Delano's pride began to be roused. Himself became reserved. But all seemed one to the Spaniard. Quitting him, therefore, Captain Delano once more went to the deck.

The ship was now within less than two miles of the sealer. The whale-boat was seen darting over the interval.

To be brief, the two vessels, thanks to the pilot's skill, ere long in neighbourly style lay anchored together.

Before returning to his own vessel, Captain Delano had intended communicating to Don Benito the smaller details of the proposed services to be rendered. But, as it was, unwilling anew to subject himself to rebuffs, he resolved, now that he had seen the *San Dominick* safely moored, immediately to quit her, without further allusion to hospitality or business. Indefinitely postponing his ulterior plans, he would regulate his future actions according to future circumstances. His boat was ready to receive him; but his host still tarried below. Well, thought Captain Delano, if he has little breeding, the more need to show mine. He descended to the cabin to bid a ceremonious, and, it may be, tacitly rebukeful adieu. But to his great satisfaction, Don Benito, as if he began to feel the weight of that treatment with which his slighted guest had, not indecorously, retaliated upon him, now supported by his servant, rose to his feet, and grasping Captain Delano's hand, stood tremulous; too much agitated to speak. But the good augury hence drawn was suddenly dashed, by his resuming all his previous reserve, with augmented gloom, as, with half-averted eyes, he silently reseated himself on his cushions. With a corresponding return of his own chilled feelings, Captain Delano bowed and withdrew.

He was hardly midway in the narrow corridor, dim as a tunnel, leading from the cabin to the stairs, when a sound, as of the tolling for execution in some jail-yard, fell on his ears. It was the echo of the ship's flawed bell, striking the hour, drearily reverberated in this subterranean vault. Instantly, by a fatality not to be withstood, his mind, responsive to the portent, swarmed with superstitious suspicions. He paused. In images far swifter than these sentences, the minutest details of all his former distrusts swept through him.

Hitherto, credulous good-nature had been too ready to furnish excuses for reasonable fears. Why was the Spaniard, so superfluously punctilious at times, now heedless of common propriety in not accompanying to the side his departing guest? Did indisposition forbid? Indisposition had not forbidden more irksome exertion that day. His last equivocal demeanour recurred. He had risen to his feet, grasped his guest's hand, motioned toward his hat; then, in an instant, all was eclipsed in sinister muteness and gloom. Did this imply one brief, repentant relenting at the final moment, from some iniquitous plot, followed by remorseless return to it? His last glance seemed to express a calamitous, yet acquiescent farewell to Captain Delano forever. Why decline the invitation to visit the sealer that evening? Or was the Spaniard less hardened than the Jew, who refrained not from supping at the board of him whom the same night he meant to betray? What imported all those day-long enigmas and contradictions, except they were intended to mystify, preliminary to some stealthy blow? Atufal, the pretended rebel, but punctual shadow, that moment lurked by the threshold without. He seemed a sentry, and more. Who, by his own confession, had stationed him there? Was the negro now lying in wait?

The Spaniard behind—his creature before: to rush from darkness to light was the involuntary choice.

The next moment, with clenched jaw and hand, he passed Atufal, and stood unharmed in the light. As he saw his trim ship lying peacefully at anchor, and almost within ordinary call; as he saw his household boat, with familiar faces in it, patiently rising and falling on the short waves by the *San Dominick's* side; and then, glancing about the decks where he stood, saw the oakum-pickers still gravely plying their fingers; and heard the low, buzzing whistle and industrious hum of the hatchet-polishers, still bestirring themselves over their endless occupation; and more than all, as he saw the benign aspect of nature, taking her innocent repose in the evening; the screened sun in the quiet camp of the west shining out like the mild light from Abraham's tent; as charmed eye and ear took in all these, with the chained figure of the black, clenched jaw and hand relaxed. Once again he smiled at the phantoms which had mocked him, and felt something like a tinge of remorse, that, by harbouring them even for a moment, he should, by implication, have betrayed an atheist doubt of the ever-watchful Providence above.

There was a few minutes' delay, while, in obedience to his orders, the boat was being hooked along to the gangway. During this interval, a sort of saddened satisfaction stole over Captain Delano, at thinking of the kindly offices he had that day discharged for a stranger. Ah, thought he, after good actions one's conscience is never ungrateful, however much so the benefitted party may be.

Presently, his foot, in the first act of descent into the boat, pressed the first round of the side-ladder, his face presented inward upon the deck. In the same moment, he heard his name courteously sounded; and, to his pleased surprise, saw Don Benito advancing—an unwonted energy in his air, as if, at the last moment, intent upon making amends for his recent discourtesy. With instinctive good feeling, Captain Delano, withdrawing his foot, turned and reciprocally advanced. As he did so, the Spaniard's nervous eagerness increased, but his vital energy failed; so that, the better to support him, the servant, placing his master's hand on his naked shoulder, and gently holding it there, formed himself into a sort of crutch.

When the two captains met, the Spaniard again fervently took the hand of the American, at the same time casting an earnest glance into his eyes, but, as before, too much overcome to speak.

I have done him wrong, self-reproachfully thought Captain Delano; his apparent coldness has deceived me; in no instance has he meant to offend.

Meantime, as if fearful that the continuance of the scene might too much unstring his master, the servant seemed anxious to terminate it. And so still presenting himself as a crutch, and walking between the two captains, he advanced with them towards the gangway; while still, as if full of kindly contrition, Don Benito would not let go the hand of Captain Delano, but retained it in his, across the black's body.

Soon they were standing by the side, looking over into the boat, whose crew turned up their curious eyes. Waiting a moment for the Spaniard to relinquish his hold, the now embarrassed Captain Delano lifted his foot, to overstep the threshold of the open gangway; but still Don Benito would not let go his hand. And yet, with an agitated tone, he said, "I can go no further; here I must bid you adieu. Adieu, my dear, dear Don Amasa. Go—go!" suddenly tearing his hand loose, "go, and God guard you better than me, my best friend."

Not unaffected, Captain Delano would now have lingered; but catching the meekly admonitory eye of the servant, with a hasty farewell he descended into his boat, followed by the continual adieus of Don Benito, standing rooted in the gangway.

Seating himself in the stern, Captain Delano, making a last salute, ordered the boat shoved off. The crew had their oars on end. The bowsman pushed the boat a sufficient distance for the oars to be lengthwise dropped. The instant that was done, Don Benito sprang over the bulwarks, falling at the feet of Captain Delano; at the same time calling towards his ship, but in tones so frenzied, that none in the boat could understand him. But, as if not equally obtuse, three sailors, from three different and distant parts of the ship, splashed into the sea, swimming after their captain, as if intent upon his rescue.

The dismayed officer of the boat eagerly asked what this meant. To which, Captain Delano, turning a disdainful smile upon the unaccountable Spaniard, answered that, for his part, he neither knew nor cared; but it seemed as if Don Benito had taken it into his head to produce the impression among his people that the boat wanted to kidnap him. "Or else—give way for your lives," he wildly added, starting at a clattering hubbub in the ship, above which rang the tocsin of the hatchet-polishers; and seizing Don Benito by the throat he added, "this plotting pirate means murder!" Here, in apparent verification of the words, the servant, a dagger in his hand, was seen on the rail overhead, poised, in the act of leaping, as if with desperate fidelity to befriend his master to the last; while, seemingly to aid the black, the three white sailors were trying to clamber into the hampered bow. Meantime, the whole host of negroes, as if inflamed at the sight of their jeopardized captain, impended in one sooty avalanche over the bulwarks.

All this, with what preceded, and what followed, occurred with such involutions of rapidity, that past, present, and future seemed one.

Seeing the negro coming, Captain Delano had flung the Spaniard aside, almost in the very act of clutching him, and by the unconscious recoil, shifting his place, with arms thrown up, so promptly grappled the servant in his descent, that with dagger presented at Captain Delano's heart, the black seemed of purpose to have leaped there as to his mark. But the weapon was wrenched away, and the assailant dashed down into the bottom of the boat, which now, with disentangled oars, began to speed through the sea.

At this juncture, the left hand of Captain Delano, on one side, again clutched the half-reclining Don Benito, heedless that he was in a speechless faint, while his right foot, on the other side, ground the prostrate negro; and his right arm pressed for added speed on the after oar, his eye bent forward, encouraging his men to their utmost.

But here, the officer of the boat, who had at last succeeded in beating off the towing sailors, and was now, with face turned aft, assisting the bowsman at his oar, suddenly called to Captain Delano, to see what the black was about; while a Portuguese oarsman shouted to him to give heed to what the Spaniard was saying.

Glancing down at his feet, Captain Delano saw the freed hand of the servant aiming with a second dagger—a small one, before concealed in his wool—with this he was snakishly writhing up from the boat's bottom, at the heart of his master, his countenance lividly vindictive, expressing the centered purpose of his soul; while the Spaniard, half-choked, was vainly shrinking away, with husky words, incoherent to all but the Portuguese.

That moment, across the long-benighted mind of Captain Delano, a flash of revelation swept, illuminating, in unanticipated clearness, his host's whole mysterious demeanor, with every enigmatic event of the day, as well as the entire past voyage of the *San Dominick*. He smote Babo's hand down, but his own heart smote him harder. With infinite pity he withdrew his hold from Don Benito. Not Captain Delano, but Don Benito, the black, in leaping into the boat, had intended to stab.

Both the black's hands were held, as, glancing up towards the *San Dominick*, Captain Delano, now with scales dropped from his eyes, saw the negroes, not in misrule, not in tumult, not as if frantically concerned for Don Benito, but with mask torn away, flourishing hatchets, and knives, in ferocious piratical revolt. Like delirious black dervishes, the six Ashantees danced on the poop. Prevented by their foes from springing into the water, the Spanish boys were hurrying up to the topmost spars, while such of the few Spanish sailors, not already in the sea, less alert, were descried, helplessly mixed in, on deck, with the blacks.

Meantime Captain Delano hailed his own vessel, ordering the ports up, and the guns run out. But by this time the cable of the *San Dominick* had been cut; and the fag-end, in lashing out, whipped away the canvas shroud about the beak, suddenly revealing, as the bleached hull swung round towards the open ocean, death for the figure-head, in a human skeleton; chalky comment on the chalked words below, "*Follow your leader.*"

At the sight, Don Benito, covering his face, wailed out:

" 'Tis he, Aranda! my murdered, unburied friend!"

Upon reaching the sealer, calling for ropes, Captain Delano bound the negro, who made no resistance, and had him hoisted to the deck. He would then have assisted the now almost helpless Don Benito up the side; but Don Benito, wan as he was, refused to move, or be moved, until the negro should have been first put below out of view. When, presently assured that it was done, he no more shrank from the ascent.

The boat was immediately dispatched back to pick up the three swimming sailors. Meantime, the guns were in readiness, though, owing to the *San Dominick* having glided somewhat astern of the sealer, only the aftermost one could be brought to bear. With this, they fired six times; thinking to cripple the fugitive ship by bringing down her spars. But only a few inconsiderable ropes were shot away. Soon the ship was beyond the gun's range, steering broad out of the bay; the blacks thickly clustering round the bowsprit, one moment with taunting cries towards the whites, the next with upthrown gestures hailing the now dusky moors of ocean—cawing crows escaped from the hand of the fowler.

The first impulse was to slip the cables and give chase. But, upon second thoughts, to pursue with whale-boat and yawl seemed more promising.

Upon inquiring of Don Benito what fire-arms they had on board the *San Dominick*, Captain Delano was answered that they had none that could be used; because, in the earlier stages of the mutiny, a cabin-passenger, since dead, had secretly put out of order the locks of what few muskets there were. But with all his remaining strength, Don Benito entreated the American not to give chase, either with ship or boat; for the negroes had already proved themselves such desperadoes, that, in case of a present assault, nothing but a total massacre of the whites could be looked for. But, regarding this warning as coming from one whose spirit had been crushed by misery, the American did not give up his design.

The boats were got ready and armed. Captain Delano ordered his men into them. He was going himself when Don Benito grasped his arm.

"What! have you saved my life, Señor, and are you now going to throw away your own?"

The officers also, for reasons connected with their interests and those of the voyage, and a duty owing to the owners, strongly objected against their commander's going. Weighing their remonstrances a moment, Captain Delano felt bound to remain; appointing his chief mate—an athletic and resolute man, who had been a privateer's-man—to head the party. The more to encourage the sailors, they were told, that the Spanish captain considered his ship good as lost; that she and her cargo, including some gold and silver, were worth more than a thousand doubloons. Take her, and no small part should be theirs. The sailors replied with a shout.

The fugitives had now almost gained an offing. It was nearly night; but the moon was rising. After hard, prolonged pulling, the boats came up on the ship's quarters, at a suitable distance laying upon their oars to discharge their muskets. Having no bullets to return, the negroes sent their yells. But, upon the second volley, Indian-like, they hurled their hatchets. One took off a sailor's fingers. Another struck the whale-boat's bow, cutting off the rope there, and remaining stuck in the gunwale like a woodman's axe. Snatching it, quivering from its lodgment, the mate hurled it back, The returned gauntlet now stuck in the ship's broken quarter-gallery, and so remained.

The negroes giving too hot a reception, the whites kept a more respectful distance. Hovering now just out of reach of the hurtling hatchets, they, with a view to the close encounter which must soon come, sought to decoy the blacks into entirely disarming themselves of their most murderous weapons in a hand-to-hand fight, by foolishly flinging them, as missiles, short of the mark, into the sea. But, ere long, perceiving the stratagem, the negroes desisted, though not before many of them had to replace their lost hatchets with handspikes; an exchange which, as counted upon, proved, in the end, favourable to the assailants.

Meantime, with a strong wind, the ship still clove the water; the boats alternately falling behind, and pulling up, to discharge fresh volleys.

The fire was mostly directed towards the stern, since there, chiefly, the negroes, at present, were clustering. But to kill or maim the negroes was not the object. To take them, with the ship, was the object. To do it, the ship must be boarded; which could not be done by boats while she was sailing so fast.

A thought now struck the mate. Observing the Spanish boys still aloft, high as they could get, he called to them to descend to the yards, and cut adrift the sails. It was done. About this time, owing to causes hereafter to be shown, two Spaniards, in the dress of sailors, and conspicuously showing themselves, were killed; not by volleys, but by deliberate marksman's shots; while, as it afterwards appeared, by one of the general discharges, Atufal, the black, and the Spaniard at the helm likewise were killed. What now, with the loss of sails, and loss of leaders, the ship became unmanageable to the negroes.

With creaking masts, she came heavily round to the wind; the prow slowly swinging into view of the boats, its skeleton gleaming in the horizontal moonlight, and casting a gigantic ribbed shadow upon the water. One extended arm of the ghost seemed beckoning the whites to avenge it.

"Follow your leader!" cried the mate; and, one on each bow, the boats boarded. Sealing-spears and cutlasses crossed hatchets and hand-spikes. Huddled upon the long-boat amidships, the negresses raised a wailing chant, whose chorus was the clash of the steel.

For a time, the attack wavered; the negroes wedging themselves to beat it back; the half-repelled sailors, as yet unable to gain a footing, fighting as troopers in the saddle, one leg sideways flung over the bulwarks, and one without, plying their cutlasses like carters' whips. But in vain. They were almost overborne, when, rallying themselves into a squad as one man, with a huzza, they sprang inboard, where, entangled, they involuntarily separated again. For a few breaths' space, there was a vague, muffled, inner sound, as of submerged sword-fish rushing hither and thither through shoals of black-fish. Soon, in a reunited band, and joined by the Spanish seamen, the whites came to the surface, irresistibly driving the negroes toward the stern. But a barricade of casks and sacks, from side to side, had been thrown up by the main-mast. Here the negroes faced about, and though scorning peace or truce, yet fain would have had respite. But, without pause, overleaping the barrier, the unflagging sailors again closed. Exhausted, the blacks now fought in despair. Their red tongues lolled, wolf-like, from their black mouths. But the pale sailor's teeth were set; not a word was spoken; and in five minutes more, the ship was won.

Nearly a score of the negroes were killed. Exclusive of those by the balls, many were mangled; their wounds—mostly inflicted by the long-edged sealing-spears, resembling those shaven ones of the English at Preston Pans, made by the poled scythes of the Highlanders. On the other side, none were killed, though several were wounded; some severely, including the mate. The surviving negroes were temporarily secured, and the ship, towed back into the harbour at midnight, once more lay anchored.

Omitting the incidents and arrangements ensuing, suffice it that, after two days spent in refitting, the ships sailed in company for Concepcion, in Chile, and thence for Lima, in Peru; where, before the vice-regal courts, the whole affair, from the beginning, underwent investigation.

Though, midway on the passage, the ill-fated Spaniard, relaxed from constraint, showed some signs of regaining health with free-will; yet, agreeably to his own foreboding, shortly before arriving at Lima, he relapsed, finally becoming so reduced as to be carried ashore in arms. Hearing of his story and plight, one of the many religious institutions of the City of Kings opened an hospitable refuge to him, where both physician and priest were his nurses, and a member of the order volunteered to be his one special guardian and consoler, by night and by day.

The following extracts, translated from one of the official Spanish documents, will, it is hoped, shed light on the preceding narrative, as well as, in the first place, reveal the true port of departure and true history of the *San Dominick*'s voyage, down to the time of her touching at the island of Santa Maria.

But, ere the extracts come, it may be well to preface them with a remark.

The document selected, from among many others, for partial translation, contains the deposition of Benito Cereno; the first taken in the case. Some disclosures therein were, at the time, held dubious for both learned and natural reasons. The tribunal inclined to the opinion that the deponent, not undisturbed in his mind by recent events, raved of some things which could never have happened. But subsequent depositions of the surviving sailors, bearing out the revelations of their captain in several of the strangest particulars, gave credence to the rest. So that the tribunal, in its final decision, rested its capital sentences upon statements which, had they lacked confirmation, it would have deemed it but duty to reject.

I, DON JOSÉ DE ABOS AND PADILLA, *His Majesty's Notary for the Royal Revenue, and Register of this Province, and Notary Public of the Holy Crusade of this Bishoprick, &c.*

Do certify and declare, as much as is requisite in law, that, in the criminal cause commenced the twenty-fourth of the month of September, in the year seventeen hundred and ninety-nine, against the negroes of the ship San Dominick, *the following declaration before me was made:*

Declaration of the first Witness, DON BENITO CERENO.

The same day and month and year, His Honour, Doctor Juan Martinez de Rozas, Councillor of the Royal Audience of this Kingdom, and learned in the law of this Intendency, ordered the captain of the ship San Dominick, *Don Benito Cereno, to appear; which he did in his litter, attended by the monk Infelez; of whom he received the oath, which he took by God, our Lord, and a Sign of the Cross; under which he promised to tell the truth of whatever he should know and should be asked;—and being interrogated agreeably to the tenor of the act, commencing the process, he said, that on the twentieth of May last, he set sail with his ship from the port of Valparaiso, bound to that of Callao; loaded with the produce of the country beside thirty cases of hardware and one hundred and sixty blacks, of both sexes, mostly belonging to Don Alexandro Aranda, gentleman, of the City of Mendoza; that the crew of the ship consisted of thirty-six men, beside the persons who went as passengers; that the negroes were in part as follows:*

[Here, in the original, follows a list of some fifty names, descriptions, and ages, compiled from certain recovered documents of Aranda's and also from recollections of the deponent, from which portions only are extracted.]

*—One, from about eighteen to nineteen years, named José, and this was the man that waited upon his master, Don Alexandro, and who speaks well the Spanish, having served him four or five years; * * * a mulatto, named Francesco, the cabin steward, of a good person and*

*voice having sung in the Valparaiso churches, native of the province of Buenos Ayres, aged about thirty-five years. * * * A smart negro, named Dago, who had been for many years a grave-digger among the Spaniards, aged forty-six years. * * * Four old negroes, born in Africa, from sixty to seventy, but sound, calkers by trade, whose names are as follows:—the first was named Mure, and he was killed (as was also his son named Diamelo); the second, Nacta; the third, Yola, likewise killed; the fourth, Ghofan; and six full-grown negroes, aged from thirty to forty-five, all raw, and born among the Ashantees—Matinqui, Yau, Lecbe, Mapenda, Yambaio, Akim, four of whom were killed; * * * a powerful negro named Atufal, who being supposed to have been a chief in Africa, his owner set great store by him. * * * And a small negro of Senegal, but some years among the Spaniards, aged about thirty, which negro's name was Babo; * * * that he does not remember the names of the others, but that still expecting the residue of Don Alexandro's papers will be found, will then take due account of them all, and remit to the court; * * * and thirty-nine women and children of all ages.*

[The catalogue over, the deposition goes on:]

** * * That all the negroes slept upon deck, as is customary in this navigation, and none wore fetters, because the owner, his friend Aranda, told him that they were all tractable; * * * that on the seventh day after leaving port, at three o'clock in the morning, all the Spaniards being asleep except two officers at the watch, who were the boat-swain, Juan Robles, and the carpenter, Juan Bautista Gayete, and the helmsman and his boy, the negroes revolted suddenly, wounded dangerously the boatswain and the carpenter, and successively killed eighteen men of those who were sleeping upon deck, some with hand-spikes and hatchets, and others by throwing them alive overboard, after tying them; that of the Spaniards upon deck, they left about seven, as he thinks, alive and tied, to manœuvre the ship, and three or four more, who hid themselves, remained also alive. Although in the act of revolt the negroes made themselves masters of the hatchway, six or seven wounded went through it to the cockpit, without any hindrance on their part; that during the act of revolt, the mate and another person, whose name he does not recollect, attempted to come up through the hatchway, but being quickly wounded, were obliged to return to the cabin; that the deponent resolved at break of day to come up the companion-way,**

where the negro Babo was, being the ringleader, and Atufal, who assisted him, and having spoken to them, exhorted them to cease committing such atrocities asking them, at the same time, what they wanted and intended to do, offering, himself, to obey their commands; that notwithstanding this, they threw, in his presence, three men, alive and tied, overboard; that they told the deponent to come up, and that they would not kill him; which having done, the negro Babo asked him whether there were in those seas any negro countries where they might be carried, and he answered them, No; that the negro Babo afterwards told him to carry them to Senegal, or to the neighbouring islands of St. Nicholas; and he answered, that this was impossible, on account of the great distance, the necessity involved of rounding Cape Horn, the bad condition of the vessel, the want of provisions, sails, and water; but that the negro Babo replied to him he must carry them in any way; that they would do and conform themselves to anything the deponent should require as to eating and drinking; that after a long conference, being absolutely compelled to please them, for they threatened to kill all the whites if they were not, at all events, carried to Senegal, he told them that what was most wanting for the voyage was water; that they would go near the coast to take it, and thence they would proceed on their course; that the negro Babo agreed to it; and the deponent steered towards the intermediate ports, hoping to meet some Spanish or foreign vessel that would save them; that within ten or eleven days they saw the land, and continued their course by it in the vicinty of Nasca; that the deponent observed that the negroes were now restless and mutinous, because he did not effect the taking in of water, the negro Babo having required, with threats, that it should be done, without fail, the following day; he told him he saw plainly that the coast was steep, and the rivers designated in the maps were not to be found, with other reasons suitable to the circumstances; that the best way would be to go to the island of Santa Maria, where they might water easily, it being a solitary island, as the foreigners did; that the deponent did not go to Pisco, that was near, nor make any other port of the coast, because the negro Babo had intimated to him several times, that he would kill all the whites the very moment he should perceive any city, town, or settlement of any kind on the shores to which they should be carried: that having determined to go to the island of Santa Maria, as the deponent had planned, for the purpose of trying whether, on the passage or near the island itself, they could find any vessel *that should favour them, or whether he could escape from it in a boat to the neighbouring coast of Arauco, to adopt the necessary means he immediately changed his course, steering for the island; that the negroes Babo and Atufal held daily conferences, in which they discussed what was necessary for their design of returning to Senegal, whether they were to kill all the Spaniards, and particularly the deponent; that eight days after parting from the coast of Nasca, the deponent being on the watch a little after day-break, and soon after the negroes had their meeting, the negro Babo came to the place where the deponent was, and told him that he had determined to kill his master, Don Alexandro Aranda, both because he and his companions could not otherwise be sure of their liberty, and that to keep the seamen in subjection, he wanted to prepare a warning of what road they should be made to take did they or any of them oppose him; and that, by means of the death of Don Alexandro, that warning would best be given; but, that what this last meant, the deponent did not at the time comprehend, nor could not, further than the death of Don Alexandro was intended; and moreover the negro Babo proposed to the deponent to call the mate Raneds, who was sleeping in the cabin, before the thing was done, for fear, as the deponent understood it, that the mate, who was a good navigator, should not be killed with Don Alexandro and the rest; that the deponent, who was the friend, from youth, of Don Alexandro, prayed and conjured, but all was useless; for the negro Babo answered him that the thing could not be prevented, and that all the Spaniards risked their death if they should attempt to frustrate his will in this matter, or any other; that, in this conflict, the deponent called the mate, Raneds, who was forced to go apart, and immediately the negro Babo commanded the Ashantee Matinqui and the Ashantee Lecbe to go and commit the murder; that those two went down with hatchets to the berth of Don Alexandro; that, yet half alive and mangled, they dragged him on deck; that they were going to throw him overboard in that state, but the negro Babo stopped them, bidding the murder be completed on the deck before him, which was done, when, by his orders, the body was carried below, forward; that nothing more was seen of it by the deponent for three days; * * * that Don Alonzo Sidonia, an old man, long resident at Valparaiso, and lately appointed to a civil office in Peru, whither he had taken passage, was at the time sleeping in the berth opposite Don Alexandro's; that awakening at his cries, surprised by them, and at the* *sight of the negroes with their bloody hatchets in their hands, he threw himself into the sea through a window which was near him, and was drowned, without it being in the power of the deponent to assist or take him up; * * * that a short time after killing Aranda, they brought upon deck his cousin-german, of middle-age, Don Francisco Masa, of Mendoza, and the young Don Joaquin, Marques de Aramboalaza, then lately from Spain, with his Spanish servant Ponce, and the three young clerks of Aranda, José Morairi, Lorenzo Bargas, and Hermenegildo Gandix, all of Cadiz; that Don Joaquin and Hermenegildo Gandix, the negro Babo, for purposes hereafter to appear, preserved alive; but Don Francisco Masa, José Morairi, and Lorenzo Bargas, with Ponce the servant, beside the boat-swain, Juan Robles, the boat-swain's mates, Manuel Viscaya and Roderigo Hurta, and four of the sailors, the negro Babo ordered to be thrown alive into the sea, although they made no resistance, nor begged for anything else but mercy; that the boat-swain Juan Robles, who knew how to swim, kept the longest above water, making acts of contrition, and, in the last words he uttered, charged this deponent to cause mass to be said for his soul to our Lady of Succour; * * * that, during the three days which followed, the deponent, uncertain what fate had befallen the remains of Don Alexandro, frequently asked the negro Babo where they were, and, if still on board, whether they were to be preserved for interment ashore, entreating him so to order it; that the negro Babo answered nothing till the fourth day, when at sunrise, the deponent coming on deck, the negro Babo showed him a skeleton, which had been substituted for the ship's proper figurehead—the image of Cristobal Colon, the discoverer of the New World; that the negro Babo asked him whose skeleton that was, and whether from its whiteness, he should not think it a white's; that, upon his covering his face the negro Babo, coming close, said words to this effect: "Keep faith with the blacks from here to Senegal, or you shall in spirit, as now in body, follow your leader," pointing to the prow; * * * that the same morning the negro Babo took by succession each Spaniard forward, and asked him whose skeleton that was, and whether, from its whiteness, he should not think it a white's; that each Spaniard covered his face; that then to each the negro Babo repeated the words in the first place said to the deponent; * * * that they (the Spaniards), being then assembled aft, the negro Babo harangued them, saying that he had now done all; that the deponent (as navigator for the negroes) might pursue his course, warning him and all of them that they should,* *soul and body, go the way of Don Alexandro, if he saw them (the Spaniards) speak or plot anything against them (the negroes), a threat which was repeated every day; that, before the events last mentioned, they had tied the cook to throw him overboard, for it is not known what thing they heard him speak, but finally the negro Babo spared his life, at the request of the deponent; that a few days after, the deponent, endeavouring not to omit any means to preserve the lives of the remaining whites, spoke to the negroes peace and tranquillity, and agreed to draw up a paper, signed by the deponent and the sailors who could write, as also by the negro Babo, for himself and all the blacks, in which the deponent obliged himself to carry them to Senegal, and they not to kill any more, and he formally to make over to them the ship, with the cargo, with which they were for that time satisfied and quieted. * * * But the next day, the more surely to guard against the sailors' escape, the negro Babo commanded all the boats to be destroyed but the long-boat, which was unseaworthy, and another, a cutter in good condition, which knowing it would yet be wanted for towing the water casks, he had it lowered down into the hold. * * **

[Various particulars of the prolonged and perplexed navigation ensuing here follow, with incidents of a calamitous calm, from which portion one passage is extracted, to wit:]

—That on the fifth day of the calm, all on board suffering much from the heat, and want of water, and five having died in fits, and mad, the negroes became irritable, and for a chance gesture, which they deemed suspicious—though it was harmless—made by the mate, Raneds, to the deponent in the act of handing a quadrant, they killed him; but that for this they afterwards were sorry, the mate being the only remaining navigator on board, except the deponent.

* * *

—That omitting other events which daily happened, and which can only serve uselessly to recall past misfortunes and conflicts, after seventy-three days' navigation, reckoned from the time they sailed from Nasca, during which they navigated under a scanty allowance of water, and were afflicted with the calms before mentioned, they at last arrived at the island of Santa Maria, on the seventeenth of the month of

August, at about six o'clock in the afternoon, at which hour they cast anchor very near the American ship, Bachelor's Delight, *which lay in the same bay, commanded by the generous Captain Amasa Delano; but at six o'clock in the morning, they had already described the port, and the negroes became uneasy, as soon as at distance they saw the ship, not having expected to see one there; that the negro Babo pacified them, assuring them that no fear need be had; that straightway he ordered the figure on the bow to be covered with canvas, as for repairs, and had the decks a little set in order; that for a time the negro Babo and the negro Atufal conferred; that the negro Atufal was for sailing away, but the negro Babo would not, and, by himself, cast about what to do; that at last he came to the deponent, proposing to him to say and do all that the deponent declares to have said and done to the American captain; * * * that the negro Babo warned him that if he varied in the least, or uttered any word, or gave any look that should give the least intimation of the past events or present state, he would instantly kill him, with all his companions, showing a dagger, which he carried hid, saying something which, as he understood it, meant that that dagger would be alert as his eyes; that the negro Babo then announced the plan to all his companions, which pleased them; that he then, the better to disguise the truth, devised many expedients, in some of them uniting deceit and defense; that of this sort was the device of the six Ashantees before named, who were his bravoes; that them he stationed on the break of the poop, as if to clean certain hatchets (in cases, which were part of the cargo), but in reality to use them, and distribute them at need, and at a given word he told them; that, among other devices, was the device of presenting Atufal, his right hand man, as chained, though in a moment the chains could be dropped; that in every particular he informed the deponent what part he was expected to enact in every device, and what story he was to tell on every occasion, always threatening him with instant death if he varied in the least: that, conscious that many of the negroes would be turbulent, the negro Babo appointed the four aged negroes who were calkers, to keep what domestic order they could on the decks; that again and again he harangued the Spaniards and his companions, informing them of his intent, and of his devices, and of the invented story that this deponent was to tell; charging them lest any of them varied from that story; that these arrangements were made and matured during the interval of two or three hours, between their first sighting the ship and the arrival on* *board of Captain Amasa Delano; that this happened about half-past seven o'clock in the morning, Captain Amasa Delano coming in his boat, and all gladly receiving him; that the deponent, as well as he could force himself, acting then the part of principal owner, and a free captain of the ship, told Captain Amasa Delano, when called upon, that he came from Buenos Ayres, bound to Lima, with three hundred negroes; that off Cape Horn, and in a subsequent fever, many negroes had died; that also, by similar casualties, all the sea-officers and the greatest part of the crew had died.*

* * *

[And so the deposition goes on, circumstantially recounting the fictitious story dictated to the deponent by Babo, and through the deponent imposed upon Captain Delano; and also recounting the friendly offers of Captain Delano, with other things, but all of which is here omitted. After the fictitious story, etc., the deposition proceeds:]

—that that generous Captain Delano remained on board all the day, till he left the ship anchored at six o'clock in the evening, deponent speaking to him always of his pretended misfortunes, under the forementioned principles, without having had it in his power to tell a single word, or give him the least hint, that he might know the truth and state of things; because the negro Babo, performing the office of an officious servant with all the appearance of submission of the humble slave, did not leave the deponent one moment; that this was in order to observe the deponent's actions and words, for the negro Babo understands well the Spanish; and besides, there were thereabouts some others who were constantly on the watch, and likewise understood the Spanish; * * * *that upon one occasion, while deponent was standing on the deck conversing with Amasa Delano, by a secret sign the negro Babo drew him (the deponent) aside, the act appearing as if originating with the deponent; that then, he being drawn aside, the negro Babo proposed to him to gain from Amasa Delano full particulars about his ship, and crew, and arms; that the deponent asked "for what?" that the negro Babo answered he might conceive; that, grieved at the prospect of what might overtake the generous Captain Amasa Delano, the deponent at first refused to ask the desired questions, and used every argument to induce the negro Babo to give up this new design; that the negro Babo*

*showed the point of his dagger; that, after the information had been obtained, the negro Babo again drew him aside, telling him that that very night he (the deponent) would be captain of two ships, instead of one, for that, great part of the American's ship's crew being to be absent fishing, the six Ashantees without any one else, would easily take it; that at this time he said other things to the same purpose; that no entreaties availed; that, before Amasa Delano's coming on board, no hint had been given touching the capture of the American ship; that to prevent this project the deponent was powerless; * * * —that in some things his memory is confused, he cannot distinctly recall every event; * * * that as soon as they had cast anchor at six o'clock in the evening, as has before been stated, the American captain took leave, to return to his vessel; that upon a sudden impulse, which the deponent believes to have come from God and his angels, he, after the farewell had been said, followed the generous Captain Amasa Delano as far as the gunwale, where he stayed, under pretence of taking leave, until Amasa Delano should have been seated in his boat; that on shoving off, the deponent sprang from the gunwale into the boat and fell into it, he knows not how, God guarding him; that—*

[Here, in the original, follows the account of what further happened at the escape, and how the *San Dominick* was retaken, and of the passage to the coast; including in the recital many expressions of "eternal gratitude" to the "generous Captain Amasa Delano." The deposition then proceeds with recapitulatory remarks, and a partial renumeration of the negroes, making record of their individual part in the past events, with a view to furnishing, according to command of the court, the data whereon to found the criminal sentences to be pronounced. From this portion is the following:]

*—That he believes that all the negroes, though not in the first place knowing to the design of revolt, when it was accomplished, approved it. * * * That the negro José, eighteen years old, and in the personal service of Don Alexandro, was the one who communicated the information to the negro Babo about the state of things in the cabin, before the revolt; that this is known, because, in the preceding nights, he used to come from his berth, which was under his master's, in the cabin, to the deck where the ringleader and his associates were, and had secret conversations with the negro Babo, in which he was several times seen by*

*the mate; that one night, the mate drove him away twice; * * * that this same negro José was the one who, without being commanded to do so by the negro Babo, as Lecbe and Martinqui were, stabbed his master, Don Alexandro, after he had been dragged halflifeless to the deck; * * * that the mulatto steward, Francesco, was of the first band of revolters, that he was, in all things, the creature and tool of the negro Babo; that, to make his court, he, just before a repast in the cabin, proposed to the negro Babo, poisoning a dish for the generous Captain Amasa Delano; this is known and believed, because the negroes have said it; but that the negro Babo, having another design, forbade Francesco; * * * that the Ashantee Lecbe was one of the worst of them; for that, on the day the ship was retaken, he assisted in the defence of her, with a hatchet in each hand, with one of which he wounded in the breast, the chief mate of Amasa Delano, in the first act of boarding; this all knew; that, in sight of the deponent, Lecbe struck with a hatchet, Don Francisco Masa, when, by the negro Babo's orders, he was carrying him to throw him overboard alive, beside participating in the murder, before mentioned, of Don Alexandro Aranda, and others of the cabin-passengers; that, owing to the fury with which the Ashantees fought in the engagement with the boats, but this Lecbe and Yau survived; that Yau was bad as Lecbe; that Yau was the man who, by Babo's command, willingly prepared the skeleton of Don Alexandro, in a way the negroes afterwards told the deponent, but which he, so long as reason is left him, can never divulge; that Yau and Lecbe were the two who, in a calm by night, riveted the skeleton to the bow; this also the negroes told him; that the negro Babo was he who traced the inscription below it; that the negro Babo was the plotter from first to last; he ordered every murder, and was the helm and keel of the revolt; that Atufal was his lieutenant in all; but Atufal, with his own hand, committed no murder; nor did the negro Babo; * * * that Atufal was shot, being killed in the fight with the boats, ere boarding; * * * that the negresses of age, were knowing to the revolt, and testified themselves satisfied at the death of their master, Don Alexandro; that, had the negroes not restrained them, they would have tortured to death, instead of simply killing, the Spaniards slain by command of the negro Babo; that the negresses used their utmost influence to have the deponent made away with; that, in the various acts of murder, they sang songs and danced—not gaily, but solemnly; and before the engagement with the boats, as well as during the action, they sang melancholy songs to the negroes,* *and that this melancholy tone was more inflaming than a different one would have been, and was so intended; that all this is believed, because the negroes have said it;—that of the thirty-six men of the crew, exclusive of the passengers (all of whom are now dead), which the deponent had knowledge of, six only remained alive, with four cabin-boys and ship-boys, not included with the crew; * * * —that the negroes broke an arm of one of the cabin-boys and gave him strokes with hatchets.*

[Then follow various random disclosures referring to various periods of time. The following are extracted:]

*—That during the presence of Captain Amasa Delano on board, some attempts were made by the sailors, and one by Hermenegildo Gandix, to convey hints to him of the true state of affairs; but that these attempts were ineffectual, owing to fear of incurring death, and furthermore, owing to the devices which offered contradictions to the true state affairs, as well as owing to the generosity and piety of Amasa Delano incapable of sounding such wickedness; * * * that Luys Galgo, a sailor about sixty years of age, and formerly of the King's navy, was one of those who sought to convey tokens to Captain Amasa Delano; but his intent, though undiscovered, being suspected, he was, on a pretence, made to retire out of sight, and at last into the hold and there was made away with. This the negroes have since said; * * * that one of the ship-boys, feeling from Captain Amasa Delano's presence, some hopes of release, and not having enough prudence, dropped some chance word respecting his expectations, which being overheard and understood by a slave-boy with whom he was eating at the time, the latter struck him on the head with a knife, inflicting a bad wound, but of which the boy is now healing; that likewise, not long before the ship was brought to anchor, one of the seamen, steering at the time, endangered himself by letting the blacks remark some expression in his countenance, arising from a similar cause to the above; but this sailor, by his heedful after conduct, escaped; * * * that these statements are made to show the court that from the beginning to the end of the revolt, it was impossible for the deponent and his men to act otherwise than they did; * * * —that the third clerk, Hermenegildo Gandix, who before had been forced to live among the seamen, wearing a seaman's habit, and in all respects appearing to be one for the time, he, Gandix, was killed by a musket ball fired through mistake from the boats before boarding;*

*having in his fright run up the mizzen-rigging, calling to the boats—"don't board," lest upon their boarding the negroes should kill him; that this inducing the Americans to believe he some way favoured the cause of the negroes, they fired two balls at him, so that he fell wounded from the rigging, and was drowned in the sea; * * * —that the young Don Joaquin, Marques de Aramboalaza, like Hermenegildo Gandix, the third clerk, was degraded to the office and appearance of a common seaman; that upon one occasion when Don Joaquin shrank, the negro Babo commanded the Ashantee Lecbe to take tar and heat it, and pour it upon Don Joaquin's hands; * * *—that Don Joaquin was killed owing to another mistake of the Americans, but one impossible to be avoided, as upon the approach of the boats, Don Joaquin, with a hatchet tied edge out and upright to his hand, was made by the negroes to appear on the bulwarks; whereupon, seen with arms in his hands, and in a questionable attitude, he was shot for a renegade seaman; * * * that on the person of Don Joaquin was found secreted a jewel, which, by papers that were discovered, proved to have been meant for the shrine of our Lady of Mercy in Lima; a votive offering, beforehand prepared and guarded, to attest his gratitude, when he should have landed in Peru, his last destination for the safe conclusion of his entire voyage from Spain; * * * —that the jewel, with the other effects of the late Don Joaquin, is in the custody of the Hospital de Sacerdotes, awaiting the disposition of the honourable court; * * * —that, owing to the condition of the deponent, as well as the haste in which the boats departed for the attack, the Americans were not forewarned that there were, among the apparent crew, a passenger and one of the clerks disguised by the negro Babo; * * * —that, beside the negroes killed in the action, some were killed after the capture and re-anchoring at night, when shackled to the ring-bolts on deck; that these deaths were committed by the sailors, ere they could be prevented. That so soon as informed of it, Captain Amasa Delano used all his authority, and in particular with his own hand, struck down Martinez Gola, who, having found a razor in the pocket of an old jacket of his, which one of the shackled negroes had on, was aiming it at the negro's throat; that the noble Captain Amasa Delano also wrenched from the hand of Bartholomew Barlo a dagger, secreted at the time of the massacre of the whites, with which he was in the act of stabbing a shackled negro, who, the same day, with another negro had thrown him down and jumped upon him; * * * —that, for all the events befalling through so long a time, during which* *the ship was in the hands of the negro Babo, he cannot here give account; but that, what he has said is the most substantial of what occurs to him at present, and is the truth under the oath which he has taken; which declaration he affirmed and ratified, after hearing it read to him.*

He said that he is twenty-nine years of age, and broken in body and mind; that when finally dismissed by the court, he shall not return home to Chile, but betake himself to the monastery on Mount Agonia without; and signed with his honour, and crossed himself, and, for the time, departed as he came, in his litter, with the monk Infelez, to the Hospital de Sacerdotes. Benito Cereno. Dr. Rozas.

If the Deposition have served as the key to fit into the lock of the complications which precede it, then, as a vault whose door has been flung back, the *San Dominick*'s hull lies open to-day.

Hitherto the nature of this narrative, besides rendering the intricacies in the beginning unavoidable, has more or less required that many things, instead of being set down in order of occurrence, should be retrospectively, or irregularly given; this last is the case with the following passages, which will conclude the account.

During the long, mild voyage to Lima, there was, as before hinted, a period during which the sufferer a little recovered his health, or, at least in some degree, his tranquillity. Ere the decided relapse which came, the two captains had many cordial conversations—their fraternal unreserve in similar contrast with former withdrawments.

Again and again it was repeated, how hard it had been to enact the part forced on the Spaniard by Babo.

"Ah, my dear friend," Don Benito once said, "at those very times you thought me so morose and ungrateful, nay, when, as you now admit, you have thought me plotting your murder, at those very times my heart was frozen; I could not look at you, thinking of what, both on board this ship and your own, hung, from other hands, over my kind benefactor. And as God lives, Don Amasa, I know not whether desire for my own safety alone could have nerved me to that leap into your boat, had it not been for the thought that, did you, unenlightened, return to your ship, you, my best friend, with all who might be with you, stolen upon, that night, in your hammocks, would never in this world have wakened again. Do but think how you walked this deck, how you sat in this cabin, every inch of ground mined into honey-combs under you. Had I dropped the least hint, made the least advance towards an understanding between us; death, explosive death—yours and mine—would have ended the scene."

"True, true," cried Captain Delano, starting, "you have saved my life, Don Benito, more than I yours; saved it, too, against my knoweldge and will."

"Nay, my friend," rejoined the Spaniard, courteous even to the point of religion, "God charmed your life, but you saved mine. To think of some things you did—those smilings and chattings, rash pointings and gesturings. For less than these, they slew my mate, Raneds; but you had the Prince of Heaven's safe-conduct through all ambuscades."

"Yes, all is owing to Providence, I know: but the temper of my mind that morning was more than commonly pleasant, while the sight of so much suffering, more apparent than real, added to my good-nature, compassion, and charity, happily interweaving the three. Had it been otherwise, doubtless, as you hint, some of my interferences might have ended unhappily enough. Besides, those feelings I spoke of enabled me to get the better of momentary distrust, at times when acuteness might have cost me my life, without saving another's. Only at the end did my suspicions get the better of me, and you know how wide of the mark they then proved."

"Wide, indeed," said Don Benito, sadly; "you were with me all day; stood with me, sat with me, talked with me, looked at me, ate with me, drank with me, and yet, your last act was to clutch for a monster, not only an innocent man, but the most pitiable of all men. To such degree may malign machinations and deceptions impose. So far may even the best man err, in judging the conduct of one with the recesses of whose condition he is not acquainted. But you were forced to it; and you were in time undeceived. Would that, in both respects, it was so ever, and with all men."

"You generalize, Don Benito; and mournfully enough. But the past is passed; why moralize upon it? Forget it. See, yon bright sun has forgotten it all, and the blue sea, and the blue sky; these have turned over new leaves."

"Because they have no memory," he dejectedly replied; "because they are not human."

"But these mild trades that now fan your cheek, do they not come with a human-like healing to you? Warm friends, steadfast friends are the trades."

"With their steadfastness they but waft me to my tomb, Señor," was the foreboding response.

"You are saved," cried Captain Delano, more and more astonished and pained; "You are saved: what has cast such a shadow upon you?"

"The negro."

There was silence, while the moody man sat, slowly and unconsciously gathering his mantle about him, as if it were a pall.

There was no more conversation that day.

But if the Spaniard's melancholy sometimes ended in muteness upon topics like the above, there were others upon which he never spoke at all; on which, indeed, all his old reserves were piled. Pass over the worst, and, only to elucidate, let an item or two of these be cited. The dress, so precise and costly, worn by him on the day whose events have been narrated, had not willingly been put on. And that silver-mounted sword, apparent symbol of despot command, was not, indeed, a sword, but the ghost of one. The scabbard, artificially stiffened, was empty.

As for the black—whose brain, not body, had schemed and led the revolt, with the plot—his slight frame, inadequate to that which it held, had at once yielded to the superior muscular strength of his captor, in the boat. Seeing all was over, he uttered no sound, and could not be forced to. His aspect seemed to say, since I cannot do deeds, I will not speak words. Put in irons in the hold, with the rest, he was carried to Lima. During the passage, Don Benito did not visit him. Nor then, nor at any time after would he look at him. Before the tribunal he refused. When pressed by the judges he fainted. On the testimony of the sailors alone rested the legal identity of Babo.

Some months after, dragged to the gibbet at the tail of a mule, the black met his voiceless end. The body was burned to ashes; but for many days, the head, that hive of subtlety, fixed on a pole in the Plaza, met, unabashed, the gaze of the whites; and across the Plaza looked towards St. Bartholomew's church, in whose vaults slept then, as now, the recovered bones of Aranda: and across the Rimac bridge looked towards the monastery, on Mount Agonia without; where, three months after being dismissed by the court, Benito Cereno, borne on the bier, did, indeed, follow his leader.

ANDRÉ GIDE

(*In the imaginative literature of our century, the name of André Gide stands out boldly for honesty and self-knowledge. Revealed in all Gide's writing is the ceaseless struggle within him of the personal antimonies, the sensual and the spiritual, which threatened to pull him apart. That he was able eventually to achieve a personal integration of the opposites which might have destroyed him is testimony to the love of life which was the mainstay of his creative endurance.*

However first undertaken, analysis of The Pastoral Symphony *cannot proceed far without calling into play such words as* innocence, self-deception, hypocrisy, love, *and* evil. *The novella is clearly one of ideas. The unbearable clash and conflict occur plainly in the arena of moral and intellectual preoccupations. The irony is in the difference between what a man thinks he is, and what he is. Gide obliges us to recognize the forces that do not allow the two images of the man to be more nearly identical.*

As readers, we may very likely be inclined to say that the pastor's tragic self-deception must be blamed upon the restrictive confines of his religious ethic. Of course, this view is all wrong; the pastor's interpretation of the scriptures is lyric and soaring far beyond that which any self-imposed joyless, punitive code could ever allow. Rather, we may be sure, Gide would want us to know first, the real reason for the man's failure and, then, the suffering that any such moral idealist can cause when he shuts his eyes to the true human condition and to his own. Freedom is not for the weak and spiritually blind. Observation of these matters may best be directed toward the disparity between the pastor's First and Second Notebooks. It can be shown that Gide probes moral cowardice; he is not concerned with an attack on institutionalized religion.

The Pastoral Symphony *is one of those rare fictionized tracts that succeed in demonstrating that ideas are more exciting than emotions. The author asks if we can afford to cling to ready-made formulas and obsolete clichés in a world in which the old illusions are fading away. But in satisfying a cerebral necessity, Gide admirably does not cease to be an artist. Mainly, the critical reader will consider how effectively the pastor as mouthpiece serves as a literary device. A less arduous exercise will be to weigh the variety of meanings attaching to the title of the novella.*

The deplorable way in which a human being may compromise or falsify his true self is, then, the subject of the story. The pastor presents a superficial, external self to the world. That image he dishonestly accepts as the real self. He is not strong enough to pierce his stiff external shell and find his true personality beneath it. His own possibilities as a human being are not realized. He becomes a destructive, a life-denying force. These are the things André Gide talks about, in just such terms. It is, therefore, too easy to say that the moral essay which is The Pastoral Symphony *demonstrates only that the road to hell is paved with good intentions.*

The Pastoral Symphony

FIRST NOTEBOOK

10 February 189–

THE SNOW has been falling continuously for the last three days and all the roads are blocked. It has been impossible for me to go to R—, where I have been in the habit of holding a service twice a month for the last fifteen years. This morning not more than thirty of my flock were gathered together in La Brévine chapel.

I will take advantage of the leisure this enforced confinement affords me to think over the past and to set down how I came to take charge of Gertrude.

I propose to write here the whole history of her formation and development, for I seem to have called up out of the night her sweet and pious soul for no other end but adoration and love. Blessed be the Lord for having entrusted me with this task!

Two years and six months ago I had just driven back one afternoon from La Chaux-de-Fonds when a little girl who was a stranger to me came up in a great hurry to take me to a place about five miles away where she said an old woman lay dying. My horse was still in the shafts, so I made the child get into the carriage and set off at once, after first providing myself with a lantern, as I thought it likely I should not be able to get back before dark.

I had supposed myself to be perfectly acquainted with the whole countryside in the neighborhood of my parish; but when we had passed La Saudraie farm, the child made me take a road that I had never ventured down before. About two miles farther on, however, I recognized on the left-hand side a mysterious little lake where I had sometimes been to skate as a young man. I had not seen it for fifteen years, for none of my pastoral duties take me that way; I could not have said where it lay and it had so entirely dropped out of my mind that when I suddenly recognized it in the golden enchantment of the rose-flecked evening sky, I felt as though I had seen it before only in a dream.

The road ran alongside the stream that falls out of the lake, cut across the extreme end of the forest, and then skirted a peat-bog. I had certainly never been there before.

The sun was setting and for a long time we had been driving in the shade when my young guide pointed out a cottage on the hillside which would have seemed uninhabited but for a tiny thread of smoke that rose from the chimney, looking blue in the shade and brightening as it reached the gold of the sky. I tied the horse up to an apple tree close by and then followed the child into the dark room where the old woman had just died.

The gravity of the landscape, the silence and solemnity of the hour had struck me to the heart. A woman still in her youth was kneeling beside the bed. The child, whom I had taken to be the deceased woman's granddaughter, but who was only her servant, lighted a smoky tallow dip and then stood motionless at the foot of the bed. During our long drive I had tried to get her to talk, but had not succeeded in extracting two words from her.

The kneeling woman rose. She was not a relation as I had first supposed, but only a neighbor, a friend, whom the servant girl had brought there when she saw her mistress's strength failing, and who now offered to watch by the dead body. The old woman, she said, had passed away painlessly. We agreed together on the arrangements for the burial and the funeral service. As often before in this out-of-the-world country, it fell to me to settle everything. I was a little uneasy, I admit, at leaving the house, in spite of the poverty of its appearance, in the sole charge of this neighbor and of the little servant girl. But it seemed very unlikely that there was any treasure hidden away in a corner of this wretched dwelling . . . and what else could I do? I inquired nevertheless whether the old woman had left any heirs.

Upon this, the woman took the candle and held it up so as to light the corner of the hearth, and I could make out crouching in the fireplace, and apparently asleep, a nondescript-looking creature, whose face was almost entirely hidden by a thick mass of hair.

"The blind girl there—she's a niece, the servant says. That's all that's left of the family, it seems. She must be sent to the poorhouse; I don't see what else can be done with her."

I was shocked to hear the poor thing's future disposed of in this way in her presence and afraid such rough words might give her pain.

"Don't wake her up," I said softly, as a hint to the woman that she should at any rate lower her voice.

"Oh, I don't think she's asleep. But she's an idiot; she can't speak or understand anything, I'm told. I have been in the room since this morning and she has hardly so much as stirred. I thought at first she was deaf; the servant thinks not, but that the old woman was deaf herself and never uttered a word to her, nor to anyone else; she hadn't opened her mouth for a long time past except to eat and drink."

"How old is she?"

"About fifteen, I suppose. But as to that, I know no more about it than you do. . . ."

It did not immediately occur to me to take charge of the poor, forlorn creature myself; but after I had prayed—or, to be more accurate, while I was still praying on my knees between the woman and the little servant girl, who were both kneeling too—it suddenly came upon me that God had set a kind of obligation in my path and that I could not shirk it without cowardice. When I rose, I had decided to take the girl away that very evening, though I had not actually asked myself what

I should do with her afterward, nor into whose charge I should put her. I stayed a few moments longer gazing at the old woman's sleeping face, with its puckered mouth, looking like a miser's purse with strings tightly drawn so as to let nothing escape. Then, turning toward the blind girl, I told the neighbor of my intention.

"Yes, it is better she should not be there tomorrow when they come to take the body away," said she. And that was all.

Many things would be easily accomplished but for the imaginary objections men sometimes take a pleasure in inventing. From our childhood upwards, how often have we been prevented from doing one thing or another we should have liked to do, simply by hearing people about us repeat: "He won't be able to . . ."!

The blind girl allowed herself to be taken away like a lifeless block. The features of her face were regular, rather fine, but utterly expressionless. I took a blanket off the mattress where she must have usually slept, in a corner under a staircase that led from the room to the loft.

The neighbor was obliging and helped me wrap her up carefully, for the night was very clear and chilly; after having lighted the carriage lamp, I started home, taking the girl with me. She sat huddled up against me—a soulless lump of flesh, with no sign of life beyond the communication of an obscure warmth. The whole way home I was thinking: "Is she asleep? And what can this black sleep be like? . . . And in what way do her waking hours differ from her sleeping? But this darkened body is surely tenanted; an immured soul is waiting there for a ray of Thy grace, O Lord, to touch it. Wilt Thou perhaps allow my love to dispel this dreadful darkness? . . ."

I have too much regard for the truth to pass over in silence the unpleasant welcome I had to encounter on my return home. My wife is a garden of virtues; and in the times of trouble we have sometimes gone through I have never for an instant had cause to doubt the stuff of which her heart is made; but it does not do to take her natural charity by surprise. She is an orderly person, careful neither to go beyond nor to fall short of her duty. Even her charity is measured, as though love were not an inexhaustible treasure. This is the only point on which we differ. . . .

Her first thoughts when she saw me bring home the girl that evening broke from her in this exclamation:

"What kind of job have you saddled yourself with now?"

As always happens when we have to come to an understanding, I began by telling the children—who were standing round, open-mouthed and full of curiosity and surprise—to leave the room. Ah, how different this welcome was from what I could have wished! Only my dear little Charlotte began to dance and clap her hands when she understood that something new, something alive, was coming out of the carriage. But the others, who have been well trained by their mother, very soon damped the child's pleasure and made her fall into step.

There was a moment of great confusion. And as neither my wife nor the children yet knew that they had to do with a blind person, they could not understand the extreme care with which I guided her footsteps. I myself was disconcerted by the odd moans the poor afflicted creature began to utter as soon as I let go her hand, which I had held in mine during the whole drive. There was nothing human in the sounds she made; they were more like the plaintive whines of a puppy. Torn away for the first time as she had been from the narrow round of customary sensations that had formed her universe, her knees now failed her; but when I pushed forward a chair, she sank on the floor in a heap, as if she were incapable of sitting down; I then led her up to the fireplace and she regained her calm a little as soon as she was able to crouch down in the same position in which I had first seen her beside the old woman's fire, leaning against the chimney-piece. In the carriage too, she had slipped off the seat and spent the whole drive huddled up at my feet. My wife, however, whose instinctive impulses are always the best, came to my help; it is her reflection that is constantly at odds with her heart and very often gets the better of it.

"What do you mean to do with *that?*" she asked when the girl had settled down.

I shivered in my soul at this use of the word *that*, and had some difficulty in restraining a movement of indignation. As I was still under the spell of my long and peaceful meditation, however, I controlled myself. Turning toward the whole party, who were standing round in a circle again, I placed my hand on the blind girl's head and said as solemnly as I could:

"I have brought back the lost sheep."

But Amélie will not admit that there can be anything unreasonable or superreasonable in the teaching of the Gospel. I saw she was going to object, and it was then I made a sign to Jacques and Sarah, who, as they are accustomed to our little conjugal differences and have not much natural curiosity (not enough, I often think), led the two younger children out of the room.

Then, as my wife still remained silent and a little irritated, I thought, by the intruder's presence: "You needn't mind speaking before her," I said. "The poor child doesn't understand."

Upon this Amélie began to protest that she had absolutely nothing to say—which is her usual prelude to the lengthiest explanations—and there was nothing for her to do but to submit, as usual, to all my most unpractical vagaries, however contrary to custom and good sense they might be. I have already said that I had not in the least made up my mind what I was going to do with the child. It had not occurred to me, or only in the vaguest way, that there was any possibility of taking her into our house permanently, and I may almost say it was Amélie herself who first suggested it to me by asking whether I didn't think there were "enough of us in the house already"? Then she declared that I always hurried on ahead without taking any thought for those who could not keep up with me, that for her part she considered five children quite enough, and that since the birth of Claude (who at that very moment set up a howl from his cradle, as if he had heard his name) she had as much as she could put up with and that she couldn't stand any more.

At the beginning of her outburst some of Christ's words rose from my heart to my lips; I kept them back, however, for I never think it becoming to allege the authority of the Holy Book as an excuse for my conduct. But when she spoke of her fatigue, I was struck with confusion, for I must admit it has more than once happened to me to let my wife suffer from the consequences of my impulsive and inconsiderate zeal. In the meantime, however, her recriminations had enlightened me as to my duty; I begged Amélie therefore, as mildly as possible, to consider whether she would not have done the same in my place and whether she could have possibly abandoned a creature who had been so obviously left without anyone to help her; I added that I was under no illusion as to the extra fatigue the charge of this new inmate would add to the cares of the household, and that I regretted I was not more often able to help her with them. In this way I pacified her as best I could, begging her at the same time not to visit her anger on the innocent girl, who had done nothing to deserve it. Then I pointed out that Sarah was now old enough to be more of a help to her and that Jacques was no longer in need of her care. In short, God put into my mouth the right words to help her accept what I am sure she would have undertaken of her own accord if the circumstances had given her time to reflect and if I had not forestalled her decision without consulting her.

I thought the cause was almost gained, and my dear Amélie was already approaching Gertrude with the kindest of intentions; but her irritation suddenly blazed up again higher than ever when, on taking up the lamp to look at the child more closely, she discovered her to be in a state of unspeakable dirt.

"Why, she's filthy!" she cried. "Go and brush yourself quickly. No, not here. Go and shake your clothes outside. Oh dear! Oh dear! The children will be covered with them. There's nothing in the world I hate so much as vermin."

It cannot be denied that the poor child was crawling with them; and I could not prevent a feeling of disgust as I thought how close I had kept her to me during our long drive.

When I came back a few minutes later, having cleaned myself as best I could, I found my wife had sunk into an armchair and with her head in her hands was giving way to a fit of sobbing.

"I did not mean to put your fortitude to such a test," I said tenderly. "In any case it is late tonight and too dark to do anything. I will sit up and keep the fire going and the child can sleep beside it. Tomorrow we will cut her hair and wash her properly. You need not attend to her until you have got over your repugnance." And I begged her not to say anything of that to the children.

It was supper time. My protégée, at whom our old Rosalie cast many a scowling glance as she waited on us, greedily devoured the plateful of soup I handed her. The meal was a silent one. I should have liked to relate my adventure, to talk to the children and touch their hearts by making them understand and feel the strangeness of such a condition of total deprivation. I should have liked to rouse their pity, their sympathy for the guest God had sent us; but I was afraid of reviving Amélie's irritation. It seemed as though the word had been passed to take no notice of what had happened and to forget all about it, though certainly not one of us can have been thinking of anything else.

I was extremely touched when, more than an hour after everyone had gone to bed and Amélie had left me, I saw my little Charlotte steal gently through the half-open door in her nightdress and bare feet; she flung her arms round my neck and hugged me fiercely.

"I didn't say good-night to you properly," she murmured.

Then, pointing with her little forefinger to the blind girl, who was now peacefully slumbering and whom she had been curious to see again before going to sleep:

"Why didn't I kiss her too?" she whispered.

"You shall kiss her tomorrow. We must let her be now. She is asleep," I said as I went with her to the door.

Then I sat down again and worked till morning, reading or preparing my next sermon.

"Certainly," I remember thinking, "Charlotte seems much more affectionate than the elder children, but when they were her age, I believe they all got round me too. My big boy Jacques, nowadays so distant and reserved . . . One thinks them tender-hearted, when really they are only coaxing and wheedling one."

27 February

The snow fell heavily again last night. The children are delighted because they say we shall soon be obliged to go out by the windows. It is a fact that this morning the front door is blocked and the only way out is by the washhouse. Yesterday I made sure the village was sufficiently provisioned, for we shall doubtless remain cut off from the rest of the world for some time to come. This is not the first winter we have been snowbound, but I cannot remember ever having seen so thick a fall. I take advantage of it to go on with the tale I began yesterday.

I have said that when I first brought home this afflicted child I had not clearly thought out what place she would take in our household. I knew the limits of my wife's powers of endurance; I knew the size of our house and the smallness of our income. I had acted, as usual, in the way that was natural to me, quite as much as on principle, and without for a moment calculating the expense into which my impulse might land me—a proceeding I have always thought contrary to the Gospels' teaching. But it is one thing to trust one's cares to God and quite another to shift them onto other people. I soon saw I had laid a heavy burden on Amélie's shoulders—so heavy that at first I felt struck with shame.

I helped her as best I could to cut the little girl's hair, and I saw that she did even that with disgust. But when it came to washing and cleaning her, I was obliged to leave it to my wife; and I realized that I perforce escaped the heaviest and most disagreeable tasks.

For the rest, Amélie ceased to make the slightest objection. She seemed to have thought things over during the night and resigned herself to her new duties; she even seemed to take some pleasure in them and I saw her smile when she had finished washing and dressing Gertrude. After her head had been shaved and I had rubbed it with ointment, a white cap was put on her; some of Sarah's old clothes and some clean linen took the place of the wretched rags Amélie threw into the fire. The name of Gertrude was chosen by Charlotte and immediately adopted by us all, in our ignorance of her real name, which the orphan girl herself was unaware of, and which I did not know how to find out. She must have been a little younger than Sarah, whose last year's clothes fitted her.

I must here confess the profound and overwhelming disappointment I felt during the first days. I had certainly built up a whole romance for myself on the subject of Gertrude's education, and the reality was a cruel disillusion. The indifference, the apathy of her countenance, or rather its total lack of expression froze my good intentions at their very source. She sat all day long by the fireside, seemingly on the defensive, and as soon as she heard our voices, still more when we came near her, her features appeared to harden; from being expressionless they became hostile; if anyone tried to attract her attention, she began to groan and grunt like an animal. This sulkiness only left her at meal times. I helped her myself and she flung herself on her food with a kind of bestial avidity that was most distressing to witness. And as love responds to love, so a feeling of aversion crept over me at this obstinate withholding of her soul. Yes truly, I confess that at the end of the first ten days I had begun to despair, and my interest in her was even so far diminished that I almost regretted my first impulse and wished I had never brought her home with me. And the absurd thing was that Amélie, being not unnaturally a little triumphant over feelings I was really unable to hide from her, seemed all the more lavish of care and kindness now that she saw Gertrude was becoming a burden to me, and that I felt her presence among us as a mortification.

This was how matters stood when I received a visit from my friend Dr. Martins, of Val Travers, in the course of one of his rounds. He was very much interested by what I told him of Gertrude's condition and was at first greatly astonished she should be so backward, considering her only infirmity was blindness; but I explained that in addition to this she had had to suffer from the deafness of the old woman who was her sole guardian, and who never spoke to her, so that the poor child had been utterly neglected. He persuaded me that in that case I was wrong to despair, but that I was not employing the proper method.

"You are trying to build," he said, "before making sure of your foundations. You must reflect that her whole mind is in a state of chaos and that even its first lineaments are as yet unformed. The first thing to be done is to make her connect together one or two sensations of touch and taste and attach a sound to them—a word—to serve as a kind of label. This you must repeat over and over again indefatigably and then try to get her to say it after you.

"Above all, don't go too quickly; take her at regular hours and never for very long at a time. . . .

"For the rest, this method," he added, after having described it to me minutely, "has nothing particularly magic about it. I did not invent it and other people have applied it. Don't you remember in the philosophy class at school, our professors told us of an analogous case apropos of Condillac and his animated statue—unless," he corrected himself "I read it later in a psychological review. . . . Never mind; I was much struck by it and I even remember the name of the poor girl, who was still more afflicted than Gertrude, for she was a deaf-mute as well as blind. She was discovered somewhere in England toward the middle of last century by a doctor who devoted himself to educating her. Her name was Laura Bridgman. The doctor kept a journal, as you ought to do, of the child's progress—or rather, in the first place, of his efforts to instruct her. For days and weeks he went on, first making her feel alternately two little objects, a pin and a pen, and then putting her fingers on the two words *pin* and *pen* printed in a Braille book for the blind. For weeks and weeks there was no result. Her body seemed quite vacant. He did not lose courage, however. 'I felt like a person,' says he, 'leaning over the edge of a deep dark well and desperately dangling a rope in the hope that a hand would catch hold of it.' For he did not for one moment doubt that someone was there at the bottom of the well and that in the end the rope would be caught hold of. And one day, at last, he saw Laura's impassive face light up with a kind of smile. I can well believe that tears of love and gratitude sprang to his eyes and that he straightway fell on his knees and gave thanks to God. Laura had understood at last what it was the doctor wanted. She was saved! From that day forward she was all attention; her progress was rapid; she was soon able to learn by herself and eventually became the head of an institution for the blind—unless that was some other person—for there have been other cases recently that the reviews and newspapers have been full of; they were all astonished—rather foolishly, in my opinion—that such creatures should be happy. For it is a fact that all these walled-up prisoners were happy, and as soon as they were able to express anything, it was their *happiness* they spoke of. The journalists of course went into ecstasies and pointed the 'moral' for people who 'enjoy' all their five senses and yet have the audacity to complain. . . ."

Here an argument arose between Martins and me, for I objected to his pessimism and could not allow what he seemed to infer—that our senses serve in the long run only to make us miserable.

"That's not what I meant," he protested; "I merely wanted to say, first, that man's spirit imagines beauty, comfort, and harmony more easily and gladly than it can the disorder and sin that everywhere tarnish, stain, degrade, and mar this world; and further, that this state of things is revealed to us by our five senses, which also help us to contribute to it. So that I feel inclined to put the words '*si sua mala nescient*' after Virgil's '*Fortunatos nimium,*' instead of '*si sua bona norint*' as we are taught. How happy men would be if they knew nothing of evil!"

Then he told me of one of Dickens's stories—which he thinks was directly inspired by Laura Bridgman's case; he promised to send it to me, and four days later I received *The Cricket on the Hearth,* which I read with the greatest pleasure. It is a rather lengthy but at times very touching tale of a little blind girl, maintained by her father, a poor toymaker, in an illusory world of comfort, wealth, and happiness. Dickens exerts all his art in representing this deception as an act of piety, but, thank Heaven, I shall not have to make use of any such falsehood with Gertrude.

The day after Martins's visit I began to put his method into practice with all the application I was capable of. I am sorry now I did not take notes, as he advised, of Gertrude's first steps along the twilit path where I myself at first was but a groping guide. During the first weeks more patience was needed than can well be believed, not only because of the amount of time an education of this kind requires, but also because of the reproaches it brought me. It is painful for me to have to say that these reproaches came from Amélie; but, for that matter, if I mention this here it is because it has not left in me the slightest trace of animosity or bitterness—I declare this most solemnly, in case these lines should come to her eyes later on. (Does not Christ's teaching of the forgiveness of injuries follow immediately after the parable of the lost sheep?) More than that—at the very moment when I most suffered from her reproaches, I could not feel angry with her for disapproving the length of time I devoted to Gertrude. What I chiefly deplored was that she failed to believe that my efforts would be at all successful. Yes, it was her want of faith that grieved me—without, however, discouraging me. How often I heard her repeat: "If only any good were to come of it all! . . ." And she remained stubbornly convinced that my work was labor lost; so that naturally she thought it wrong of me to devote the time to Gertrude's education which she always declared would have been better employed otherwise. And whenever I was occupied with Gertrude, she managed to make out that I was wanted at that moment for someone or something else, and that I was giving her time that ought to have been given to others. In fact, I think she felt a kind of maternal jealousy, for she more than once said to me: "You never took so much pains with any of your own children"—which was true; for though I am very fond of my children, I have never thought it my business to take much pains with them.

It has often been my experience that the parable of the lost sheep is one of the most difficult of acceptance for certain people, who yet believe themselves to be profoundly Christian at heart. That each single sheep of the flock should be in turn more precious in the eyes of the shepherd than the rest of the flock as a whole is beyond and above their power of conception. And the words: "If a man have a hundred sheep and one of them be gone astray, doth he not leave the ninety and nine and goeth into the mountains and seeketh that which is gone astray?"—words all aglow with charity, such persons would, if they dared speak frankly, declare to be abominably unjust.

Gertrude's first smiles consoled me for everything and repaid me for my pains a hundredfold. For "and if so be that he find it, verily I say unto you, he rejoiceth more of that sheep than of the ninety and nine which went not astray." Yes, verily, the smile that dawned for me one morning on that marble face of hers, when she seemed suddenly touched to understanding and interest by what I had been trying for so many days to teach her, flooded my heart with a more seraphic joy than was ever given me by any child of my own.

5 March

I noted this date as if it had been a birthday. It was not so much a smile as a transfiguration. Her features flashed into life—a sudden illumination, like the crimson glow that precedes dawn in the high Alps, thrilling the snowy peak on which it lights and calling it up out of darkness—such a flood it seemed, of mystic color; and I thought too of the pool of Bethesda at the moment the angel descends to stir the slumbering water. A kind of ecstasy rapt me at sight of the angelic expression that came over Gertrude's face so suddenly, for it was clear to me that this heavenly visitor was not so much intelligence as love. And in a very transport of gratitude I kissed her forehead and felt that I was offering thanks to God.

The progress she made after this was as rapid as the first steps had been slow. It is only with an effort that I can now recall our manner of proceeding; it seemed to me sometimes that Gertrude advanced by leaps and bounds, as though in defiance of all method. I can remember that at first I dwelt more on the qualities of objects than on their variety—hot, cold, sweet, bitter, rough, soft, light—and then on actions: to pick up, to put down, to remove, to approach, to tie, to cross, to assemble, to disperse, etc. And very soon I abandoned all attempt at method and began to talk to her without troubling much whether her mind was always able to follow me; but I went slowly, inviting and provoking her questions as she seemed inclined. Certainly her mind was at work during the hours I left her to herself; for every time I came back to her after an absence, it was to find with fresh surprise that the wall of darkness that separated us had grown less thick. After all, I said to myself, it is so that the warmth of the air and the insistence of spring gradually triumph over winter. How often have I wondered at the melting of the snow! Its white cloak seems to wear thin from underneath, while to all appearance it remains unchanged. Every winter Amélie falls into the trap: "The snow is as thick as ever," she declares. And indeed it still seems so, when all at once there comes a break and suddenly, in patches here and there, life once more shows through.

Fearing that Gertrude might become peaky if she continued to sit beside the fire like an old woman, I had begun to make her go out. But she refused to do this unless she held my arm. I realized from her surprise and fear when she first left the house, and before she was able to tell me so in words, that she had never as yet ventured out of doors. In the cottage where I had found her no one had cared for her further than to give her food and prevent her from dying—for I cannot say that anyone helped her to live. Her little universe of darkness was bounded by the walls of the single room she never left; she scarcely ventured on summer days as far as the threshold, when the door stood open to the great universe of light. She told me later that when she heard the birds' song she used to suppose it was simply the effect of light, like the gentle warmth which she felt on her cheeks and hands, and that, without precisely thinking about it, it seemed to her quite natural that the warm air should begin to sing, just as the water begins to boil on the fire. The truth is she did not trouble to think; she took no interest in anything and lived in a state of frozen numbness till the day I took charge of her. I remember her inexhaustible delight when I told her that the little voices came from living creatures, whose sole function apparently was to express the joy that lies broadcast throughout all nature. (It was from that day that she began to say: "I am as joyful as a bird.") And yet the idea that these songs proclaim the splendor of a spectacle she could not behold had begun by making her melancholy.

"Is the world really as beautiful as the birds say?" she would ask. "Why do people not tell us so oftener? Why do *you* never tell me so? Is it for fear of grieving me because I cannot see it? That would be wrong. I listen so attentively to the birds; I think I understand everything they say."

"People who can see do not hear them as well as you do, my Gertrude," I said, hoping to comfort her.

"Why don't other animals sing?" she went on. Sometimes her questions surprised me and left me perplexed for a moment, for she forced me to reflect on things I had hitherto taken for granted. It was thus it occurred to me for the first time that the closer an animal lives to the ground and the heavier its weight, the duller it is. I tried to make her understand this; and I told her of the squirrel and its gambols.

She asked me if the birds were the only animals that flew.

"There are butterflies too," I told her.

"And do they sing?"

"They have another way of telling their joy. It is painted on their wings. . . ." And I described the rainbow colors of the butterfly.

28 February

Now let me turn back a little, for yesterday I allowed myself to be carried away.

In order to teach Gertrude, I had had to learn the Braille alphabet myself; but she was soon able to read much quicker than I could; I had some difficulty in deciphering the writing; and besides found it easier to follow with my eyes than with my fingers. For that matter, I was not the only one to give her lessons. And at first I was glad to be helped in this respect, for I have a great deal to do in the parish, the houses being so widely scattered that my visits to the poor and the sick sometimes oblige me to go far afield. Jacques had managed to break his arm while skating during the Christmas holidays, which he was spending with us; for during term time he goes to Lausanne, where he received his early education, and where he is studying at the theological school. The fracture was not serious and Martins, whom I at once sent for, was easily able to set it without the help of a surgeon; but it was considered advisable for Jacques to keep indoors for some time. He now suddenly began to take an interest in Gertrude, to whom he had hitherto paid no attention, and occupied himself with helping me to teach her to read. His assistance only lasted the time of his convalescence—about three weeks—but during those weeks Gertrude's progress was very marked. She was now fired with extraordinary zeal. Her young intelligence, but yesterday so benumbed and torpid, its first steps hardly taken, and scarcely able to walk, seemed now already preparing to run. I wondered at the ease with which she succeeded in formulating her thoughts and at the rapidity with which she learned to express herself —not childishly, but at once correctly, conveying her ideas by the help of images, taken in the most delightful and unexpected way from the objects we had just taught her to recognize, or from others we described to her, when we could not actually put them within her grasp; for she always used things she could touch or feel in order to explain what was beyond her reach, after the method of land-surveyors measuring distances.

But I think it is unnecessary to note here all the first steps of her education, doubtless the same in the early education of all blind people. I suppose too that in each case the teacher must have been plunged into a similar perplexity by the question of colors. (And this subject led me to the reflection that there is nowhere any mention of colors in the

Gospels.) I do not know how other people set about it; for my part, I began by naming the colors of the prism to her in the order in which they occur in the rainbow; but then a confusion was immediately set up in her mind between color and brightness; and I realized that her imagination was unable to draw any distinction between the *quality* of the shade and what painters, I believe, call its "*value*." She had the greatest difficulty in understanding that every color in its turn might be more or less dark and that they might be mixed one with another to an unlimited extent. It puzzled her exceedingly, and she came back to the subject again and again.

About this time the opportunity was given me of taking her to a concert at Neuchâtel. The part played by each instrument in the symphony suggested to me the idea of recurring to this question of colors. I bade Gertrude observe the different resonances of the brasses, the strings, and the wood instruments, and that each of them was able in its own way to produce the whole series of sounds, from the lowest to the highest, with varying intensity. I asked her to imagine the colors of nature in the same way—the reds and oranges analogous to the sounds of the horns and trombones; the yellows and greens like those of the violins, cellos, and double basses; the violets and blue suggested by the clarinets and oboes. A sort of inner rapture now took place of all her doubts and uncertainties.

"How beautiful it must be!" she kept on repeating.

Then suddenly she added: "But the white? I can't understand now what the white can be like."

And I at once saw how insecure my comparison was.

"White," I tried however to explain, "is the extreme treble limit where all the tones are blended into one, just as black is the bass or dark limit."

But this did not satisfy me any more than it did her; and she pointed out at once that the wood instruments, the brasses, and the violins remain distinct in the bass as well as in the treble parts. How often I have been obliged to remain puzzled and silent, as I did then, searching about for some comparison I might appeal to.

"Well," said I at last, "imagine white as something absolutely pure, something in which color no longer exists, but only light; and black, on the contrary, something so full of color that it has become dark. . . ."

I recall this fragment of dialogue merely as an example of the difficulties I encountered only too often. Gertrude had this good point, that she never pretended to understand, as people so often do, thus filling their minds with inaccurate or false statements, which in the end vitiate all their reasoning. So long as she could not form a clear idea of any notion, it remained a cause of anxiety and discomfort to her.

As regards what I have just related, the difficulty was increased by the fact that the notion of light and that of heat began by being closely associated with each other in her mind, and I had the greatest trouble afterward in disconnecting them.

Thus, through these experiments with her, it was constantly brought home to me how greatly the visual world differs from the world of sound, and that any comparison between the two must necessarily be a lame one.

29 February

I have been so full of my comparisons that I have not yet said what immense pleasure the Neuchâtel concert gave Gertrude. It was actually the *Pastoral Symphony* that was being played. I say *actually* because, as will be easily understood, there is no work I could have more wished her to hear. For a long time after we had left the concert-room, Gertrude remained silent, as though lost in ecstasy.

"Is what you see really as beautiful as that?" she asked at last.

"As beautiful as what, dear child?"

"As that 'scene on the bank of a stream'?"

I did not answer at once, for I was reflecting that those ineffable harmonies painted the world as it might have been, as it would be without evil and without sin, rather than the world as it really was. And I have never yet ventured to speak to Gertrude of evil and sin and death.

"Those who have eyes," I said at last, "do not know their happiness."

"But I who have not," she cried, "*I* know the happiness of hearing."

She pressed up against me as she walked and hung on to my arm in the way small children do.

"Pastor, do you feel how happy I am? No, no, I don't say so to please you. Look at me. Can't you see on people's faces whether they are speaking the truth? I always know by their voices. Do you remember the day you answered me that you weren't crying when my aunt" (that is what she called my wife) "had reproached you with being no help to her? And I cried out: 'Pastor, that's not true!' Oh, I felt at once from your voice that you weren't telling me the truth; there was no need for me to feel your cheeks to know that you had been crying." And she repeated very loud: "No, there was no need for me to feel your cheeks" —which made me turn red, for we were still in the town and the passers-by turned round to look at us. She went on, however:

"You mustn't try to deceive me, you know. First of all, because it would be very mean to try to deceive a blind person . . . and then because you wouldn't succeed," she added, laughing. "Tell me, pastor, you aren't unhappy, are you?"

I put her hand to my lips, as though to make her feel, without having to confess it, that part of my happiness came from her, and answered as I did so.

"No, Gertrude, I am not unhappy. How should I be unhappy?"

"And yet you cry sometimes?"

"I have cried sometimes."

"Not since that time?"

"No, I have not cried again since then."

"And you have not felt inclined to cry?"

"No, Gertrude."

"And tell me—have you felt inclined since then not to speak the truth to me?"

"No, dear child."

"Can you promise never to try to deceive me?"

"I promise."

"Well, tell me quickly, then—am I pretty?"

This sudden question dumbfounded me, all the more because I had studiously avoided up to then taking any notice of Gertrude's undeniable beauty; and moreover I considered it perfectly unnecessary that she should be informed of it herself.

"What can it matter to you?" I said.

"I am anxious," she went on, "I should like to know whether I do not—how shall I put it?—make too much of a discord in the symphony. Whom else should I ask, pastor?"

"It is not a pastor's business to concern himself with the beauty of people's faces," said I, defending myself as best I could.

"Why not?"

"Because the beauty of their souls suffices him."

"You had rather I thought myself ugly," was her reply with a charming pout; so that, giving up the struggle, I exclaimed:

"Gertrude, you know quite well you are pretty."

She was silent and her face took on an expression of great gravity, which did not leave her until we got home.

On our return Amélie at once managed to make me feel she disapproved of the way I had been spending my day. She might have told me so before; but she had let Gertrude and me start without a word, according to her habit of letting people do things and of reserving to herself the right to blame them afterward. For that matter, she did not actually reproach me; but her very silence was accusing; for surely it would have been natural to have inquired what we had heard, since she knew I was taking Gertrude to the concert. Would not the child's pleasure have been increased if she had felt that the smallest interest had been taken in it? But Amélie did not remain entirely silent—she merely seemed to put a sort of affectation into avoiding any but the most indifferent topics; and it was not till evening, when the little ones had gone to bed, and after I had asked her in private and with some severity if she was vexed with me for taking Gertrude to the concert, that I got the following answer:

"You do things for her you would never have done for any of your own children."

So it was always the same grievance, and the same refusal to understand that the feast is prepared for the child who returns to us, not for those who have stayed at home, as the parable shows us. It grieved me too to see that she took no account of Gertrude's infirmity—poor Gertrude, who could hope for no other kind of pleasure. And if I providentially happened to be free that afternoon—I, who am as a rule so much in request—Amélie's reproach was all the more unfair, because she knew perfectly well that the other children were busy or occupied in one way or other, and that she herself did not care for music, so that even if she had all the time in the world, it would never enter her head to go to a concert, not even if it were given at our very door.

What distressed me still more was that Amélie had actually said this in front of Gertrude; for though I had taken my wife on one side, she had raised her voice so much that Gertrude heard her. I felt not so much sad as indignant, and a few moments later, when Amélie had left us, I went up to Gertrude and, taking her frail little hand in mine, I lifted it to my face. "You see," I said, "this time I am not crying."

"No," answered she, trying to smile, "this time it is my turn." And as she looked up at me, I suddenly saw her face was flooded with tears.

8 March

The only pleasure I can give Amélie is to refrain from doing the things she dislikes. These very negative signs of love are the only ones she allows me. The degree to which she has already narrowed my life is a thing she cannot realize. Oh, would to Heaven she would demand something difficult of me! How gladly I would undertake a rash, a dangerous task for her! But she seems to have a repugnance for everything that is not usual; so that for her, progress in life consists merely in adding like days to like days. She does not desire—she will not even accept—any new virtue, nor even an increase of the old ones. When it is not with disapproval, it is with mistrust that she views every effort of the soul to find in Christianity something other than the domestication of our instincts.

I must confess that I entirely forgot, that afternoon at Neuchâtel, to go and pay our haberdasher's bill and to bring her back some spools of thread she wanted. But I was more vexed with myself for this than she could have been; especially as I had been quite determined not to forget her commissions, being very well aware that "he that is faithful in that which is least is faithful also in much," and being afraid too of the conclusions she might draw from my forgetfulness. I should even have been glad if she had reproached me with it, for I certainly deserved reproaches. But, as often happens, the imaginary grievance outweighed the definite charge. Ah, how beautiful life would be and how bearable our wretchedness if we were content with real evils without opening the doors to the phantoms and monsters of our imagination! . . . But I am straying here into observations that would do better as the subject of a sermon (Luke xii, 29: "Neither be ye of doubtful mind"). It is the history of Gertrude's intellectual and moral development that I purposed tracing here and I must now return to it.

I had hoped to follow its course step by step in this book and had begun to tell the story in detail. Not only, however, do I lack time to note all its phases with minuteness, but I find it extremely difficult at the present moment to remember their exact sequence. Carried away by my tale, I began by setting down remarks of Gertrude's and conversations with her that are far more recent; a person reading these pages would no doubt be astonished at hearing her express herself so justly and reason so judiciously in such a little while. The fact is her progress was amazingly rapid; I often wondered at the promptness with which her mind fastened on the intellectual food I offered it, and indeed on everything it could catch hold of, absorbing it all by a constant process of assimilation and maturation. The way in which she forestalled my thoughts and outstripped them was a continual surprise to me, and often from one lesson to another I ceased to recognize my pupil.

At the end of a very few months there was no appearance of her intelligence having lain dormant for so long. Even at this early stage she showed more sense and judgment than the generality of young girls, distracted as they are by the outside world and prevented from giving their best attention by a multitude of futile preoccupations. She was moreover a good deal older, I think, than we had at first supposed. Indeed, it seemed as though she were determined to profit by her blindness, so that I actually wondered whether this infirmity was not in many ways an advantage. In spite of myself I compared her with Charlotte, so easily distracted by the veriest trifles, so that many a time while hearing the child say over her lessons, as I sometimes did, I found myself thinking: "Dear me, how much better she would listen if only she could not see!"

Needless to say, Gertrude was a very eager reader, but as I wished as far as possible to keep in touch with the development of her mind, I preferred her not to read too much—or at any rate not much without me—and especially not the Bible—which may seem very strange for a Protestant. I will explain myself; but before touching on a question so important, I wish to relate a small circumstance that is connected with music and should be placed, as far I can remember shortly after the concert at Neuchâtel.

Yes, the concert, I think, took place three weeks before the summer vacation, which brought Jacques home. In the meantime I had often sat with Gertrude at the little harmonium of our chapel, which is usually played by Mlle de la M., with whom Gertrude is at present staying. Louise de la M. had not yet begun to give Gertrude music lessons. Notwithstanding my love for music, I do not know much about it, and I felt very little able to teach her anything when I sat beside her at the keyboard.

"No," she had said after the first gropings, "you had better leave me. I had rather try by myself."

And I left her all the more willingly that the chapel did not seem to me a proper place in which to be shut up alone with her, as much out of respect for the sanctity of the place as for fear of gossip—though as a rule I endeavor to disregard it; in this case, however, it is a matter that concerns not only me but her. So when a round of visits called me in that direction, I would take her to the church and leave her there, often for long hours together, and then would go to fetch her on my return. In this way she spent her time patiently hunting out harmonies, and I would find her again toward evening pondering over some concord of sounds that had plunged her into a long ecstasy.

On one of the first days of August, barely more than six months ago, it so happened that I had gone to visit a poor widow in need of consolation and had not found her in. I therefore returned at once to fetch Gertrude from the church, where I had left her; she was not expecting me back so soon, and I was extremely surprised to find Jacques with her. Neither of them heard me come in, for the little noise I made was covered by the sound of the organ. It is not in my nature to play the spy, but everything that touches Gertrude touches me; so stepping as softly as I could, I stole up the few steps that lead to the gallery—an excellent post of observation. I must say that during the whole time I was there I did not hear a word from either of them that they might not have said before me. But he sat very close to her, and several time I saw him take her hand in order to guide her fingers over the keys. Was it not in itself strange that she should accept instructions and guidance from him when she had previously refused them from me, preferring, she said, to practice by herself? I was more astonished, more pained, than I liked to own and was just on the point of intervening when I saw Jacques suddenly take out his watch.

"I must leave you now," he said; "my father will be coming back in a moment."

I saw him lift her unresisting hand to his lips; then he left. A few moments later I went noiselessly down the stairs and opened the church door so that she might hear me and think I had only just arrived.

"Well, Gertrude! Are you ready to go home? How is the organ getting on?"

"Very well," she answered in the most natural tone; "I have really made some progress today."

A great sadness filled my heart, but we neither of us made any allusion to the episode I have just described.

I was impatient to find myself alone with Jacques. My wife, Gertrude, and the children used as a rule to go to bed rather early after supper, while we two sat on late over our studies. I was waiting for this moment. But before speaking to him I felt my heart bursting with such a mixture of feelings that I could not—or dared not—begin on the subject that was tormenting me. And it was he who abruptly broke the silence by announcing his intention of spending the rest of the vacation with us. Now, a few days earlier he had spoken to us about a tour he wanted to make in the high Alps—a plan my wife and I heartily approved of; I knew his friend T., who was to be his traveling companion, was counting on him; it was therefore quite obvious to me that this sudden change of plan was not unconnected with the scene I had just come upon. I was at first stirred by violent indigation, but was afraid to give way to it lest it should put an end to my son's confidence altogether; I was afraid too of pronouncing words I should afterward regret; so making a great effort over myself, I said as naturally as I could:

"I thought T. was counting on you."

"Oh," he answered, "not absolutely, and besides he will have no difficulty in finding someone else to go with him. I can rest here quite as well as in the Oberland, and I really think I can spend my time better than mountaineering."

"In fact," I said, "you have found something to occupy you at home."

He noticed some irony in the tone of my voice and looked at me, but being unable as yet to guess the motive of it, went on unconcernedly:

"You know I have always liked reading better than climbing."

"Yes, my dear boy," said I, returning his glance with one as searching; "but are not lessons in harmonium-playing even more attractive than reading?"

No doubt he felt himself blush, for he put his hand to his forehead, as though to shade his eyes from the lamplight; but he recovered himself almost immediately and went on in a voice I could have wished less steady:

"Do not blame me too much, Father. I did not mean to hide anything from you and you have only forestalled by a very little the confession I was preparing to make you."

He spoke deliberately, as if he were reading the words out of a book, finishing his sentences with as much calm, it seemed, as if it were a matter in which he had no concern. The extraordinary self-possession he showed brought my exasperation to a climax. Feeling that I was about to interrupt him, he raised his hand, as much as to say: "No, you can speak afterward; let me finish first." But I seized his arm and shook it.

"Oh," I exclaimed impetuously, "I would rather never see you again than have you trouble the purity of Gertrude's soul. I don't want your confessions! To abuse infirmity, innocence, candor—what abominable cowardice! I should never have thought you capable of it. And to speak of it with such cold-blooded unconcern! . . . Understand me: it is I who have charge of Gertrude and I will not suffer you to speak to her, to touch her, to see her for one single day more."

"But, Father," he went on as calmly as ever, driving me almost beside myself, "you may be sure that I respect Gertrude as much as you can. You are making a strange mistake if you think there is anything reprehensible—I don't say in my conduct, but in my intentions and in my secret heart. I love Gertrude and respect her, I tell you, as much as I love her. The idea of troubling her, of abusing her innocence, is as abominable to me as to you."

Then he protested that what he wanted was to be her help, her friend, her husband; that he had thought he ought not to speak to me about it until he had made up his mind to marry her; that Gertrude herself did not know of his intention and that he had wanted to speak to me about it first.

"This is the confession I had to make to you," he wound up; "and I have nothing else to confess, believe me."

These words filled me with stupor. As I listened, I felt my temples throbbing. I had been prepared with nothing but reproaches, and the fewer grounds he gave me for indignation, the more at a loss I felt, so that at the end of his speech I had nothing left to say.

"Let us go to bed," I said at last, after some moments of silence. I got up and put my hand on his shoulder. "Tomorrow I will tell you what I think about it all."

"Tell me at any rate that you aren't still angry with me."

"I must have the night to think it over."

When I saw Jacques again the next morning, I seemed to be looking at him for the first time. I suddenly realized that my son was no longer a child but a young man; so long as I thought of him as a child, the love that I had accidentally discovered might appear monstrous. I had passed the whole night persuading myself that on the contrary it was perfectly natural and normal. Why was it that my dissatisfaction only became keener still? It was not till later that this became clear to me. In the meantime I had to speak to Jacques and tell him my decision. Now an instinct as sure as the voice of conscience warned me that this marriage must be prevented at all costs.

I took Jacques down to the bottom of the garden.

"Have you said anything to Gertrude?" I began by asking him.

"No," he answered; "perhaps she feels I love her, but I have not yet told her so."

"Then you must promise me not to speak of it yet awhile."

"I am determined to obey you, Father; but may I not know your reasons?"

I hesitated to give them, feeling doubtful whether those that first came into my mind were the wisest to put forward. To tell the truth, conscience rather than reason dictated my conduct.

"Gertrude is too young," I said at last. "You must reflect that she has not yet been confirmed. You know she was unhappily not like other children and did not begin to develop till very late. She is so trustful that she would no doubt be only too easily touched by the first words of love she heard; that is why it is of importance not to say them. To take possession of what is defenseless is cowardice; I know that you are not a coward. Your feelings, you say, are in no way reprehensible; I say they are wrong because they are premature. It is our duty to be prudent for Gertrude till she is able to be prudent for herself. It is a matter of conscience."

Jacques has one excellent point—that the simple words I often used to him as a child: "I appeal to your conscience," have always been sufficient to check him. Meanwhile, as I looked at him, I thought that if Gertrude were able to see, she could not fail to admire the tall slender figure, so straight and yet so lithe, the smooth forehead, the frank look, the face, so childlike still, though now, as it were, overshadowed by a sudden gravity. He was bareheaded, and his fair hair, which was rather long at that time, curled a little at the temples and half hid his ears.

"There is another thing I want to ask you," I went on, rising from the bench where we had been sitting. "You had intended, you said, to go away the day after tomorrow; I beg you not to put off your leaving.

You were to remain away a whole month at least; I beg you not to shorten your absence by a single day. Is that agreed?"

"Very well, Father, I will obey."

I thought he turned extremely pale—so pale that the color left even his lips. But I persuaded myself that such prompt submission argued no very great love, and I felt inexpressively relieved. I was touched besides by his obedience.

"That's the child I love," I said gently. And drawing him to me, I put my lips to his forehead. There was a slight recoil on his part, but I refused to feel hurt by it.

10 March

Our house is so small that we are obliged to live more or less on top of one another, which is sometimes very inconvenient for my work, although I keep a little room for myself upstairs where I can receive my visitors in private—and especially inconvenient when I want to speak to one of the family in private, without such an air of solemnity as would be the case if the interview took place in this little parlor of mine, which the children call my "sanctum" and into which they are forbidden to enter. On that particular morning, however, Jacques had gone to Neuchâtel to buy a pair of boots for his mountaineering, and as it was very fine, the children had gone out after lunch with Gertrude, whom they take charge of, while she at the same time takes charge of them. (It is a pleasure for me to note that Charlotte is particularly attentive to her.) At tea, then, a meal we always take in the common sitting-room, I was quite naturally left alone with Amélie. This was just what I wanted, for I was longing to speak to her. It happens to me so rarely to have a tête-à-tête with her, that I felt almost shy, and the importance of what I had to say agitated me as much as if it had been a question, not of Jacques's affairs, but of my own. I felt too, before I began to speak, how two people who love each other and live practically the same life can yet remain (or become) as much of an enigma to each other as if they lived behind stone walls. Words in this case—those spoken or those heard—have the pathetic sound of vain knocking against the resistance of that dividing barrier, which, unless watch be kept, will grow more and more impenetrable. . . .

"Jacques was speaking to me last night and again this morning," I began as she poured out the tea; and my voice was as faltering as

Jacques's had been steady the day before. "He told me he loved Gertrude."

"It was quite right of him to tell you," said she without looking at me and continuing her housewifely task, as if I had said the most natural thing in the world—or rather as if I had said nothing she did not already know.

"He told me he wanted to marry her; he is resolved to—"

"It was only to be expected," she murmured with a slight shrug of her shoulders.

"Then you suspected it?" I asked in some vexation.

"I've seen it coming on for a long while. But that's the kind of thing men never notice."

It would have been no use to protest, and besides there was perhaps some truth in her rejoinder, so "In that case," I simply objected, "you might have warned me."

She gave me the little crooked smile with which she sometimes accompanies and screens her reticences, and then, with a sideways nod of her head, "If I had to warn you," she said, "of everything you can't see for yourself, I should have my work cut out for me!"

What did she mean by this insinuation? I did not know or care to know, and went on, without attending to it:

"Well, but I want to hear what you think about it."

She sighed. Then: "You know, my dear, that I never approved of that child's staying with us."

I found it difficult not to be irritated by her harking back in this way to the past.

"Gertrude's staying with us is not what we are discussing," I said, but Amélie went on:

"I have always thought it would lead to no good."

With a strong desire to be conciliatory, I caught at her phrase:

"Then you think it would be no good if it led to such a marriage? That's just what I wanted to hear you say. I am glad we are of the same opinion." Then I added that Jacques had submitted quietly to the reasons I had given him, so that there was no need for her to be anxious; that it had been agreed he was to leave the next day for his trip and stay away a whole month.

"As I have no more wish than you that he should find Gertrude here when he comes back," I wound up, "I think the best thing would be to hand her over to the care of Mademoiselle de la M. and I could continue to see her there; for there's no denying that I have very serious obligations to her. I have just been to sound our friend and she is quite ready to oblige us. In this way you will be rid of a presence that is painful to you. Louise de la M. will look after Gertrude; she seemed delighted with the arrangement; she is looking forward already to giving her harmony lessons."

Amélie seemed determined to remain silent, so that I went on:

"As we shall not want Jacques to see Gertrude there, I think it would be a good thing to warn Mademoiselle de la M. of the state of affairs, don't you?"

I hoped by putting this question to get something out of her; but she kept her lips tightly shut, as if she had sworn not to speak. And I went on—not that I had anything more to add, but because I could not endure her silence:

"For that matter, perhaps Jacques will have got over his love by the time he gets back. At his age one hardly knows what one wants."

"And even later one doesn't always know," said she at last, rather oddly.

Her enigmatical and slightly oracular way of speaking irritated me, for I am too frank by nature to put up easily with mystery-making. Turning toward her, I begged her to explain what she meant to imply by that.

"Nothing, my dear," she answered sadly. "I was only thinking that a moment ago you were wishing to be warned of the things you didn't notice yourself."

"Well?"

"Well, I was thinking that it's not always easy to warn people."

I have said that I hate mysteries and I object on principle to hints and double meanings.

"When you want me to understand you, perhaps you will explain yourself more clearly," I replied, rather brutally, perhaps, and I was sorry as soon as I had said it; for I saw her lips tremble a moment. She turned her head aside, then got up and took a few hesitating, almost tottering steps about the room.

"But, Amélie," I cried, "why do you go on being unhappy now that everything is all right again?"

I felt that my eyes embarrassed her, and it was with my back turned and my elbows on the table, resting my head in my hands, that I went on to say:

"I spoke to you unkindly just now. Forgive me."

At that I heard her come up behind me; then I felt her lay her fingers gently on my head as she said tenderly and in a voice trembling with tears:

"My poor dear!"

Then she left the room quickly.

Amélie's words, which I then thought so mysterious, became clear to me soon after this; I have written them down as they struck me at the moment; and that day I only understood that it was time Gertrude should leave.

12 *March*

I had imposed on myself the duty of devoting a little time daily to Gertrude—a few hours or a few minutes, according to the occupations in hand. The day after this conversation with Amélie, I had some free time, and as the weather was inviting, I took Gertrude with me through the forest to that fold in the Jura where in the clear weather one can see, through a curtain of branches and across an immense stretch of land at one's feet, the wonder of the snowy Alps emerging from a thin veil of mist. The sun was already declining on the left when we reached our customary seat. A meadow of thick, closely cropped grass sloped downwards at our feet; farther off, a few cows were grazing; each of them among these mountain herds wears a bell at its neck.

"They outline the landscape," said Gertrude as she listened to their tinkling.

She asked me, as she does every time we go for a walk, to describe the places where we had stopped.

"But you know it already," I told her; "on the fringe of the forest, where one can see the Alps."

"Can one see them clearly today?"

"Yes, in all their splendor."

"You told me they were a little different every day."

"What shall I compare them to this afternoon? To a thirsty midsummer's day. Before evening they will have melted into the air."

"I should like you to tell me if there are any lilies in the big meadows before us."

"No, Gertrude, lilies do not grow on these heights, or only a few rare species."

"Not even the lilies called the lilies of the field?"

"There are no lilies of the field."

"Not even in the fields round Neuchâtel?"

"There are no lilies of the field."

"Then why did our Lord say: 'Consider the lilies of the field'?"

"There were some in his day, no doubt, for him to say so; but they have disappeared before men and their plows."

"I remember you have often told me that what this world most needs is confidence and love. Don't you think that with a little more confidence men would see them again? When I listen to His word, I assure you I see them. I will describe them to you, shall I? They are like bells of flame—great bells of azure, filled with the perfume of love and swinging in the evening breeze. Why do you say there are none there before us? I feel them! I see the meadow filled with them."

"They are not more beautiful than you see them, my Gertrude."

"Say they are not less beautiful."

"They are as beautiful as you see them."

" 'And yet I say unto you that even Solomon in all his glory was not arrayed like one of these,' " said she, quoting Christ's words; and when I heard her melodious voice, I felt I was listening to them for the first time. " 'In all his glory,' " she repeated thoughtfully, and was silent for a time. I went on:

"I have told you, Gertrude, that it is those who have eyes who cannot see." And a prayer rose from the bottom of my heart: "I thank Thee, O Lord, that Thou revealest to the humble what Thou hidest from the wise."

"If you knew," she exclaimed in a rapture of delight, "if you knew how easily I imagine it all! Would you like me to describe the landscape to you? . . . Behind us, above us, and around us are the great fir trees, with their scent of resin and ruddy trunks, stretching out their long dark horizontal branches and groaning as the wind tries to bend them. At our feet, like an open book on the sloping desk of the mountain, lies the broad green meadow, shot with shifting colors—blue in the shade, golden in the sun, and speaking in clear words of flowers—gentians, pulsatillas, ranunculus, and Solomon's beautiful lilies; the cows come and spell them out with their bells; and the angels come and read them—for you say that the eyes of men are closed. Below the book I see a great smoky, misty river of milk, hiding abysses of mystery—an immense river, whose only shore is the beautiful, dazzling Alps far, far away in the distance. . . . That's where Jacques is going. Tell me, is he really starting tomorrow?"

"He is to start tomorrow. Did he tell you so?"

"He didn't tell me so, but I guessed it. Will he be away long?"

"A month. . . . Gertrude, I want to ask you something. Why didn't you tell me that he used to meet you in the church?"

"He came twice. Oh, I don't want to hide anything from you; but I was afraid of making you unhappy."

"It would make me unhappy if you didn't tell me."

Her hand sought mine.

"He was sad at leaving."

"Tell me, Gertrude—did he say he loved you?"

"He didn't say so, but I can feel it without being told. He doesn't love me as much as you do."

"And you, Gertrude, does it make you unhappy that he should go away?"

"I think it is better he should go. I couldn't respond."

"But tell me, does it make you unhappy that he should go?"

"You know, pastor, that it's you I love. . . . Oh, why do you take your hand away? I shouldn't speak so if you weren't married. But no one marries a blind girl. Then why shouldn't we love each other? Tell me, pastor, do you think there's anything wrong in it?"

"It's never in love that the wrong lies."

"I feel there is nothing but good in my heart. I don't want to make Jacques suffer. I don't want to make anyone suffer. . . . I only want to give happiness."

"Jacques was thinking of asking you to marry him."

"Will you let me speak to him before he goes? I should like to make him understand that he must give up loving me. Pastor, you understand, don't you, that I can't marry anyone? You'll let me speak to him, won't you?"

"This evening."

"No, tomorrow; just before he leaves. . . ."

The sun was setting in majestic splendor. The evening air was warm. We had risen and, talking as we went, we turned back along the somber homeward path.

SECOND NOTEBOOK

25 April

I HAVE BEEN obliged to put this book aside for some time.

The snow melted at last and as soon as the roads were passable, there were a great many things to be done that I had been obliged to put off all the long while our village was isolated from the outer world. It was only yesterday I was able for the first time to find a few moments' leisure again.

Last night I read over everything I had written here. . . .

Now that I dare call by its name the feeling that so long lay unacknowledged in my heart, it seems almost incomprehensible that I should have mistaken it until this very day—incomprehensible that those words of Amélie's that I recorded should have appeared mysterious—that even after Gertrude's naïve declarations, I could still have doubted that I loved her. The fact is that I would not then allow that any love outside marriage could be permissible, nor at the same time would I allow that there could be anything whatever forbidden in the feeling that drew me so passionately to Gertrude.

The innocence of her avowals, their very frankness, reassured me. I told myself she was only a child. Real love would not go without confusion and blushes. As far as I was concerned, I persuaded myself I loved her as one loves an afflicted child. I tended her as one tends a sick person—and so I made a moral obligation, a duty, of what was really a passionate inclination. Yes, truly, on the very evening she spoke to me in the way I have described, so happy was I, so light of heart, that I misunderstood my real feelings, and even as I transcribed our talk, I misunderstood them still. For I should have considered love reprehensible, and my conviction was that everything reprehensible must lie heavy on the soul; therefore, as I felt no weight on my soul, I had no thought of love.

These conversations not only were set down just as they occurred, but were also written while I was in the same frame of mind as when they took place; to tell the truth, it was only when I reread them last night that I understood. . . .

As soon as Jacques had gone (I had allowed Gertrude to speak to him before he left, and when he returned for the last few days of his vacation, he affected either to avoid her altogether or to speak to her only in my presence), our life slipped back into its usual peaceful course. Gertrude, as had been arranged, went to stay at Mlle Louise's, where I visited her every day. But, again in my fear of love, I made a point of not talking to her of anything likely to agitate us. I spoke to her only as a pastor and for the most part in Louise's presence, occupying myself chiefly with her religious instruction and with preparing her for Holy Communion, which she has just partaken of this Easter.

I too communicated on Easter Day.

This was a fortnight ago. To my surprise, Jacques, who was spending a week's holiday with us, did not accompany us to the Lord's Table. And I greatly regret having to say that Amélie also abstained—for the first time since our marriage. It seemed as though the two of them had come to an understanding and resolved by their abstention from this solemn celebration to throw a shadow over my joy. Here again I congratulated myself that Gertrude could not see and that I was left to bear the weight of this shadow alone. I know Amélie too well not to be aware of all the blame she wished indirectly to convey by her conduct. She never openly disapproves of me, but she makes a point of showing her displeasure by leaving me in a sort of isolation.

I was profoundly distressed that a grievance of this kind—such a one, I mean, as I shrink from contemplating—should have so affected Amélie's soul as to turn her aside from her higher interests. And when I came home I prayed for her in all sincerity of heart.

As for Jacques's abstention, it was due to quite another motive, as a conversation I had with him a little later on made clear.

3 May

Gertrude's religious instruction has led me to reread the Gospels with a fresh eye. It seems to me more and more that many of the notions that constitute our Christian faith originate not from Christ's own words but from St. Paul's commentaries.

This was, in fact, the subject of the discussion I have just had with Jacques. By disposition he is somewhat hard and rigid, and his mind is not sufficiently nourished by his heart; he is becoming traditionalist and dogmatic. He reproaches me with choosing out of the Christian doctrine "what pleases me." But I do not pick and choose among Christ's words. I simply, between Christ and St. Paul, choose Christ.

He, on the contrary, for fear of finding them in opposition, refuses to dissociate them, refuses to feel any difference of inspiration between them, and makes objections when I say that in one case it is a man I hear, while in the other it is God. The more he argues, the more persuaded I am he does not feel that Christ's slightest word has a divine accent that is unique.

I search the Gospels, I search in vain for commands, threats, prohibitions. . . . All of these come from St. Paul. And it is precisely because they are not to be found in the words of Christ that Jacques is disturbed. Souls like his think themselves lost as soon as they are deprived of their props, their handrails, their fences. And besides they cannot endure others to enjoy a liberty they have resigned, and want to obtain by compulsion what would readily be granted by love.

"But, Father," he said, "I too desire the soul's happiness."

"No, my friend, you desire its submission."

"It is in submission that happiness lies."

I leave him the last word because I dislike arguing; but I know that happiness is endangered when one seeks to obtain it by what should on the contrary be the effect of happiness—and if it is true that the loving soul rejoices in a willing submission, nothing is farther from happiness than submission without love.

For the rest, Jacques reasons well, and if I were not distressed at seeing so much doctrinal harshness in so young a mind, I should no doubt admire the quality of his arguments and his unbending logic. It often seems to me that I am younger than he is—younger today than I was yesterday—and I repeat to myself the words:

"Except ye become as little children, ye shall not enter into the kingdom of heaven."

Do I betray Christ, do I slight, do I profane the Gospels when I see in them above all a *method for attaining the life of blessedness?* The state of joy, which our doubt and the hardness of our hearts prevent, is an obligation laid upon every Christian. Every living creature is more or less capable of joy. Every living creature ought to tend to joy. Gertrude's smile alone teaches me more in this respect than all my lessons teach her.

And these words of Christ's stood out before my eyes in letters of light: "If ye were blind ye should have no sin." Sin is that which darkens the soul—which prevents its joy. Gertrude's perfect happiness, which shines forth from her whole being, comes from the fact that she does not know sin. There is nothing in her but light and love.

I have put into her vigilant hands the four Gospels, the Psalms, the Apocalypse, and the three Epistles of St. John, so that she may read: "God is light, and in him is no darkness at all," as in the Gospel she has already heard the Saviour say: "I am the light of the world." I will not give her the Epistles of St. Paul, for if, being blind, she knows not sin, what is the use of troubling her by letting her read: "sin by the commandment might become exceeding sinful" (Romans vii, 13) and the whole of the dialectic that follows, admirable as it may be.

8 May

Dr. Martins came over yesterday from Chaux-de-Fonds. He examined Gertrude's eyes for a long time with the ophthalmoscope. He told me he had spoken about Gertrude to Dr. Roux, the Lausanne specialist, and is to report his observations to him. They both have an idea that Gertrude might be operated on with success. But we have agreed to say nothing about it to her as long as things are not more certain. Martins is to come and let me know what they think after they have consulted. What would be the good of raising Gertrude's hopes if there is any risk of their being immediately extinguished? And besides is she not happy as she is? . . .

10 May

At Easter Jacques and Gertrude saw each other again in my presence —at least, Jacques saw Gertrude and spoke to her, but only about trifles. He seemed less agitated than I feared; and I persuade myself afresh that if his love had really been very ardent, he would not have got over it so easily, even though Gertrude had told him last year before he went away that it was hopeless. I noticed that he no longer says "thou" to Gertrude, but calls her "you" which is certainly preferable; however, I had not asked him to do so and I am glad it was his own idea. There is undoubtedly a great deal of good in him.

I suspect, however, that this submission of Jacques's was not arrived at without a struggle. The unfortunate thing is that the constraint he has been obliged to impose on his feelings now seems to him good in itself; he would like to see it imposed on everyone; I felt this in the discussion I had with him that I have recorded farther back. Is it not La Rochefoucauld who says that the mind is often the dupe of the heart? I need not say that, knowing Jacques as I do, I did not venture to point this out to him there and then, for I take him to be one of those people who are only made more obstinate by argument; but the very same evening I found what furnished me with a reply—and from St. Paul himself (I could only beat him with his own weapons)—and left a little note in his room, in which I wrote out the text: "Let not him which eateth not judge him that eateth: for God hath received him" (Romans xiv, 3).

I might as well have copied out what follows: "I know, and am persuaded by the Lord Jesus, that there is nothing unclean of itself: but to him that esteemeth any thing to be unclean, to him it is unclean." But I did not dare to, for I was afraid that Jacques might proceed to suspect me of some wrongful interpretation with regard to Gertrude—a suspicion that must not so much as cross his imagination for a second. Evidently it is here a question of food; but in how many passages of the Scriptures are we not called on to give the words a double and triple meaning? ("If thine eye . . ." and the multiplication of the loaves, the miracle of Cana, etc.) This is not a matter of logic-chopping; the meaning of this text is wide and deep: the restriction must not be dictated by the law but by love, and St. Paul exclaims immediately afterward: "But if thy brother be grieved with thy meat, now walkest thou not charitably." It is where love fails that the chink in our armor lies. That is where the Evil One attacks us. Lord, remove from my heart all that does not belong to love. . . . For I was wrong to provoke Jacques: the next morning I found on my table the same note on which I had written out the text; Jacques had simply written on the back of it another text from the same chapter: "Destroy not him with thy meat for whom Christ died" (Roman xiv, 15).

I have reread the whole chapter. It is the starting-point for endless discussion. And is Gertrude to be tormented with these perplexities? Is the brightness of her sky to be darkened with these clouds? Am I not nearer Christ, do I not keep her nearer to Him, when I teach her, when I let her believe, that the only sin is that which hurts the happiness of others or endangers our own?

Alas! There are some souls to whom happiness is uncongenial; they cannot, they do not know how to avail themselves of it. . . . I am thinking of my poor Amélie. I never cease imploring her, urging her—I wish

I could force her to be happy. Yes, I wish I could lift everyone up to God. But she will none of it; she curls up like certain flowers that never open to the sun. Everything she sees causes her uneasiness and distress.

"What's the good, my dear?" she answered me the other day, "we can't all be blind."

Ah, how her irony grieves me! And what courage I need not to be disturbed by it! And yet it seems to me she ought to understand that this allusion to Gertrude's infirmity is particularly painful to me. She makes me feel, indeed, that what I admire above all in Gertrude is her infinite mildness; I have never heard her express the slightest resentment against anyone. It is true I do not allow her to hear anything that might hurt her.

And as the soul that is happy diffuses happiness around it by the radiation of love, so everything in Amélie's neighborhood becomes gloomy and morose. Amiel would say that her soul gives out black rays. When, after a harassing day of toil—visits to the sick, the poor, the afflicted—I come in at nightfall, tired out and with a heart longing for rest, affection, warmth, it is to find, more often than not, worries, recriminations and quarrels, which I dread a thousand times more than the cold, the wind, and the rain out of doors. I know well enough that our old Rosalie invariably wants her own way, but she is not always in the wrong, nor Amélie always in the right when she tries to make her give in. I know that Gaspard and Charlotte are horribly unruly; but would not Amélie get better results if she scolded them less loudly and less constantly? So much nagging, so many reprimands and expostulations, lose their edge like pebbles on the seashore; they are far less disturbing to the children than to me. I know that Claude is teething (at least that is what his mother declares every time he sets up a howl), but does it not encourage him to howl for her or Sarah to run and pick him up and be forever petting him? I am convinced he would not howl so often if he was left to howl once or twice to his heart's content when I am not there. But I know that is the very time they spoil him most.

Sarah is like her mother, and for that reason I should have wished to send her to school. She is not, alas, what her mother was at her age, when we were first engaged, but what the material cares of life have made her—I was going to say that *cultivation* of the cares of life, for Amélie certainly does cultivate them. I find it indeed very difficult to recognize in her today the angel of those early times who smiled encouragement on every high-minded impulse of my heart, who I dreamed would be the sharer of my every hope and fear, and whom I looked on as my guide and leader along the path to heaven—or did love blind me in those days? . . . I cannot see that Sarah has any interests that are not vulgar; like her mother, she allows herself to be entirely taken up with paltry household matters; the very features of her face, unilluminated as they are by any inward flame, look dull and almost hard. She has no taste for poetry or for reading in general; I never overhear any conversation between her and her mother in which I have any inclination to take part, and I feel my isolation even more painfully when I am with them than when I retire to my study, as it is becoming my custom to do more and more often.

And I have also fallen into the habit this autumn, encouraged by the shortness of the days, of taking tea at Mlle de la M.'s whenever my rounds permit it—that is, whenever I can get back early enough. I have not yet mentioned that since last November Louise de la M. has extended her hospitality to three little blind girls, entrusted to her care by Martins. Gertrude is teaching them to read and to work at sundry little tasks over which they have already begun to be quite clever.

How restful, how comforting I find its warm friendly atmosphere every time I re-enter the Grange, and how much I miss it if I am obliged to let two or three days pass without going there. Mlle de la M., it is hardly necessary to say, has sufficient means to take in and provide for Gertrude and the three little borders without putting herself out in any way; three maidservants help her with the greatest devotion and save her all fatigue. Can one imagine fortune and leisure better bestowed? Louise de la M. has always interested herself in the poor; she is a profoundly religious woman and seems hardly to belong to this earth or to live for anything but love; though her hair is already silvery under its lace cap, nothing can be more childlike than her laugh, nothing more harmonious than her movements, nothing more musical than her voice. Gertrude has caught her manners, her way of speaking, almost the intonation, not only of her voice, but of her mind, of her whole being—a likeness upon which I tease them both, but which neither of them will admit. How sweet it is, when I can find the time, to linger in their company, to see them sitting beside each other, Gertrude either leaning her head on her friend's shoulder or clasping one of her hands in hers, while I read them some lines out of Lamartine or

Hugo; how sweet to behold the beauties of such poetry reflected in the mirror of their limpid souls! Even the little pupils are touched by it. These children, in this atmosphere of peace and love, develop astonishingly and make remarkable progress. I smiled at first when Mlle Louise spoke of teaching them to dance—for their health's sake as much as for their amusement—but now I admire the rhythmic grace to which they have attained, though they themselves, alas, are unable to appreciate it. And yet Louise de la M. has persuaded me that though they cannot see, they do physically perceive the harmony of their movements. Gertrude takes part in their dances with the most charming grace and sweetness, and moreover seems to take the keenest pleasure in them. Or sometimes it is Louise de la M. who directs the little girls' movements, and then Gertrude seats herself at the piano. Her progress in music has been astonishing; she plays the organ in chapel now every Sunday and preludes short improvisations to the singing of the hymns. Every Sunday she comes to lunch with us; my children are delighted to see her, notwithstanding that their tastes are growing more and more divergent. Amélie is not too irritable and we get through the meal without a hitch. After lunch the whole family goes back with Gertrude to the Grange and has tea there. It is a treat for my children, and Louise enjoys spoiling them and loading them with delicacies. Amélie, who is far from being insensible to attentions of this kind, unbends at last and looks ten years younger. I think she would find it difficult now to do without this halt in the wearisome round of her daily life.

18 May

Now that the fine weather has returned, I have been able to go out again with Gertrude—a thing I had not done for a long time (for there have been fresh falls of snow quite recently and the roads have been in a terrible state until only a few days ago), and it is a long time too since I have found myself alone with her.

We walked quickly; the sharp air colored her cheeks and kept blowing her fair hair over her face. As we passed alongside a peat-bog, I picked one or two rushes that were in flower and slipped their stalks under her béret; then I twined them into her hair so as to keep them in place.

We had scarcely spoken to each other as yet in the astonishment of finding ourselves alone together, when Gertrude turned her sightless face toward me and asked abruptly:

"Do you think Jacques still loves me?"

"He has made up his mind to give you up," I replied at once.

"But do you think he knows you love me?" she went on.

Since the conversation I have related above, more than six months had gone by without (strange to say) the slightest word of love having passed between us. We were never alone, as I have said, and it was better so. . . . Gertrude's question made my heart beat so fast that I was obliged to slacken our pace a little.

"My dear Gertrude, everyone knows I love you," I cried. But she was not to be put off.

"No, no; you have not answered my question."

And after a moment's silence she went on, with lowered head:

"Aunt Amélie knows it; and *I* know it makes her sad."

"She would be sad anyway," I protested with an unsteady voice. "It is her nature to be sad."

"Oh, you always try to reassure me," she answered with some impatience. "But I don't want to be reassured. There are a great many things, I feel sure, you don't tell me about for fear of troubling or grieving me; a great many things I don't know, so that sometimes—"

Her voice dropped lower and lower; she stopped as if for want of breath. And when, taking up her last words, I asked:

"So that sometimes—?"

"So that sometimes," she continued sadly, "I think all the happiness I owe you is founded upon ignorance."

"But, Gertrude—"

"No, let me say this: I don't want a happiness of that kind. You must understand that I don't—I don't care about being happy. I would rather know. There are a great many things—sad things assuredly—that I can't see, but you have no right to keep them from me. I have reflected a great deal during these winter months; I am afraid, you know, that the whole world is not as beautiful as you have made me believe, pastor—and, in fact, that it is very far from it."

"It is true that man has often defaced it," I argued timidly, for the rush of her thoughts frightened me and I tried to turn it aside, though without daring to hope I should succeed. She seemed to be waiting for these words, for she seized on them at once as though they were the missing link in the chain.

"Exactly!" she cried; "I want to be sure of not adding to the evil."

For a long time we walked on very quickly and in silence. Everything I might have said was checked beforehand by what I felt she was thinking; I dreaded to provoke some sentence that might set both our fates trembling in the balance. And as I thought of what Martins had said about the possibility of her regaining her sight, a dreadful anxiety gripped my heart.

"I wanted to ask you," she went on at last "—but I don't know how to say it. . . ."

Certainly she needed all her courage to speak, just as I needed all mine to listen. But how could I have foreseen the question that was tormenting her?

"Are the children of a blind woman necessarily born blind?"

I don't know which of us this conversation weighed down more, but it was necessary for us to go on.

"No, Gertrude," I said, "except in very special cases. There is in fact no reason why they should be."

She seemed extremely reassured. I should have liked in my turn to ask her why she wanted to know this; I had not the courage and went on clumsily:

"But, Gertrude, to have children, one must be married."

"Don't tell me that, pastor. I know it's not true."

"I have told you what it was proper for me to tell you," I protested. "But it is true, the laws of nature do allow what is forbidden by the laws of man and of God."

"You have often told me the laws of God were the laws of love."

"But such love as that is not the same that also goes by the name of charity."

"Is it out of charity you love me?"

"No, my Gertrude, you know it is not."

"Then you admit our love is outside the laws of God?"

"What do you mean?"

"Oh, you know well enough, and I ought not to be the one to say so."

I sought in vain for some way of evasion; the beating of my heart set all my arguments flying in confusion.

"Gertrude," I exclaimed wildly, "—you think your love wrong?"

She corrected me:

"*Our love.* . . . I say to myself I ought to think so."

"And then—?"

I heard what sounded like a note of supplication in my voice, while without waiting to take breath she went on:

"But that I cannot stop loving you."

All this happened yesterday. I hesitated at first to write it down.... I have no idea how our walk came to an end. We hurried along as if we were being pursued, while I held her arm tightly pressed against me. My soul was so absent from my body that I felt as if the smallest pebble in the path might send us both rolling to the ground.

19 May

Martins came back this morning. Gertrude's is a case for operation. Roux is certain of it and wishes to have her under his care for a time. I cannot refuse and yet, such is my cowardice, that I asked to be allowed to reflect. I asked to have time to prepare her gently.... My heart should leap for joy, but it feels inexpressibly heavy, weighed down by a sick misgiving. At the thought of having to tell Gertrude her sight may be restored to her, my heart fails me altogether.

19 May. Night

I have seen Gertrude and I have not told her. At the Grange this evening there was no one in the drawing-room; I went upstairs to her room. We were alone.

I held her long in my arms pressed to my heart. She made no attempt to resist, and as she raised her face to mine our lips met....

21 May

O Lord, is it for us Thou has clothed the night with such depth and such beauty? Is it for me? The air is warm and the moon shines in at my open window as I sit listening to the vast silence of the skies. Oh, from all creation rises a blended adoration that bears my heart along, lost in an ecstasy that knows no words. I cannot—I cannot pray with calm. If there is any limitation to love, it is set by man and not by Thee, my God. However guilty my love may appear in the eyes of men, oh, tell me that in Thine it is sacred.

I try to rise above the idea of sin; but sin seems to me intolerable, and I will not give up Christ. No, I will not admit that I sin in loving Gertrude. I could only succeed in tearing this love from my heart if I tore my heart out with it, and for what? If I did not already love her, it would be my duty to love her for pity's sake; to cease to love her would be to betray her; she needs my love. . . .

Lord, I know not. . . . I know nothing now but Thee. Be Thou my guide. Sometimes I feel that darkness is closing round me and that it is I who have been deprived of the sight that is to be restored to her.

Gertrude went into the Lausanne nursing-home yesterday and is not to come out for three weeks. I am expecting her return with extreme apprehension. Martins is to bring her back. She has made me promise not to try to see her before then.

22 *May*

A letter from Martins: the operation has been successful. God be thanked!

24 *May*

The idea that she who loved me without seeing me must now see me causes me intolerable discomfort. Will she know me? For the first time in my life I consult the mirror. If I feel her eyes are less indulgent than her heart and less loving, what will become of me? O Lord, I sometimes think I have need of her love in order to love Thee!

8 *June*

An unusual amount of work has enabled me to get through these last days with tolerable patience. Every occupation that takes me out of myself is a merciful one; but all day long and through all that happens her image is with me.

She is coming back tomorrow. Amélie, who during these last weeks has shown only the best side of herself and seems endeavoring to distract my thoughts, is preparing a little festivity with the children to welcome her return.

9 *June*

Gaspard and Charlotte have picked what flowers they could find in the woods and fields. Old Rosalie has manufactured a monumental cake, which Sarah is decorating with gilt paper ornaments. We are expecting her this morning for lunch.

I am writing to fill in the time of waiting. It is eleven o'clock. Every moment I raise my head and look out at the road along which Martins's carriage will come. I resist the temptation to go and meet them; it is better—especially for Amélie's sake—that I should not welcome her apart from the others. My heart leaps. . . . Ah, here they are!

9 June. Evening

Oh, in what abominable darkness I am plunged!

Pity, Lord, pity! I renounce loving her, but do Thou not let her die!

How right my fears were! What has she done? What did she want to do? Amélie and Sarah tell me they went with her as far as the door of the Grange, where Mlle de la M. was expecting her. So she must have gone out again. . . . What happened?

I try to put my thoughts into some sort of order. The accounts they give are incomprehensible or contradictory. My mind is utterly confused . . . Mlle de la M.'s gardener has just brought her back to the Grange unconscious; he says he saw her walking by the river, then she crossed the garden bridge, then stooped and disappeared; but as he did not at first realize that she had fallen, he did not run to her help as he should have done; he found her at the little sluice, where she had been carried by the stream. When I saw her soon afterward, she had not recovered consciousness; or at least had lost it again, for she came to for a moment, thanks to the prompt measures that were taken. Martins, who, thank Heaven, had not yet left, cannot understand the kind of stupor and lassitude in which she is now sunk. He has questioned her in vain; she seems either not to hear or else to be determined not to speak. Her breathing is very labored and Martins is afraid of pneumonia; he has ordered sinapisms and cupping and has promised to come again tomorrow. The mistake was leaving her too long in her wet clothes while they were trying to bring her round; the water of the river is icy. Mlle de la M., who is the only person who has succeeded in getting a few words from her, declares she wanted to pick some of the forget-me-nots that grow in abundance on this side of the river, and that, being still unaccustomed to measure distances, or else mistaking the floating carpet of flowers for solid ground, she suddenly lost her footing. . . . If I could only believe it! If I could only persuade myself it was nothing but an accident, what a dreadful load would be lifted from my heart! During the whole meal, though it was so gay, the strange smile that never left her face made me uneasy; a forced smile, which I had never seen her wear before, but which I tried my utmost to believe was the smile of her newly born sight; a smile that seemed to stream from her eyes onto her face like tears, and beside which the vulgar mirth of the others seemed to me offensive. She did not join in the mirth; I felt as if she had discovered a secret she would surely have confided to me if we had been alone. She hardly spoke; but no one was surprised at that, because she is often silent when she is with others, and all the more so when their merriment grows noisy.

Lord, I beseech Thee, let me speak to her. I must know or how can I continue to live? . . . And yet if she really wished to end her life, is it just because she *knew*? Knew what? Dear, what horrible thing can you have learned? What did I hide from you that was so deadly? What can you so suddenly have seen?

I have been spending two hours at her bedside, my eyes never leaving her forehead, her pale cheeks, her delicate eyelids, shut down over some unspeakable sorrow, her hair still wet and like seaweed as it lies spread round her on the pillow—listening to her difficult, irregular breathing.

10 June

Mlle Louise sent for me this morning just as I was starting to go to the Grange. After a fairly quiet night Gertrude has at last emerged from her torpor. She smiled when I went into the room and motioned to me to come and sit by her bedside. I did not dare question her, and no doubt she was dreading my questions, for she said immediately, as though to forestall anything emotional:

"What do you call those little blue flowers that I wanted to pick by the river? Flowers the color of the sky. Will you be cleverer than I and pick me a bunch of them? I should like to have them here beside my bed. . . ."

The false cheerfulness of her voice was dreadful to me; and no doubt she was aware of it, for she added more gravely:

"I can't speak to you this morning; I am too tired. Go and pick those flowers for me, will you? You can come back again later."

And when an hour later I brought her the bunch of forget-me-nots, Mlle Louise told me that Gertrude was resting and could not see me before evening.

I saw her again this evening. She was lying—almost sitting up in bed —propped against a pile of pillows. Her hair was now fastened up, with the forget-me-nots I had brought her twisted into the plaits above her forehead.

She was obviously very feverish and drew her breath with great difficulty. She kept the hand I put out to her in her burning hand; I remained standing beside her.

"I must confess something to you, pastor; because this evening I am afraid of dying," she said. "What I told you this morning was a lie. It was not to pick flowers. . . . Will you forgive me if I say I wanted to kill myself?"

I fell on my knees beside the bed, still keeping her frail hand in mine; but she disengaged it and began to stroke my forehead, while I buried my face in the sheets so as to hide my tears and stifle my sobs.

"Do you think it was very wrong?" she went on tenderly; then, as I answered nothing:

"My friend, my friend," she said, "you must see that I take up too much room in your heart and in your life. When I came back to you, that was what struck me at once—or, at any rate, that the place I took belonged to another and that it made her unhappy. My crime is that I did not feel it sooner; or rather—for indeed I knew it all along—that I allowed you to love me in spite of it. But when her face suddenly appeared to me, when I saw such unhappiness on her poor face, I could not bear the idea that that unhappiness was my work. . . . No, no, don't blame yourself for anything; but let me go, and give her back her joy."

The hand ceased stroking my forehead; I seized it and covered it with kisses and tears. But she drew it away impatiently and began to toss in the throes of some fresh emotion.

"That is not what I wanted to say to you; no, it's not that I want to say," she kept repeating, and I saw the sweat on her damp forehead. Then she closed her eyes and kept them shut for a time, as though to concentrate her thoughts or to recover her former state of blindness; and in a voice that at first was trailing and mournful, but that soon, as she reopened her eyes, grew louder, grew at last animated even to vehemence:

"When you gave me back my sight," she began, "my eyes opened on a world more beautiful than I had ever dreamed it could be; yes, truly, I had never imagined the daylight so bright, the air so brilliant, the sky so vast. But I had never imagined men's faces so full of care either; and when I went into your house, do you know what it was that struck me first? . . . Oh, it can't be helped, I must tell you: what I saw first of all was our fault, our sin. No, don't protest. You remember Christ's words: 'If ye were blind ye should have no sin.' But now I see. . . . Get up, pastor. Sit there, beside me. Listen to me without interrupting. During the time I spent in the nursing-home I read—or rather I had read to me some verses of the Bible I did not know—some you had never read me. I remember a text of St. Paul's which I repeated to myself all one day: 'For I was alive without the law once; but when the commandment came, sin revived, and I died.' "

She spoke in a state of extreme excitement and in a very loud voice, almost shouting the last words, so that I was made uncomfortable by the idea that they might be heard outside the room; then she shut her eyes and repeated in a whisper, as though for herself alone:

"Sin revived—and *I* died."

I shivered and my heart froze in a kind of terror. I tried to turn aside her thoughts.

"Who read you those texts?" I asked.

"Jacques," she said, opening her eyes and looking at me fixedly. "Did you know he was converted?"

It was more than I could bear; I was going to implore her to stop, but she had already gone on:

"My friend, I am going to grieve you very much; but there must be no falsehood between us now. When I saw Jacques, I suddenly realized it was not you I loved—but him. He had your face—I mean the face I imagined you had. . . . Ah! why did you make me refuse him? I might have married him. . . ."

"But, Gertrude, you still can," I cried with despair in my heart.

"He is entering the priesthood," she said impetuously. Then, shaken by sobs: "Oh, I want to confess to him," she moaned in a kind of ecstasy. . . . "You see for yourself there's nothing left me but to die. I am thirsty. Please call someone. I can't breathe. Leave me. I want to be alone. Ah! I had hoped that speaking to you would have brought me more relief. You must say good-by. We must say good-by. I cannot bear to be with you any more."

I left her. I called Mlle de la M. to take my place beside her; her extreme agitation made me fear the worst, but I could not help seeing that my presence did her harm. I begged that I might be sent for if there was a change for the worse.

11 June

Alas! I was never to see her again alive. She died this morning after a night of delirium and exhaustion. Jacques, who at Gertrude's dying request was telegraphed for by Mlle de la M., arrived a few hours after the end. He reproached me cruelly for not having called in a priest while there was yet time. But how could I have done so when I was still unaware that during her stay at Lausanne, and evidently urged by him, Gertrude had abjured the Protestant faith? He told me in the same breath of his own conversion and Gertrude's. And so they both left me at the same time; it seemed as if, separated by me during their lifetime, they had planned to escape me here and be united to each other in God. But I tell myself that Jacques's conversion is more a matter of the head than the heart.

"Father," he said, "it is not fitting for me to make accusations against you; but it was the example of your error that guided me."

After Jacques had left again, I knelt down beside Amélie and asked her to pray for me, as I was in need of help. She simply repeated "Our Father . . ." but after each sentence she left long pauses, which we filled with our supplication.

I would have wept, but I felt my heart more arid than the desert.

JAMES JOYCE

Though James Joyce was once admired only by a narrow cult of readers, he now commands a broad ascendancy in all of modern literature. As one of the great literary masterpieces of our century, his novel Ulysses *has influenced most of the important prose fiction that has come after it. But* Ulysses' *destiny is not to be read, loved, and appreciated by all people. Since the Joyce name has become a permanent part of our artistic vocabulary, however, and since every syllable he wrote is now faithfully studied, interest in the novel has spilled over to his stories, which are admirable in themselves and not as difficult to comprehend. It is the student's pleasurable assignment, therefore, to assess* The Dead *on its own literary merits. Those merits are great, and as the reader may judge, they are merits of art as technique rather than vision.*

As a technical virtuoso of the craft of writing, Joyce is one to watch. Judiciously assembling significant details of sight and sound, he carefully puts together his script. There is talk, there is movement—the screen comes alive. The images, although created with a roving camera eye, are firm and clear. A kind of parabolic simplicity can be perceived in the story surface.

Gabriel Conroy is the focus of interest. With uncompromising honesty, Joyce shows the discrepancy between Gabriel's life and that of everyone else, including his wife. Through Gabriel—at variance with the rest of the world and with his inner self—the reader is given a merciless view of the deadening influence of Irish life upon Irish spirit. An interesting catalogue could be made of the different kinds of "dead" the story exhibits with unusually sharp reportorial accuracy.

Our author, of course, is concerned with the living or, more precisely, with the effects of the dead upon the living. The American reader cannot easily escape comparing Joyce's satire on the provincial aspects of Irish civilization with similar satiric exposés of American life by some of our own writers of fiction. Whatever tenderness Joyce allows is best seen in his irrepressible touches of humor and poetry, but one is conscious that these subserve the main satiric intention and are not mere embellishments. There is never any doubt concerning the writer's serious moral purpose. The reader knows Joyce is making a direct statement about life; the artist is not simply relating a series of incidents without comment.

What then, exactly, does Joyce accomplish in this story? Immediately, the student is obliged to set in motion the machinery of critical method: What did the writer seem to be trying to do? How did he go about trying to do it? Did he succeed or fail in what he set for himself? Was it worth doing?

The Dead

LILY, the caretaker's daughter, was literally run off her feet. Hardly had she brought one gentleman into the little pantry behind the office on the ground floor and helped him off with his overcoat than the wheezy hall-door bell clanged again and she had to scamper along the bare hallway to let in another guest. It was well for her she had not to attend to the ladies also. But Miss Kate and Miss Julia had thought of that and had converted the bathroom upstairs into a ladies' dressing-room. Miss Kate and Miss Julia were there, gossiping and laughing and fussing, walking after each other to the head of the stairs, peering down over the banisters and calling down to Lily to ask her who had come.

It was always a great affair, the Misses Morkan's annual dance. Everybody who knew them came to it, members of the family, old friends of the family, the members of Julia's choir, any of Kate's pupils that were grown up enough, and even some of Mary Jane's pupils too. Never once had it fallen flat. For years and years it had gone off in splendid style, as long as anyone could remember; ever since Kate and Julia, after the death of their brother Pat, had left the house in Stoney

---

From *Dubliners* by James Joyce. Published by B. W. Heubsch, Inc., in 1916.

Batter and taken Mary Jane, their only niece, to live with them in the dark, gaunt house on Usher's Island, the upper part of which they had rented from Mr. Fulham, the corn-factor on the ground floor. That was a good thirty years ago if it was a day. Mary Jane, who was then a little girl in short clothes, was now the main prop of the household, for she had the organ in Haddington Road. She had been through the Academy and gave a pupils' concert every year in the upper room of the Antient Concert Rooms. Many of her pupils belonged to the better-class families on the Kingstown and Dalkey line. Old as they were, her aunts also did their share. Julia, though she was quite grey, was still the leading soprano in Adam and Eve's, and Kate, being too feeble to go about much, gave music lessons to beginners on the old square piano in the back room. Lily, the caretaker's daughter, did housemaid's work for them. Though their life was modest, they believed in eating well; the best of everything: diamond-bone sirloins, three-shilling tea and the best bottled stout. But Lily seldom made a mistake in the orders, so that she got on well with her three mistresses. They were fussy, that was all. But the only thing they would not stand was back answers.

Of course, they had good reason to be fussy on such a night. And then it was long after ten o'clock and yet there was no sign of Gabriel and his wife. Besides they were dreadfully afraid that Freddy Malins might turn up screwed. They would not wish for worlds that any of Mary Jane's pupils should see him under the influence; and when he was like that it was sometimes very hard to manage him. Freddy Malins always came late, but they wondered what could be keeping Gabriel: and that was what brought them every two minutes to the banisters to ask Lily had Gabriel or Freddy come.

"O, Mr. Conroy," said Lily to Gabriel when she opened the door for him, "Miss Kate and Miss Julia thought you were never coming. Good-night, Mrs. Conroy."

"I'll engage they did," said Gabriel, "but they forget that my wife here takes three mortal hours to dress herself."

He stood on the mat, scraping the snow from his goloshes, while Lily led his wife to the foot of the stairs and called out:

"Miss Kate, here's Mrs. Conroy."

Kate and Julia came toddling down the dark stairs at once. Both of them kissed Gabriel's wife, said she must be perished alive, and asked was Gabriel with her.

"Here I am as right as the mail, Aunt Kate! Go on up. I'll follow," called out Gabriel from the dark.

He continued scraping his feet vigorously while the three women went upstairs, laughing, to the ladies' dressing-room. A light fringe of snow lay like a cape on the shoulders of his overcoat and like toecaps on the toes of his goloshes; and, as the buttons of his overcoat slipped with a squeaking noise through the snow-stiffened frieze, a cold, fragrant air from out-of-doors escaped from crevices and folds.

"Is it snowing again, Mr. Conroy?" asked Lily.

She had preceded him into the pantry to help him off with his overcoat. Gabriel smiled at the three syllables she had given his surname and glanced at her. She was a slim, growing girl, pale in complexion and with hay-coloured hair. The gas in the pantry made her look still paler. Gabriel had known her when she was a child and used to sit on the lowest step nursing a rag doll.

"Yes, Lily," he answered, "and I think we're in for a night of it."

He looked up at the pantry ceiling, which was shaking with the stamping and shuffling of feet on the floor above, listened for a moment to the piano and then glanced at the girl, who was folding his overcoat carefully at the end of a shelf.

"Tell me, Lily," he said in a friendly tone, "do you still go to school?"

"O no, sir," she answered. "I'm done schooling this year and more."

"O, then," said Gabriel gaily, "I suppose we'll be going to your wedding one of these fine days with your young man, eh?"

The girl glanced back at him over her shoulder and said with great bitterness:

"The men that is now is only all palaver and what they can get out of you."

Gabriel coloured, as if he felt he had made a mistake and, without looking at her, kicked off his goloshes and flicked actively with his muffler at his patent-leather shoes.

He was a stout, tallish young man. The high colour of his cheeks pushed upwards even to his forehead, where it scattered itself in a few formless patches of pale red; and on his hairless face there scintillated restlessly the polished lenses and the bright gilt rims of the glasses which screened his delicate and restless eyes. His glossy black hair was parted in the middle and brushed in a long curve behind his ears where it curled slightly beneath the groove left by his hat.

When he had flicked lustre into his shoes he stood up and pulled his waistcoat down more tightly on his plump body. Then he took a coin rapidly from his pocket.

"O Lily," he said, thrusting it into her hands, "it's Christmas-time, isn't it? Just . . . here's a little. . . ."

He walked rapidly towards the door.

"O no, sir!" cried the girl, following him. "Really, sir, I wouldn't take it."

"Christmas-time! Christmas-time!" said Gabriel, almost trotting to the stairs and waving his hand to her in deprecation.

The girl, seeing that he had gained the stairs, called out after him:

"Well, thank you, sir."

He waited outside the drawing-room door until the waltz should finish, listening to the skirts that swept against it and to the shuffling of feet. He was still discomposed by the girl's bitter and sudden retort. It had cast a gloom over him which he tried to dispel by arranging his cuffs and the bows of his tie. He then took from his waistcoat pocket a little paper and glanced at the headings he had made for his speech. He was undecided about the lines from Robert Browning, for he feared they would be above the heads of his hearers. Some quotation that they would recognise from Shakespeare or from Melodies would be better. The indelicate clacking of the men's heels and the shuffling of their soles reminded him that their grade of culture differed from his. He would only make himself ridiculous by quoting poetry to them which they could not understand. They would think that he was airing his superior education. He would fail with them just as he had failed with the girl in the pantry. He had taken up a wrong tone. His whole speech was a mistake from first to last, an utter failure.

Just then his aunts and his wife came out of the ladies' dressing-room. His aunts were two small, plainly dressed old women. Aunt Julia was an inch or so the taller. Her hair, drawn low over the tops of her ears, was grey; and grey also, with darker shadows, was her large flaccid face. Though she was stout in build and stood erect, her slow eyes and parted lips gave her the appearance of a woman who did not know where she was or where she was going. Aunt Kate was more vivacious. Her face, healthier than her sister's, was all puckers and creases, like a shrivelled red apple, and her hair, braided in the same old-fashioned way, had not lost its ripe nut colour.

They both kissed Gabriel frankly. He was their favourite nephew, the son of their dead elder sister, Ellen, who had married T. J. Conroy of the Port and Docks.

"Gretta tells me you're not going to take a cab back to Monkstown tonight, Gabriel," said Aunt Kate.

"No," said Gabriel, turning to his wife, "we had quite enough of that last year, hadn't we? Don't you remember, Aunt Kate, what a cold Gretta got out of it? Cab windows rattling all the way, and the east wind blowing in after we passed Merion. Very jolly it was. Gretta caught a dreadful cold."

Aunt Kate frowned severely and nodded her head at every word.

"Quite right, Gabriel, quite right," she said. "You can't be too careful."

"But as for Gretta there," said Gabriel, "she'd walk home in the snow if she were let."

Mrs. Conroy laughed.

"Don't mind him, Aunt Kate," she said. "He's really an awful bother, what with green shades for Tom's eyes at night and making him do the dumb-bells, and forcing Eva to eat the stirabout. The poor child! And she simply hates the sight of it! . . . O, but you'll never guess what he makes me wear now!"

She broke out into a peal of laughter and glanced at her husband, whose admiring and happy eyes had been wandering from her dress to her face and hair. The two aunts laughed heartily, too, for Gabriel's solicitude was a standing joke with them.

"Goloshes!" said Mrs. Conroy. "That's the latest. Whenever it's wet underfoot I must put on my goloshes. Tonight even, he wanted me to put them on, but I wouldn't. The next thing he'll buy me will be a diving suit."

Gabriel laughed nervously and patted his tie reassuringly, while Aunt Kate nearly doubled herself, so heartily did she enjoy the joke. The smile soon faded from Aunt Julia's face and her mirthless eyes were directed towards her nephew's face. After a pause she asked:

"And what are goloshes, Gabriel?"

"Goloshes, Julia!" exclaimed her sister. "Goodness me, don't you know what goloshes are? You wear them over your . . . over your boots, Gretta, isn't it?"

"Yes," said Mrs. Conroy. "Guttapercha things. We both have a pair now. Gabriel says everyone wears them on the Continent."

"O, on the Continent," murmured Aunt Julia, nodding her head slowly.

Gabriel knitted his brows and said, as if he were slightly angered:

"It's nothing very wonderful, but Gretta thinks it very funny because she says the word reminds her of Christy Minstrels."

"But tell me, Gabriel," said Aunt Kate, with a brisk tact. "Of course, you've seen about the room. Gretta was saying . . ."

"O, the room is all right," replied Gabriel. "I've taken one in the Gresham."

"To be sure," said Aunt Kate, "by far the best thing to do. And the children, Gretta, you're not anxious about them?"

"O, for one night," said Mrs. Conroy. "Besides, Bessie will look after them."

"To be sure," said Aunt Kate again. "What a comfort it is to have a girl like that, one you can depend on! There's that Lily, I'm sure I don't know what has come over her lately. She's not the girl she was at all."

Gabriel was about to ask his aunt some questions on this point, but she broke off suddenly to gaze after her sister, who had wandered down the stairs and was craning her neck over the banisters.

"Now, I ask you," she said almost testily, "where is Julia going? Julia! Julia! Where are you going?"

Julia, who had gone half way down one flight, came back and announced blandly:

"Here's Freddy."

At the same moment a clapping of hands and a final flourish of the pianist told that the waltz had ended. The drawing-room door was opened from within and some couples came out. Aunt Kate drew Gabriel aside hurriedly and whispered into his ear:

"Slip down, Gabriel, like a good fellow and see if he's all right, and don't let him up if he's screwed. I'm sure he's screwed. I'm sure he is."

Gabriel went to the stairs and listened over the banisters. He could hear two persons talking in the pantry. Then he recognised Freddy Malins' laugh. He went down the stairs noisily.

"It's such a relief," said Aunt Kate to Mrs. Conroy, "that Gabriel is here. I always feel easier in my mind when he's here. . . . Julia, there's Miss Daly and Miss Power will take some refreshment. Thanks for your beautiful waltz, Miss Daly. It made lovely time."

A tall wizen-faced man, with stiff grizzled moustache and swarthy skin, who was passing out with his partner, said:

"And may we have some refreshment, too, Miss Morkan?"

"Julia," said Aunt Kate summarily, "and here's Mr. Browne and Miss Furlong. Take them in, Julia, with Miss Daly and Miss Power."

"I'm the man for the ladies," said Mr. Browne, pursing his lips until his moustache bristled and smiling in all his wrinkles. "You know, Miss Morkan, the reason they are so fond of me is——"

He did not finish his sentence, but, seeing that Aunt Kate was out of earshot, at once led the three young ladies into the back room. The middle of the room was occupied by two square tables placed end to end, and on these Aunt Julia and the caretaker were straightening and smoothing a large cloth. On the sideboard were arrayed dishes and plates, and glasses and bundles of knives and forks and spoons. The top of the closed square piano served also as a sideboard for viands and sweets. At a smaller sideboard in one corner two young men were standing, drinking hop-bitters.

Mr. Brownc led his charges thither and invited them all, in jest, to some ladies' punch, hot, strong and sweet. As they said they never took anything strong, he opened three bottles of lemonade for them. Then he asked one of the young men to move aside, and, taking hold of the decanter, filled out for himself a goodly measure of whiskey. The young men eyed him respectfully while he took a trial sip.

"God help me," he said, smiling, "it's the doctor's orders."

His wizened face broke into a broader smile, and the three young ladies laughed in musical echo to his pleasantry, swaying their bodies to and fro, with nervous jerks of their shoulders. The boldest said:

"O, now, Mr. Browne, I'm sure the doctor never ordered anything of the kind."

Mr. Browne took another sip of his whiskey and said, with sidling mimicry:

"Well, you see, I'm like the famous Mrs. Cassidy, who is reported to have said: 'Now, Mary Grimes, if I don't take it, make me take it, for I feel I want it.' "

His hot face had leaned forward a little too confidentially and he had assumed a very low Dublin accent so that the young ladies, with one instinct, received his speech in silence. Miss Furlong, who was one of Mary Jane's pupils, asked Miss Daly what was the name of the pretty waltz she had played; and Mr. Browne, seeing that he was ignored, turned promptly to the two young men who were more appreciative.

A red-faced young woman, dressed in pansy, came into the room, excitedly clapping her hands and crying:

"Quadrilles! Quadrilles!"

Close on her heels came Aunt Kate, crying:

"Two gentlemen and three ladies, Mary Jane!"

"O, here's Mr. Bergin and Mr. Kerrigan," said Mary Jane. "Mr. Kerrigan, will you take Miss Power? Miss Furlong, may I get you a partner, Mr. Bergin. O, that'll just do now."

"Three ladies, Mary Jane," said Aunt Kate.

The two young gentlemen asked the ladies if they might have the pleasure, and Mary Jane turned to Miss Daly.

"O, Miss Daly, you're really awfully good, after playing for the last two dances, but really we're so short of ladies tonight."

"I don't mind in the least, Miss Morkan."

"But I've a nice partner for you, Mr. Bartell D'Arcy, the tenor. I'll get him to sing later on. All Dublin is raving about him."

"Lovely voice, lovely voice!" said Aunt Kate.

As the piano had twice begun the prelude to the first figure Mary Jane led her recruits quickly from the room. They had hardly gone when Aunt Julia wandered slowly into the room, looking behind her at something.

"What is the matter, Julia?" asked Aunt Kate anxiously. "Who is it?"

Julia, who was carrying in a column of table-napkins, turned to her sister and said, simply, as if the question had surprised her:

"It's only Freddy, Kate, and Gabriel with him."

In fact right behind her Gabriel could be seen piloting Freddy Malins across the landing. The latter, a young man of about forty, was of Gabriel's size and build, with very round shoulders. His face was fleshy and pallid, touched with colour only at the thick hanging lobes of his ears and at the wide wings of his nose. He had coarse features, a blunt nose, a convex and receding brow, tumid and protruded lips. His heavy-lidded eyes and the disorder of his scanty hair made him look sleepy. He was laughing heartily in a high key at a story which he had been telling Gabriel on the stairs and at the same time rubbing the knuckles of his left fist backwards and forwards into his left eye.

"Good-evening, Freddy," said Aunt Julia.

Freddy Malins bade the Misses Morkan good-evening in what seemed an offhand fashion by reason of the habitual catch in his voice and then, seeing that Mr. Browne was grinning at him from the sideboard, crossed the room on rather shaky legs and began to repeat in an undertone the story he had just told to Gabriel.

"He's not so bad, is he?" said Aunt Kate to Gabriel.

Gabriel's brows were dark but he raised them quickly and answered:

"O, no, hardly noticeable."

"Now, isn't he a terrible fellow!" she said. "And his poor mother made him take the pledge on New Year's Eve. But come on, Gabriel, into the drawing-room."

Before leaving the room with Gabriel she signalled to Mr. Browne by frowning and shaking her forefinger in warning to and fro. Mr. Browne nodded in answer and, when she had gone, said to Freddy Malins:

"Now, then, Teddy, I'm going to fill you out a good glass of lemonade just to buck you up."

Freddy Malins, who was nearing the climax of his story, waved the offer aside impatiently but Mr. Browne, having first called Freddy Malins' attention to a disarray in his dress, filled out and handed him a full glass of lemonade. Freddy Malins' left hand accepted the glass mechanically, his right hand being engaged in the mechanical readjustment of his dress. Mr. Browne, whose face was once more wrinkling with mirth, poured out for himself a glass of whisky while Freddy Malins exploded, before he had well reached the climax of his story, in a kink of high-pitched bronchitic laughter and, setting down his untasted and overflowing glass, began to rub the knuckles of his left fist backwards and forwards into his left eye, repeating words of his last phrase as well as his fit of laughter would allow him.

Gabriel could not listen while Mary Jane was playing her Academy piece, full of runs and difficult passages, to the hushed drawing-room. He liked music but the piece she was playing had no melody for him and he doubted whether it had any melody for the other listeners, though they had begged Mary Jane to play something. Four young men, who had come from the refreshment-room to stand in the doorway at the sound of the piano, had gone away quietly in couples after a few minutes. The only persons who seemed to follow the music were Mary Jane herself, her hands racing along the key-board or lifted from it at the pauses like those of a priestess in momentary imprecation, and Aunt Kate standing at her elbow to turn the page.

Gabriel's eyes, irritated by the floor, which glittered with beeswax under the heavy chandelier, wandered to the wall above the piano. A picture of the balcony scene in *Romeo and Juliet* hung there and beside it was a picture of the two murdered princes in the Tower which Aunt Julia had worked in red, blue and brown wools when she was a girl. Probably in the school they had gone to as girls that kind of work had been taught for one year. His mother had worked for him as a birthday present a waistcoat of purple tabinet, with little foxes' heads upon it, lined with brown satin and having round mulberry buttons. It was strange that his mother had had no musical talent though Aunt Kate used to call her the brains carrier of the Morkan family. Both she and Julia had always seemed a little proud of their serious and matronly sister. Her photograph stood before the pierglass. She held an open book on her knees and was pointing out something in it to Constantine who, dressed in a man-o'-war suit, lay at her feet. It was she who had chosen the name of her sons for she was very sensible of the dignity of family life. Thanks to her, Constantine was now senior curate in Balbriggan and, thanks to her, Gabriel himself had taken his degree in the Royal University. A shadow passed over his face as he remembered her sullen opposition to his marriage. Some slighting phrases she had used still rankled in his memory; she had once spoken of Gretta as being country cute and that was not true of Gretta at all. It was Gretta who had nursed her during all her last long illness in their house at Monkstown.

He knew that Mary Jane must be near the end of her piece for she was playing again the opening melody with runs of scales after every bar and while he waited for the end the resentment died down in his heart. The piece ended with a trill of octaves in the treble and a final deep octave in the bass. Great applause greeted Mary Jane as, blushing and rolling up her music nervously, she escaped from the room. The most vigorous clapping came from the four young men in the doorway who had gone away to the refreshment-room at the beginning of the piece but had come back when the piano had stopped.

Lancers were arranged. Gabriel found himself partnered with Miss Ivors. She was a frank-mannered talkative young lady, with a freckled face and prominent brown eyes. She did not wear a low-cut bodice and the large brooch which was fixed in the front of her collar bore on it an Irish device and motto.

When they had taken their places she said abruptly:

"I have a crow to pluck with you."

"With me?" said Gabriel.

She nodded her head gravely.

"What is it?" asked Gabriel, smiling at her solemn manner.

"Who is G. C.?" answered Miss Ivors, turning her eyes upon him.

Gabriel coloured and was about to knit his brows, as if he did not understand, when she said bluntly:

"O, innocent Amy! I have found out that you write for *The Daily Express*. Now, aren't you ashamed of yourself?"

"Why should I be ashamed of myself?" asked Gabriel, blinking his eyes and trying to smile.

"Well, I'm ashamed of you," said Miss Ivors frankly. "To say you'd write for a paper like that. I didn't think you were a West Briton."

A look of perplexity appeared on Gabriel's face. It was true that he wrote a literary column every Wednesday in *The Daily Express*, for which he was paid fifteen shillings. But that did not make him a West Briton surely. The books he received for review were almost more welcome than the paltry cheque. He loved to feel the covers and turn over the pages of newly printed books. Nearly every day when his teaching in the college was ended he used to wander down the quays to the second-hand booksellers, to Hickey's on Bachelor's Walk, to Web's or Massey's on Aston's Quay, or to O'Clohissey's in the by-street. He did not know how to meet her charge. He wanted to say that literature was above politics. But they were friends of many years' standing and their careers had been parallel, first at the University and then as teachers: he could not risk a grandiose phrase with her. He continued blinking his eyes and trying to smile and murmured lamely that he saw nothing political in writing reviews of books.

When their turn to cross had come he was still perplexed and inattentive. Miss Ivors promptly took his hand in a warm grasp and said in a soft friendly tone:

"Of course, I was only joking. Come, we cross now."

When they were together again she spoke of the University question and Gabriel felt more at ease. A friend of hers had shown her his review of Browning's poems. That was how she had found out the secret: but she liked the review immensely. Then she said suddenly:

"O, Mr. Conroy, will you come for an excursion to the Aran Isles this summer? We're going to stay there a whole month. It will be splendid out in the Atlantic. You ought to come. Mr. Clancy is coming, and Mr. Kilkelly and Kathleen Kearney. It would be splendid for Gretta too if she'd come. She's from Connacht, isn't she?"

"Her people are," said Gabriel shortly.

"But you will come, won't you?" said Miss Ivors, laying her warm hand eagerly on his arm.

"The fact is," said Gabriel, "I have just arranged to go——"

"Go where?" asked Miss Ivors.

"Well, you know, every year I go for a cycling tour with some fellows and so——"

"But where?" asked Miss Ivors.

"Well, we usually go to France or Belgium or perhaps Germany," said Gabriel awkwardly.

"And why do you go to France and Belgium," said Miss Ivors, "instead of visiting your own land?"

"Well," said Gabriel, "it's partly to keep in touch with the languages and partly for a change."

"And haven't you your own language to keep in touch with—Irish?" asked Miss Ivors.

"Well," said Gabriel, "if it comes to that, you know, Irish is not my language."

Their neighbors had turned to listen to the cross-examination. Gabriel glanced right and left nervously and tried to keep his good humour under the ordeal which was making a blush invade his forehead.

"And haven't you your own land to visit," continued Miss Ivors, "that you know nothing of, your own people, and your own country?"

"O, to tell you the truth," retorted Gabriel suddenly, "I'm sick of my own country, sick of it!"

"Why?" asked Miss Ivors.

Gabriel did not answer for his retort had heated him.

"Why?" repeated Miss Ivors.

They had to go visiting together and, as he had not answered her, Miss Ivors said warmly:

"Of course, you've no answer."

Gabriel tried to cover his agitation by taking part in the dance with great energy. He avoided her eyes for he had seen a sour expression on her face. But when they met in the long chain he was surprised to feel his hand firmly pressed. She looked at him from under her brows for a moment quizzically until he smiled. Then, just as the chain was about to start again, she stood on tiptoe and whispered into his ear:

"West Briton!"

When the lancers were over Gabriel went away to a remote corner of the room where Freddy Malins' mother was sitting. She was a stout feeble old woman with white hair. Her voice had a catch in it like her son's and she stuttered slightly. She had been told that Freddy had come and that he was nearly all right. Gabriel asked her whether she had had a good crossing. She lived with her married daughter in Glasgow and came to Dublin on a visit once a year. She answered placidly that she had had a beautiful crossing and that the captain had been most attentive to her. She spoke also of the beautiful house her daughter kept in Glasgow, and of all the friends they had there. While her tongue rambled on Gabriel tried to banish from his mind all memory of the unpleasant incident with Miss Ivors. Of course the girl or woman, or whatever she was, was an enthusiast but there was a time for all things. Perhaps he ought not to have answered her like that. But she had no right to call him a West Briton before people, even in joke. She had tried to make him ridiculous before people, heckling him and staring at him with her rabbit's eyes.

He saw his wife making her way towards him through the waltzing couples. When she reached him she said into his ear:

"Gabriel, Aunt Kate wants to know won't you carve the goose as usual. Miss Daly will carve the ham and I'll do the pudding."

"All right," said Gabriel.

"She's sending in the younger ones first as soon as this waltz is over so that we'll have the table to ourselves."

"Were you dancing?" asked Gabriel.

"Of course I was. Didn't you see me? What row had you with Molly Ivors?"

"No row. Why? Did she say so?"

"Something like that. I'm trying to get that Mr. D'Arcy to sing. He's full of conceit, I think."

"There was no row," said Gabriel moodily, "only she wanted me to go for a trip to the west of Ireland and I said I wouldn't."

His wife clasped her hands excitedly and gave a little jump.

"O, do go, Gabriel," she cried. "I'd love to see Galway again."

"You can go if you like," said Gabriel coldly.

She looked at him for a moment, then turned to Mrs. Malins and said:

"There's a nice husband for you, Mrs. Malins."

While she was threading her way back across the room Mrs. Malins, without adverting to the interruption, went on to tell Gabriel what beautiful places there were in Scotland and beautiful scenery. Her son-in-law brought them every year to the lakes and they used to go fishing. Her son-in-law was a splendid fisher. One day he caught a beautiful big fish and the man in the hotel cooked it for their dinner.

Gabriel hardly heard what she said. Now that supper was coming near he began to think again about his speech and about the quotation. When he saw Freddy Malins coming across the room to visit his mother Gabriel left the chair free for him and retired into the embrasure of the window. The room had already cleared and from the back room came the clatter of plates and knives. Those who still remained in the drawing-room seemed tired of dancing and were conversing quietly in little groups. Gabriel's warm trembling fingers tapped the cold pane of the window. How cool it must be outside! How pleasant it would be to walk out alone, first along by the river and then through the park! The snow would be lying on the branches of the trees and forming a bright cap on the top of the Wellington Monument. How much more pleasant it would be there than at the supper-table!

He ran over the headings of his speech: Irish hospitality, sad memories, the Three Graces, Paris, the quotation from Browning. He repeated to himself a phrase he had written in his review: "One feels that one is listening to a thought-tormented music." Mis Ivors had praised the review. Was she sincere? Had she really any life of her own behind all her propagandism? There had never been any ill-feeling between them until that night. It unnerved him to think that she would be at the supper-table, looking up at him while he spoke with her critical quizzing eyes. Perhaps she would not be sorry to see him fail in his speech. An idea came into his mind and gave him courage. He would say, alluding to Aunt Kate and Aunt Julia: "Ladies and Gentlemen, the generation which is now on the wane among us may have had its faults but for my part I think it had certain qualities of hospitality, of humour, of humanity, which the new and very serious and hypereducated generation that is growing up around us seems to me to lack." Very good: that was one for Miss Ivors. What did he care that his aunts were only two ignorant old women?

A murmur in the room attracted his attention. Mr. Browne was advancing from the door, gallantly escorting Aunt Julia, who leaned upon his arm, smiling and hanging her head. An irregular musketry of applause escorted her also as far as the piano and then, as Mary Jane seated herself on the stool, and Aunt Julia, no longer smiling, half turned so as to pitch her voice fairly into the room, gradually ceased. Gabriel recognised the prelude. It was that of an old song of Aunt Julia's—*Arrayed for the Bridal.* Her voice, strong and clear in tone, attacked with great spirit the runs which embellish the air and though she sang very rapidly she did not miss even the smallest of the grace notes. To follow the voice, without looking at the singer's face, was to feel and share the excitement of swift and secure flight. Gabriel applauded loudly with all the others at the close of the song and loud applause was borne in from the invisible supper-table. It sounded so genuine that a little colour struggled into Aunt Julia's face as she bent to replace in the music-stand the old leather-bound songbook that had her initials on the cover. Freddy Malins, who had listened with his head perched sideways to hear her better, was still applauding when everyone else had ceased and talking animatedly to his mother who nodded her head gravely and slowly in acquiescence. At last, when he could clap no more, he stood up suddenly and hurried across the room to Aunt Julia whose hand he seized and held in both his hands, shaking it when words failed him or the catch in his voice proved too much for him.

"I was just telling my mother," he said, "I never heard you sing so well, never. No, I never heard your voice so good as it is tonight. Now! Would you believe that now? That's the truth. Upon my word and honour that's the truth. I never heard your voice sound so fresh and so . . . so clear and fresh, never."

Aunt Julia smiled broadly and murmured something about compliments as she released her hand from his grasp. Mr. Browne extended his open hand towards her and said to those who were near him in the manner of a showman introducing a prodigy to an audience:

"Miss Julia Morkan, my latest discovery!"

He was laughing very heartily at this himself when Freddy Malins turned to him and said:

"Well, Browne, if your're serious you might make a worse discovery. All I can say is I never heard her sing half so well as long as I am coming here. And that's the honest truth."

"Neither did I," said Mr. Browne. "I think her voice has greatly improved."

Aunt Julia shrugged her shoulders and said with meek pride:

"Thirty years ago I hadn't a bad voice as voices go."

"I often told Julia," said Aunt Kate emphatically, "that she was simply thrown away in that choir. But she never would be said by me."

She turned as if to appeal to the good sense of the others against a refractory child while Aunt Julia gazed in front of her, a vague smile of reminiscence playing on her face.

"No," continued Aunt Kate, "she wouldn't be said or led by anyone, slaving there in that choir night and day, night and day. Six o'clock on Christmas morning! And all for what?"

"Well, isn't it for the honour of God, Aunt Kate?" asked Mary Jane, twisting round on the piano-stool and smiling.

Aunt Kate turned fiercely on her niece and said:

"I know all about the honour of God, Mary Jane, but I think it's not at all honourable for the pope to turn out the women out of the choirs that have slaved there all their lives and put little whipper-snappers of boys over their heads. I suppose it is the good of the Church if the pope does it. But it's not just, Mary Jane, and it's not right."

She had worked herself into a passion and would have continued in defence of her sister for it was a sore subject with her but Mary Jane, seeing that all the dancers had come back, intervened pacifically:

"Now, Aunt Kate, you're giving scandal to Mr. Browne who is of the other persuasion."

Aunt Kate turned to Mr. Browne, who was grinning at this allusion to his religion, and said hastily:

"O, I don't question the pope's being right. I'm only a stupid old woman and I wouldn't presume to do such a thing. But there's such a thing as common everyday politeness and gratitude. And if I were in Julia's place I'd tell that Father Healey straight up to his face . . ."

"And besides, Aunt Kate," said Mary Jane, "we really are all hungry and when we are hungry we are all very quarrelsome."

"And when we are thirsty we are also quarrelsome," added Mr. Browne.

"So that we had better go to supper," said Mary Jane, "and finish the discussion afterwards."

On the landing outside the drawing-room Gabriel found his wife and Mary Jane trying to persuade Miss Ivors to stay for supper. But Miss Ivors, who had put on her hat and was buttoning her cloak, would not stay. She did not feel in the least hungry and she had already overstayed her time.

"But only for ten minutes, Molly," said Mrs. Conroy. "That won't delay you."

"To take a pick itself," said Mary Jane, "after all your dancing."

"I really couldn't," said Miss Ivors.

"I am afraid you didn't enjoy yourself at all," said Mary Jane hopelessly.

"Ever so much, I assure you," said Miss Ivors, "but you really must let me run off now."

"But how can you get home?" asked Mrs. Conroy.

"O, it's only two steps up the quay."

Gabriel hesitated a moment and said:

"If you will allow me, Miss Ivors, I'll see you home if you are really obliged to go."

But Miss Ivors broke away from them.

"I won't hear of it," she cried. "For goodness' sake go in to your suppers and don't mind me. I'm quite well able to take care of myself."

"Well, you're the comical girl, Molly," said Mrs. Conroy frankly.

"*Beannacht libh*," cried Miss Ivors, with a laugh, as she ran down the staircase.

Mary Jane gazed after her, a moody puzzled expression on her face, while Mrs. Conroy leaned over the banisters to listen for the hall-door. Gabriel asked himself was he the cause of her abrupt departure. But she did not seem to be in ill humour: she had gone away laughing. He stared blankly down the staircase.

At the moment Aunt Kate came toddling out of the supper-room, almost wringing her hands in despair.

"Where is Gabriel?" she cried. "Where on earth is Gabriel? There's everyone waiting in there, stage to let, and nobody to carve the goose!"

"Here I am, Aunt Kate!" cried Gabriel, with sudden animation, "ready to carve a flock of geese, if necessary."

A fat brown goose lay at one end of the table and at the other end, on a bed of creased paper strewn with sprigs of parsley, lay a great ham, stripped of its outer skin and peppered over with crust crumbs, a neat paper frill round its shin and beside this was a round of spiced beef. Between these rival ends ran parallel lines of side-dishes: two little minsters of jelly, red and yellow; a shallow dish full of blocks of blancmange and red jam, a large green leaf-shaped dish with a stalk-shaped handle, on which lay bunches of purple raisins and peeled almonds, a companion dish on which lay a solid rectangle of Smyrna figs, a dish of custard topped with grated nutmeg, a small bowl full of chocolates and sweets wrapped in gold and silver papers and a glass vase in which stood some tall celery stalks. In the centre of the table there stood, as sentries to a fruit-stand which upheld a pyramid of oranges and American apples, two squat old-fashioned decanters of cut glass, one containing port and the other dark sherry. On the closed square piano a pudding in a huge yellow dish lay in waiting and behind it were three squads of bottles of stout and ale and minerals, drawn up according to the colours of their uniforms, the first two black, with brown and red labels, the third and smallest squad white, with transverse green sashes.

Gabriel took his seat boldly at the head of the table and, having looked to the edge of the carver, plunged his fork firmly into the goose. He felt quite at ease now for he was an expert carver and liked nothing better than to find himself at the head of a well-laden table.

"Miss Furlong, what shall I send you?" he asked. "A wing or a slice of the breast?"

"Just a small slice of the breast."

"Miss Higgins, what for you?"

"O, anything at all, Mr. Conroy."

While Gabriel and Miss Daly exchanged plates of goose and plates of ham and spiced beef Lily went from guest to guest with a dish of hot floury potatoes wrapped in a white napkin. This was Mary Jane's idea and she had also suggested apple sauce for the goose but Aunt Kate had said that plain roast goose without any apple sauce had always been good enough for her and she hoped she might never eat worse. Mary Jane waited on her pupils and saw that they got the best slices and Aunt Kate and Aunt Julia opened and carried across from the piano bottles of stout and ale for the gentlemen and bottles of minerals for the ladies. There was a great deal of confusion and laughter and noise, the noise of orders and counter-orders, of knives and forks, of corks and glass-stoppers. Gabriel began to carve second helpings as soon as he had finished the first round without serving himself. Everyone protested loudly so that he compromised by taking a long draught of stout for he had found the carving hot work. Mary Jane settled down quietly to her supper but Aunt Kate and Aunt Julia were still toddling round the table, walking on each other's heels, getting in each other's way and giving each other unheeded orders. Mr. Browne begged of them to sit down and eat their suppers and so did Gabriel but they said there was time enough, so that, at last, Freddy Malins stood up and, capturing Aunt Kate, plumped her down on her chair amid general laughter.

When everyone had been well served Gabriel said, smiling:

"Now, if anyone wants a little more of what vulgar people call stuffing let him or her speak."

A chorus of voices invited him to begin his own supper and Lily came forward with three potatoes which she had reserved for him.

"Very well," said Gabriel amiably, as he took another preparatory draught, "kindly forget my existence, ladies and gentlemen, for a few minutes."

He set to his supper and took no part in the conversation with which the table covered Lily's removal of the plates. The subject of talk was the opera company which was then at the Theatre Royal. Mr. Bartell D'Arcy, the tenor, a dark-complexioned young man with a smart moustache, praised very highly the leading contralto of the company but Miss Furlong thought she had a rather vulgar style of production. Freddy Malins said there was a Negro chieftain singing in the second part of the Gaiety pantomime who had one of the finest tenor voices he had ever heard.

"Have you heard him?" he asked Mr. Bartell D'Arcy across the table.

"No," answered Mr. Bartell D'Arcy carelessly.

"Because," Freddy Malins explained, "now I'd be curious to hear your opinion of him. I think he has a grand voice."

"It takes Teddy to find out the really good things," said Mr. Browne familiarly to the table.

"And why couldn't he have a voice too?" asked Freddy Malins sharply. "Is it because he's only a black?"

Nobody answered this question and Mary Jane led the table back to the legitimate opera. One of her pupils had given her a pass for *Mignon.* Of course it was very fine, she said, but it made her think of poor Georgina Burns. Mr. Browne could go back farther still, to the old Italian companies that used to come to Dublin—Tietjens, Ilma de Murzka, Companini, the great Trebelli, Giuglini, Ravelli, Aramburo. Those were the days, he said, when there was something like singing to be heard in Dublin. He told too of how the top gallery of the old Royal used to be packed night after night, of how one night an Italian tenor had sung five encores to *Let me like a Soldier fall,* introducing a high

C every time, and of how the gallery boys would sometimes in their enthusiasm unyoke the horses from the carriage of some great *prima donna* and pull her themselves through the streets to her hotel. Why did they never play the grand old operas now, he asked, *Dinorah, Lucrezia Borgia*? Because they could not get the voices to sing them: that was why.

"Oh, well," said Mr. Bartell D'Arcy, "I presume there are as good singers today as there were then."

"Where are they?" asked Mr. Browne defiantly.

"In London, Paris, Milan," said Mr. Bartell D'Arcy warmly. "I suppose Caruso, for example, is quite as good, if not better than any of the men you have mentioned."

"Maybe so," said Mr. Browne. "But I may tell you I doubt it strongly."

"O, I'd give anything to hear Caruso sing," said Mary Jane.

"For me," said Aunt Kate, who had been picking a bone, "there was only one tenor. To please me, I mean. But I suppose none of you ever heard of him."

"Who was he, Miss Morkan?" asked Mr. Bartell D'Arcy politely.

"His name," said Aunt Kate, "was Parkinson. I heard him when he was in his prime and I think he had then the purest tenor voice that was ever put into a man's throat."

"Strange," said Mr. Bartell D'Arcy. "I never even heard of him."

"Yes, yes, Miss Morkan is right," said Mr. Browne. "I remember hearing of old Parkinson but he's too far back for me."

"A beautiful, pure, sweet, mellow English tenor," said Aunt Kate with enthusiasm.

Gabriel having finished, the huge pudding was transferred to the table. The clatter of forks and spoons began again. Gabriel's wife served out spoonfuls of the pudding and passed the plates down the table. Midway down they were held up by Mary Jane, who replenished them with raspberry or orange jelly or with blancmange and jam. The pudding was of Aunt Jula's making and she received praises for it from all quarters. She herself said that it was not quite brown enough.

"Well, I hope, Miss Morkan," said Mr. Browne, "that I'm brown enough for you because, you know, I'm all brown."

All the gentlemen, except Gabriel, ate some of the pudding out of compliment to Aunt Julia. As Gabriel never ate sweets the celery had been left for him. Freddy Malins also took a stalk of celery and ate it with his pudding. He had been told that celery was a capital thing for the blood and he was just then under doctor's care. Mrs. Malins, who had been silent all through the supper, said that her son was going down to Mount Melleray in a week or so. The table then spoke of Mount Melleray, how bracing the air was down there, how hospitable the monks were and how they never asked for a penny-piece from their guests.

"And do you mean to say," asked Mr. Browne incredulously, "that a chap can go down there and put up there as if it were a hotel and live on the fat of the land and then come away without paying anything?"

"O, most people give some donation to the monastery when they leave," said Mary Jane.

"I wish we had an institution like that in our Church," said Mr. Browne candidly.

He was astonished to hear that the monks never spoke, got up at two in the morning and slept in their coffins. He asked what they did it for.

"That's the rule of the order," said Aunt Kate firmly.

"Yes, but why?" asked Mr. Browne.

Aunt Kate repeated that it was the rule, that was all. Mr. Browne still seemed not to understand. Freddy Malins explained to him, as best he could, that the monks were trying to make up for the sins committed by all the sinners in the outside world. The explanation was not very clear for Mr. Browne grinned and said:

"I like that idea very much but wouldn't a comfortable spring bed do them as well as a coffin?"

"The coffin," said Mary Jane, "is to remind them of their last end."

As the subject had grown lugubrious it was buried in a silence of the table during which Mrs. Malins could be heard saying to her neighbour in an indistinct undertone:

"They are very good men, the monks, very pious men."

The raisins and almonds and figs and apples and oranges and chocolates and sweets were now passed about the table and Aunt Julia invited all the guests to have either port or sherry. At first Mr. Bartell D'Arcy refused to take either but one of his neighbours nudged him and whispered something to him upon which he allowed his glass to be filled. Gradually as the last glasses were being filled the conversation ceased. A pause followed, broken only by the noise of the wine and by unsettlings of chairs. The Misses Morkan, all three, looked down at the tablecloth. Someone coughed once or twice and then a few gentlemen patted the table gently as a signal for silence. The silence came and Gabriel pushed back his chair and stood up.

The patting at once grew louder in encouragement and then ceased altogether. Gabriel leaned his ten trembling fingers on the tablecloth and smiled nervously at the company. Meeting a row of upturned faces he raised his eyes to the chandelier. The piano was playing a waltz tune and he could hear the skirts sweeping against the drawing-room door. People, perhaps, were standing in the snow on the quay outside, gazing up at the lighted windows and listening to the waltz music. The air was pure there. In the distance lay the park where the trees were weighted with snow. The Wellington Monument wore a gleaming cap of snow that flashed westward over the white field of Fifteen Acres.

He began:

"Ladies and Gentlemen,

"It has fallen to my lot this evening, as in years past, to perform a very pleasing task but a task for which I am afraid my poor powers as a speaker are all too inadequate."

"No, no!" said Mr. Browne.

"But, however that may be, I can only ask you tonight to take the will for the deed and to lend me your attention for a few moments while I endeavour to express to you in words what my feelings are on this occasion.

"Ladies and Gentlemen, it is not the first time that we have gathered together under this hospitable roof, around this hospitable board. It is not the first time that we have been the recipients—or perhaps, I had better say, the victims—of the hospitality of certain good ladies."

He made a circle in the air with his arm and paused. Everyone laughed or smiled at Aunt Kate and Aunt Julia and Mary Jane who all turned crimson with pleasure. Gabriel went on more boldly:

"I feel more strongly with every recurring year that our country has no tradition which does it so much honour and which it should guard so jealously as that of its hospitality. It is a tradition that is unique as far as my experience goes (and I have visited not a few places abroad) among the modern nations. Some would say, perhaps, that with us it is rather a failing than anything to be boasted of. But granted even that, it is, to my mind, a princely failing, and one that I trust will long be cultivated among us. Of one thing, at least, I am sure. As long as this one roof shelters the good ladies aforesaid—and I wish from my heart it may do so for many and many a long year to come—the tradition of genuine warm-hearted courteous Irish hospitality, which our forefathers have handed down to us and which we in turn must hand down to our descendants, is still alive among us."

A hearty murmur of assent ran round the table. It shot through Gabriel's mind that Miss Ivors was not there and that she had gone away discourteously: and he said with confidence in himself:

"Ladies and Gentlemen,

"A new generation is growing up in our midst, a generation actuated by new ideas and new principles. It is serious and enthusiastic for these new ideas and its enthusiasm, even when it is misdirected, is, I believe, in the main sincere. But we are living in a sceptical and, if I may use the phrase, a thought-tormented age: and sometimes I fear that this new generation, educated or hypereducated as it is, will lack those qualities of humanity, of hospitality, of kindly humour which belonged to an older day. Listening tonight to the names of all those great singers of the past it seemed to me, I must confess, that we were living in a less spacious age. Those days might, without exaggeration, be called spacious days: and if they are gone beyond recall let us hope, at least, that in gatherings such as this we shall still speak of them with pride and affection, still cherish in our hearts the memory of those dead and gone great ones whose fame the world will not willingly let die."

"Hear, hear!" said Mr. Browne loudly.

"But yet," continued Gabriel, his voice falling into a softer inflection, "there are always in gatherings such as this sadder thoughts that will recur to our minds: thoughts of the past, of youth, of changes, of absent faces that we miss here tonight. Our path through life is strewn with many such sad memories: and were we to brood upon them always we could not find the heart to go on bravely with our work among the living. We have all of us living duties and living affections which claim, and rightly claim, our strenuous endeavours.

"Therefore, I will not linger on the past. I will not let any gloomy moralising intrude upon us here tonight. Here we are gathered together for a brief moment from the bustle and rush of our everyday routine. We are met here as friends, in the spirit of good-fellowship, as colleagues, also to a certain extent, in the true spirit of *camaraderie*, and as the guests of—what shall I call them?—the Three Graces of the Dublin musical world."

The table burst into applause and laughter at this allusion. Aunt Julia vainly asked each of her neighbours in turn to tell her what Gabriel had said.

"He says we are the Three Graces, Aunt Julia," said Mary Jane.

Aunt Julia did not understand but she looked up, smiling, at Gabriel, who continued in the same vein:

"Ladies and Gentlemen,

"I will not attempt to play tonight the part that Paris played on another occasion. I will not attempt to choose between them. The task would be an invidious one and one beyond my poor powers. For when I view them in turn, whether it be our chief hostess herself, whose good heart, whose too good heart, has become a byword with all who know her, or her sister, who seems to be gifted with perennial youth and whose singing must have been a surprise and a revelation to us all tonight, or, last but not least, when I consider our youngest hostess, talented, cheerful, hard-working and the best of nieces, I confess, Ladies and Gentlemen, that I do not know to which of them I should award the prize."

Gabriel glanced down at his aunts and, seeing the large smile on Aunt Julia's face and the tears which had risen to Aunt Kate's eyes, hastened to his close. He raised his glass of port gallantly, while every member of the company fingered a glass expectantly, and said loudly:

"Let us toast them all three together. Let us drink to their health, wealth, long life, happiness and prosperity and may they long continue to hold the proud and self-won position which they hold in their profession and the position of honour and affection which they hold in our hearts."

All the guests stood up, glass in hand, and turning towards the three seated ladies, sang in unison, with Mr. Browne as leader:

For they are jolly gay fellows,
For they are jolly gay fellows,
For they are jolly gay fellows,
Which nobody can deny.

Aunt Kate was making frank use of her handkerchief and even Aunt Julia seemed moved. Freddy Malins beat time with his pudding-fork and the singers turned towards one another, as if in melodious conference, while they sang with emphasis:

Unless he tells a lie,
Unless he tells a lie,

Then, turning once more towards their hostesses, they sang:

For they are jolly gay fellows,
For they are jolly gay fellows,
For they are jolly gay fellows,
Which nobody can deny.

The acclamation which followed was taken up beyond the door of the supper-room by many of the other guests and renewed time after time, Freddy Malins acting as officer with his fork on high.

The piercing morning air came into the hall where they were standing so that Aunt Kate said:

"Close the door, somebody. Mrs. Malins will get her death of cold."

"Browne is out there, Aunt Kate," said Mary Jane.

"Browne is everywhere," said Aunt Kate, lowering her voice.

Mary Jane laughed at her tone.

"Really," she said archly, "he is very attentive."

"He has been laid on here like the gas," said Aunt Kate in the same tone, "all during the Christmas."

She laughed herself this time good-humouredly and then added quickly:

"But tell him to come in, Mary Jane, and close the door. I hope to goodness he didn't hear me."

At that moment the hall-door was opened and Mr. Browne came in from the doorstep, laughing as if his heart would break. He was dressed in a long green overcoat with mock astrakhan cuffs and collar and wore on his head an oval fur cap. He pointed down the snow-covered quay from where the sound of shrill prolonged whistling was borne in.

"Teddy will have all the cabs in Dublin out," he said.

Gabriel advanced from the little pantry behind the office, struggling into his overcoat and, looking round the hall, said:

"Gretta not down yet?"

"She's getting on her things, Gabriel," said Aunt Kate.

"Who's playing up there?" asked Gabriel.

"Nobody. They're all gone."

"O no, Aunt Kate," said Mary Jane. "Bartell D'Arcy and Miss O'Callaghan aren't gone yet."

"Someone is fooling at the piano anyhow," said Gabriel.

Many Jane glanced at Gabriel and Mr. Browne and said with a shiver:

"It makes me feel cold to look at you two gentlemen muffled up like that. I wouldn't like to face your journey home at this hour."

"I'd like nothing better this minute," said Mr. Browne stoutly, "than a rattling fine walk in the country or a fast drive with a good spanking goer between the shafts."

"We used to have a very good horse and trap at home," said Aunt Julia sadly.

"The never-to-be forgotten Johnny," said Mary Jane, laughing.

Aunt Kate and Gabriel laughed too.

"Why, what was wonderful about Johnny?" asked Mr. Browne.

"The late lamented Patrick Morkan, our grandfather, that is," explained Gabriel, "commonly known in his later years as the old gentleman, was a glue-boiler."

"O, now, Gabriel," said Aunt Kate, laughing, "he had a starch mill."

"Well, glue or starch," said Gabriel, "the old gentleman had a horse by the name of Johnny. And Johnny used to work in the old gentleman's mill, walking round and round in order to drive the mill. That was all very well; but now comes the tragic part about Johnny. One fine day the old gentleman thought he'd like to drive out with the quality to a military review in the park."

"The Lord have mercy on his soul," said Aunt Kate compassionately.

"Amen," said Gabriel. "So the old gentleman, as I said, harnessed Johnny and put on his very best tall hat and his very best stock collar and drove out in grand style from his ancestral mansion somewhere near Back Lane, I think."

Everyone laughed, even Mrs. Malins, at Gabriel's manner and Aunt Kate said:

"O, now, Gabriel, he didn't live in Back Lane, really. Only the mill was there."

"Out from the mansion of his forefathers," continued Gabriel, "he drove with Johnny. And everything went on beautifully until Johnny came in sight of King Billy's statue: and whether he fell in love with the horse King Billy sits on or whether he thought he was back again in the mill, anyhow he began to walk round the statue."

Gabriel paced in a circle round the hall in his goloshes amid the laughter of the others.

"Round and round he went," said Gabriel, "and the old gentleman, who was a very pompous old gentleman, was highly indignant. 'Go on, sir? What do you mean, sir? Johnny! Johnny! Most extraordinary conduct! Can't understand the horse!' "

The peal of laughter which followed Gabriel's imitation of the incident was interrupted by a resounding knock at the hall door. Mary Jane ran to open it and let in Freddy Malins. Freddy Malins, with his hat well back on his head and his shoulders humped with cold, was puffiing and steaming after his exertions.

"I could only get one cab," he said.

"O, we'll find another along the quay," said Gabriel.

"Yes," said Aunt Kate. "Better not keep Mrs. Malins standing in the draught."

Mrs. Malins was helped down the front steps by her son and Mr. Browne and, after many manœuvres, hoisted into the cab. Freddy Malins clambered in after her and spent a long time settling her on the seat, Mr. Browne helping him with advice. At last she was settled comfortably and Freddy Malins invited Mr. Browne into the cab. There was a good deal of confused talk, and then Mr. Browne got into the cab. The cabman settled his rug over his knees, and bent down for the address. The confusion grew greater and the cabman was directed differently by Freddy Malins and Mr. Browne, each of whom had his head out through a window of the cab. The difficulty was to know where to drop Mr. Browne along the route, and Aunt Kate, Aunt Julia and Mary Jane helped the discussion from the doorstep with cross-directions and contradictions and abundance of laughter. As for Freddy he was speechless with laughter. He popped his head in and out of the window every moment to the great danger of his hat, and told his mother how the discussion was progressing, till at last Mr. Browne shouted to the bewildered cabman above the din of everybody's laughter:

"Do you know Trinity College?"

"Yes, sir,"said the cabman.

"Well, drive bang up against Trinity College gates," said Mr. Browne, and then we'll tell you where to go. You understand now."

"Yes, sir," said the cabman.

"Make like a bird for Trinity College."

"Right, sir," said the cabman.

The horse was whipped up and the cab rattled off along the quay amid a chorus of laughter and adieus.

Gabriel had not gone to the door with the others. He was in a dark part of the hall gazing up the staircase. A woman was standing near the top of the first flight, in the shadow also. He could not see her face but he could see the terra-cotta and salmon-pink panels of her skirt which the shadow made appear black and white. It was his wife. She was leaning on the banisters, listening to something. Gabriel was surprised at her stillness and strained his ear to listen also. But he could hear little save the noise of laughter and dispute on the front steps, a few chords struck on the piano and a few notes of a man's voice singing.

He stood still in the gloom of the hall, trying to catch the air that the voice was singing and gazing up at his wife. There was grace and mystery in her attitude as if she were a symbol of something. He asked himself what is a woman standing on the stairs in the shadow, listening to distant music, a symbol of. If he were a painter he would paint her in that attitude. Her blue felt hat would show off the bronze of her hair against the darkness and the dark panels of her skirt would show off the light ones. *Distant Music* he would call the picture if he were a painter.

The hall-door was closed; and Aunt Kate, Aunt Julia and Mary Jane came down the hall, still laughing.

"Well, isn't Freddy terrible?" said Mary Jane. "He's really terrible."

Gabriel said nothing but pointed up the stairs towards where his wife was standing. Now that the hall-door was closed the voice and the piano could be heard more clearly. Gabriel held up his hand for them to be silent. The song seemed to be in the old Irish tonality and the singer seemed uncertain both of his words and of his voice. The voice, made plaintive by distance and by the singer's hoarseness, faintly illuminated the cadence of the air with words expressing grief:

O, the rain falls on my heavy locks
And the dew wets my skin,
My babe lies cold . . .

"O," exclaimed Mary Jane. "It's Bartell D'Arcy singing and he wouldn't sing all the night. O, I'll get him to sing a song before he goes."

"O, do, Mary Jane," said Aunt Kate.

Mary Jane brushed past the others and ran to the staircase, but before she reached it the singing stopped and the piano was closed abruptly.

"O, what a pity!" she cried. "Is he coming down, Gretta?"

Gabriel heard his wife answer yes and saw her come down towards them. A few steps behind her were Mr. Bartell D'Arcy and Miss O'Callaghan.

"O, Mr. D'Arcy," cried Mary Jane, "it's downright mean of you to break off like that when we were all in raptures listening to you."

"I have been at him all the evening," said Miss O'Callaghan, "and Mrs. Conroy, too, and he told us he had a dreadful cold and couldn't sing."

"O, Mr. D'Arcy," said Aunt Kate, "now that was a great fib to tell."

"Can't you see that I'm as hoarse as a crow?" said Mr. D'Arcy roughly.

He went into the pantry hastily and put on his overcoat. The others, taken aback by his rude speech, could find nothing to say. Aunt Kate wrinkled her brows and made signs to the others to drop the subject. Mr. D'Arcy stood swathing his neck carefully and frowning.

"It's the weather," said Aunt Julia, after a pause.

"Yes, everybody has colds," said Aunt Kate readily, "everybody."

"They say," said Mary Jane, "we haven't had snow like it for thirty years; and I read this morning in the newspapers that the snow is general all over Ireland."

"I love the look of snow," said Aunt Julia sadly.

"So do I," said Miss O'Callaghan. "I think Christmas is never really Christmas unless we have the snow on the ground."

"But poor Mr. D'Arcy doesn't like the snow," said Aunt Kate, smiling.

Mr. D'Arcy came from the pantry, fully swathed and buttoned, and in a repentant tone told them the history of his cold. Everyone gave him advice and said it was a great pity and urged him to be very careful of his throat in the night air. Gabriel watched his wife, who did not join in the conversation. She was standing right under the dusty fanlight and the flame of the gas lit up the rich bronze of her hair, which he had seen her drying at the fire a few days before. She was in the same attitude and seemed unaware of the talk about her. At last she turned towards them and Gabriel saw that there was colour on her cheeks and that her eyes were shining. A sudden tide of joy went leaping out of his heart.

"Mr. D'Arcy," she said, "what is the name of that song you were singing?"

"It's called *The Lass of Aughrim,*" said Mr. D'Arcy, "but I couldn't remember it properly. Why? Do you know it?"

"*The Lass of Aughrim,*" she repeated. "I couldn't think of the name."

"It's a very nice air," said Mary Jane. "I'm sorry you were not in voice tonight."

"Now, Mary Jane," said Aunt Kate, "don't annoy Mr. D'Arcy. I won't have him annoyed."

Seeing that all were ready to start she shepherded them to the door, where good-night was said:

"Well, good-night, Aunt Kate, and thanks for the pleasant evening."

"Good-night, Gabriel. Good-night, Gretta!"

"Good-night, Aunt Kate, and thanks ever so much. Good-night, Aunt Julia."

"O, good-night, Gretta, I didn't see you."

"Good-night, Mr. D'Arcy. Good-night, Miss O'Callaghan."

"Good-night, Miss Morkan."

"Good-night, again."

"Good-night, all. Safe home."

"Good-night. Good night."

The morning was still dark. A dull, yellow light brooded over the houses and the river; and the sky seemed to be descending. It was slushy underfoot; and only streaks and patches of snow lay on the roofs, on the parapets of the quay and on the area railings. The lamps were still burning redly in the murky air and, across the river, the palace of the Four Courts stood out menacingly against the heavy sky.

She was walking on before him with Mr. Bartell D'Arcy, her shoes in a brown parcel tucked under one arm and her hands holding her skirt up from the slush. She had no longer any grace of attitude, but Gabriel's eyes were still bright with happiness. The blood went bounding along his veins; and the thoughts went rioting through his brain, proud, joyful, tender, valorous.

She was walking on before him so lightly and so erect that he longed to run after her noiselessly, catch her by the shoulders and say something foolish and affectionate into her ear. She seemed to him so frail that he longed to defend her against something and then to be alone with her. Moments of their secret life together burst like stars upon his memory. A heliotrope envelope was lying beside his breakfast-cup and he was caressing it with his hand. Birds were twittering in the ivy and the sunny web of the curtain was shimmering along the floor: he could not eat for happiness. They were standing on the crowded platform and he was placing a ticket inside the warm palm of her glove. He was standing with her in the cold, looking in through a grated window at a man making bottles in a roaring furnace. It was very cold. Her face, fragrant in the cold air, was quite close to his; and suddenly he called out to the man at the furnace:

"Is the fire hot, sir?"

But the man could not hear with the noise of the furnace. It was just as well. He might have answered rudely.

A wave of yet more tender joy escaped from his heart and went coursing in warm flood along his arteries. Like the tender fire of stars moments of their life together, that no one knew of or would ever know of, broke upon and illumined his memory. He longed to recall to her those moments, to make her forget the years of their dull existence together and remember only their moments of ecstasy. For the years, he felt, had not quenched his soul or hers. Their children, his writing, her household cares had not quenched all their souls' tender fire. In one letter that he had written to her then he had said: "Why is it that words like these seem to me so dull and cold? Is it because there is no word tender enough to be your name?"

Like distant music these words that he had written years before were borne towards him from the past. He longed to be alone with her. When the others had gone away, when he and she were in the room in the hotel, then they would be alone together. He would call her softly:

"Gretta!"

Perhaps she would not hear at once: she would be undressing. Then something in his voice would strike her. She would turn and look at him....

At the corner of Winetavern Street they met a cab. He was glad of its rattling noise as it saved him from conversation. She was looking out of the window and seemed tired. The others spoke only a few words, pointing out some building or street. The horse galloped along wearily under the murky morning sky, dragging his old rattling box after his heels, and Gabriel was again in a cab with her, galloping to catch the boat, galloping to their honeymoon.

As the cab drove across O'Connell Bridge Miss O'Callaghan said:

"They say you never cross O'Connell Bridge without seeing a white horse."

"I see a white man this time," said Gabriel.

"Where?" asked Mr. Bartell D'Arcy.

Gabriel pointed to the statue, on which lay patches of snow. Then he nodded familiarly to it and waved his hand.

"Good-night, Dan," he said gaily.

When the cab drew up before the hotel, Gabriel jumped out and, in spite of Mr. Bartell D'Arcy's protest, paid the driver. He gave the man a shilling over his fare. The man saluted and said:

"A prosperous New Year to you, sir."

"The same to you," said Gabriel cordially.

She leaned for a moment on his arm in getting out of the cab and while standing at the curbstone, bidding the others good-night. She leaned lightly on his arm, as lightly as when she had danced with him a few hours before. He had felt proud and happy then, happy that she was his, proud of her grace and wifely carriage. But now, after the kindling again of so many memories, the first touch of her body, musical and strange and perfumed, sent through him a keen pang of lust. Under cover of her silence he pressed her arm closely to his side; and, as they stood at the hotel door, he felt that they had escaped from their lives and duties, escaped from home and friends and run away together with wild and radiant hearts to a new adventure.

An old man was dozing in a great hooded chair in the hall. He lit a candle in the office and went before them to the stairs. They followed him in silence, their feet falling in soft thuds on the thickly carpeted stairs. She mounted the stairs behind the porter, her head bowed in ascent, her frail shoulders curved as with a burden, her skirt girt tightly about her. He could have flung his arms about her hips and held her still, for his arms were trembling with desire to seize her and only the stress of his nails against the palms of his hands held the wild impulse of his body in check. The porter halted on the stairs to settle his guttering candle. They halted, too, on the steps below him. In silence Gabriel could hear the falling of the molten wax into the tray and the thumping of his own heart against his ribs.

The porter led them along a corridor and opened a door. Then he set his unstable candle down on a toilet-table and asked at what hour they were to be called in the morning.

"Eight," said Gabriel.

The porter pointed to the tap of the electric-light and began a muttered apology, but Gabriel cut him short.

"We don't want any light. We have light enough from the street. And I say," he added, pointing to the candle, "you might remove that handsome article, like a good man."

The porter took up his candle again, but slowly, for he was surprised by such a novel idea. Then he mumbled good-night and went out. Gabriel shot the lock to.

A ghastly light from the street lay in a long shaft from one window to the door. Gabriel threw his overcoat and hat on a couch and crossed the room towards the window. He looked down into the street in order that his emotion might calm a little. Then he turned and leaned against a chest of drawers with his back to the light. She had taken off her hat and cloak and was standing before a large swinging mirror, unhooking her waist. Gabriel paused for a few moments, watching her, and then said:

"Gretta!"

She turned away from the mirror slowly and walked along the shaft of light towards him. Her face looked so serious and weary that the words would not pass Gabriel's lips. No, it was not the moment yet.

"You looked tired," he said.

"I am a little," she answered.

"You don't feel ill or weak?"

"No, tired: that's all."

She went on to the window and stood there, looking out. Gabriel waited again and then, fearing that diffidence was about to conquer him, he said abruptly:

"By the way, Gretta!"

"What is it?"

"You know that poor fellow Malins?" he said quickly.

"Yes. What about him?"

"Well, poor fellow, he's a decent sort of chap, after all," continued Gabriel in a false voice. "He gave me back that sovereign I lent him, and I didn't expect it, really. It's a pity he wouldn't keep away from that Browne, because he's not a bad fellow, really."

He was trembling now with annoyance. Why did she seem so abstracted? He did not know how he could begin. Was she annoyed, too, about something? If she would only turn to him or come to him of her own accord! To take her as she was would be brutal. No, he must see some ardour in her eyes first. He longed to be master of her strange mood.

"When did you lend him the pound?" she asked, after a pause.

Gabriel strove to restrain himself from breaking out into brutal language about the sottish Malins and his pound. He longed to cry to her from his soul, to crush her body again his, to overmaster her. But he said:

"O, at Christmas, when he opened that little Christmas-card shop in Henry Street."

He was in such a fever of rage and desire that he did not hear her come from the window. She stood before him for an instant, looking at him strangely. Then, suddenly raising herself on tiptoe and resting her hands lightly on his shoulders, she kissed him.

"You are a very generous person, Gabriel," she said.

Gabriel, trembling with delight at her sudden kiss and at the quaintness of her phrase, put his hands on her hair and began smoothing it back, scarcely touching it with his fingers. The washing had made it fine and brilliant. His heart was brimming over with happiness. Just when he was wishing for it she had come to him of her own accord. Perhaps her thought had been running with his. Perhaps she had felt the impetuous desire that was in him, and then the yielding mood had come upon her. Now that she had fallen to him so easily, he wondered why he had been so diffident.

He stood, holding her head between his hands. Then, slipping one arm swiftly about her body and drawing her towards him, he said softly:

"Gretta, dear, what are you thinking about?"

She did not answer nor yield wholly to his arm. He said again, softly:

"Tell me what it is, Gretta. I think I know what is the matter. Do I know?"

She did not answer at once. Then she said in an outburst of tears:

"O, I am thinking about that song, *The Lass of Aughrim*."

She broke loose from him and ran to the bed and, throwing her arms across the bed-rail, hid her face. Gabriel stood stock-still for a moment in astonishment and then followed her. As he passed in the way of the cheval-glass he caught sight of himself in full length, his broad, well-filled shirt-front, the face whose expression always puzzled him when he saw it in a mirror, and his glimmering gilt-rimmed eyeglasses. He halted a few paces from her and said:

"What about the song? Why does that make you cry?"

She raised her head from her arms and dried her eyes with the back of her hand like a child. A kinder note than he had intended went into his voice.

"Why, Gretta?" he asked.

"I am thinking about a person long ago who used to sing that song.

"And who was the person long ago?" asked Gabriel, smiling.

"It was a person I used to know in Galway when I was living with my grandmother," she said.

The smile passed away from Gabriel's face. A dull anger began to gather again at the back of his mind and the dull fires of his lust began to glow angrily in his veins.

"Someone you were in love with?" he asked ironically.

"It was a young boy I used to know," she answered, "named Michael Furey. He used to sing that song, *The Lass of Aughrim*. He was very delicate."

Gabriel was silent. He did not wish her to think that he was interested in this delicate boy.

"I can see him so plainly," she said, after a moment. "Such eyes as he had: big, dark eyes! And such an expression in them—an expression!"

"O, then, you are in love with him?" said Gabriel.

"I used to go out walking with him," she said, "when I was in Galway."

A thought flew across Gabriel's mind.

"Perhaps that was why you wanted to go to Galway with that Ivors girl?" he said coldly.

She looked at him and asked in surprise:

"What for?"

Her eyes made Gabriel feel awkward. He shrugged his shoulders and said:

"How do I know? To see him, perhaps?"

She looked away from him along the shaft of light towards the window in silence.

"He is dead," she said at length. "He died when he was only seventeen. Isn't it a terrible thing to die so young as that?"

"What was he?" asked Gabriel, still ironically.

"He was in the gasworks," she said.

Gabriel felt humiliated by the failure of his irony and by the evocation of this figure from the dead, a boy in the gasworks. While he had been full of memories of their secret life together, full of tenderness and joy and desire, she had been comparing him in her mind with another. A shameful consciousness of his own person assailed him. He saw himself as a ludicrous figure, acting as a pennyboy for his aunts, a nervous, well-meaning sentimentalist, orating to vulgarians and idealising his own clownish lusts, the pitiable fatuous fellow he had caught a glimpse of in the mirror. Instinctively he turned his back more to the light lest she might see the shame that burned upon his forehead.

He tried to keep up his tone of cold interrogation, but his voice when he spoke was humble and indifferent.

"I suppose you were in love with this Michael Furey, Gretta," he said.

"I was great with him at that time," she said.

Her voice was veiled and sad. Gabriel, feeling now how vain it would be to try to lead her whither he had purposed, caressed one of her hands and said, also sadly:

"And what did he die of so young, Gretta? Consumption, was it?"

"I think he died for me," she answered.

A vague terror seized Gabriel at this answer, as if, at that hour when he had hoped to triumph, some impalpable and vindictive being was coming against him, gathering forces against him in its vague world. But he shook himself free of it with an effort of reason and continued to caress her hand. He did not question her again, for he felt that she would tell him of herself. Her hand was warm and moist: it did not respond to his touch, but he continued to caress it just as he had caressed her first letter to him that spring morning.

"It was in the winter," she said, "about the beginning of the winter when I was going to leave my grandmother's and come up here to the convent. And he was ill at the time in his lodgings in Galway and wouldn't be let out, and his people in Oughterard were written to. He was in decline, they said, or something like that. I never knew rightly."

She paused for a moment and sighed.

"Poor fellow," she said. "He was very fond of me and he was such a gentle boy. We used to go out together, walking, you know, Gabriel, like the way they do in the country. He was going to study singing only for his health. He had a very good voice, poor Michael Furey."

"Well; and then?" asked Gabriel.

"And then when it came to the time for me to leave Galway and come up to the convent he was much worse and I wouldn't be let see him so I wrote him a letter saying I was going up to Dublin and would be back in the summer, and hoping he would be better then."

She paused for a moment to get her voice under control, and then went on:

"Then the night before I left, I was in my grandmother's house in Nuns' Island, packing up, and I heard gravel thrown up against the window. The window was so wet I couldn't see, so I ran downstairs as I was and slipped out the back into the garden and there was the poor fellow at the end of the garden, shivering."

"And did you not tell him to go back?" asked Gabriel.

"I implored of him to go home at once and told him he would get his death in the rain. But he said he did not want to live. I can see his eyes as well as well! He was standing at the end of the wall where there was a tree."

"And did he go home?" asked Gabriel.

"Yes, he went home. And when I was only a week in the convent he died and he was buried in Oughterard, where his people came from. O, the day I heard that, that he was dead!"

She stopped, choking with sobs, and, overcome by emotion, flung herself face downward on the bed, sobbing in the quilt. Gabriel held her hand for a moment longer, irresolutely, and then, shy of intruding on her grief, let it fall gently and walked quietly to the window.

She was fast asleep.

Gabriel, leaning on his elbow, looked for a few moments unresentfully on her tangled hair and half-open mouth, listening to her deep-drawn breath. So she had had that romance in her life: a man had died for her sake. It hardly pained him now to think how poor a part he, her husband, had played in her life. He watched her while she slept, as though he and she had never lived together as man and wife. His curious eyes rested long upon her face and on her hair: and, as he thought of what she must have been then, in that time of her first girlish beauty, a strange, friendly pity for her entered his soul. He did not like to say even to himself that her face was no longer beautiful, but he knew that it was no longer the face for which Michael Furey had braved death.

Perhaps she had not told him all the story. His eyes moved to the chair over which she had thrown some of her clothes. A petticoat string dangled to the floor. One boot stood upright, its limp upper fallen down: the fellow of it lay upon its side. He wondered at his riot of emotions of an hour before. From what had it proceeded? From his aunt's supper, from his own foolish speech, from the wine and dancing, the merry-making when saying good-night in the hall, the pleasure of the walk along the river in the snow. Poor Aunt Julia! She, too, would soon be a shade with the shade of Patrick Morkan and his horse. He had caught that haggard look upon her face for a moment when she was singing *Arrayed for the Bridal.* Soon, perhaps, he would be sitting in that same drawing-room, dressed in black, his silk hat on his knees. The blinds would be drawn down and Aunt Kate would be sitting beside him, crying and blowing her nose and telling him how Julia had died. He would cast about in his mind for some words that might console her, and would find only lame and useless ones. Yes, yes: that would happen very soon.

The air of the room chilled his shoulders. He stretched himself cautiously along under the sheets and lay down beside his wife. One by one, they were all becoming shades. Better pass boldly into that other world, in the full glory of some passion, than fade and wither dismally with age. He thought of how she who lay beside him had locked in her heart for so many years that image of her lover's eyes when he had told her that he did not wish to live.

Generous tears filled Gabriel's eyes. He had never felt like that himself towards any woman, but he knew that such a feeling must be love. The tears gathered more thickly in his eyes and in the partial darkness he imagined he saw the form of a young man standing under a dripping tree. Other forms were near. His soul had approached that region where dwell the vast hosts of the dead. He was conscious of, but could not apprehend, their wayward and flickering existence. His own identity was fading out into a grey impalpable world: the solid world itself, which these dead had one time reared and lived in, was dissolving and dwindling.

A few light taps upon the pane made him turn to the window. It had begun to snow again. He watched sleepily the flakes, silver and dark, falling obliquely against the lamplight. The time had come for him to set out on his journey westward. Yes, the newspapers were right: snow was general all over Ireland. It was falling on every part of the dark central plain, on the treeless hills, falling softly upon the Bog of Allen and, farther westward, softly falling into the dark mutinous

Shannon waves. It was falling, too, upon every part of the lonely churchyard on the hill where Michael Furey lay buried. It lay thickly drifted on the crooked crosses and headstones, on the spears of the little gate, on the barren thorns. His soul swooned slowly as he heard the snow falling faintly through the universe and faintly falling, like the descent of their last end, upon all the living and the dead.

KATHERINE MANSFIELD

The art of Katherine Mansfield was sufficiently intense to burn an ineffaceable mark upon twentieth-century fiction. Some forty years after her death, however, casual readers have a difficult time appreciating her work. If her writing is impatiently passed over, the reason may be the diminished capacity of today's reader to enjoy poetry. A Mansfield tale is much like a poem—carefully wrought, compact, intense, and often lyric. And like a poem, a Mansfield story can be fully savored only through the sympathetic, imaginative participation of the reader. Her desire was to reveal the meaning of an individual life, or group of lives, within the illuminating flash of a particular moment in time. That she would have this aim, and often realize it, suggests both why her contribution is so great and why she may challenge new readers.

Knowing that Katherine Mansfield is often chiefly desirous of exposing the inner quality of life should settle at once the complaint that "nothing ever happens." Any man or woman, during a sharp moment of recall, may experience a shattering spiritual crisis while quietly seated in an easy chair. The stories here are no more than a series of vignettes. Prelude *and* At the Bay *do not have logical plots with artfully interconnected episodes leading to a planned ending. Nothing is clearly resolved. A New Zealand family moves from a house in town and settles into a larger house in the country. The same family, in another sequence, spends a day at their summer beach house. Outwardly, not much more happens. Not even unusual are the overt responses of the characters to these commonplace experiences. Discovery is in recognizing that although the family is warmly united, each member dwells in his own isolated world. Spiritual aloneness is artfully unveiled. The inner life of the characters is solitary, passionate, intense, and filled with private reverie.*

To say that all the family (including the maid) are fully self-absorbed is, as close reading shows, not so. Discerning the essential difference in the grandmother, for instance, is one of the subtle rewards of close scrutiny. Old Mrs. Fairchild is a superbly revealed character. But so are the children uncannily real. Facets of the many lives are exposed with delicacy—Kezia's nostalgia upon moving, Linda's ambivalence toward Stanley, Beryl's restless sexuality. The careful reader welcomes the variety, the richness, the kaleidoscopic view of life eventually afforded him.

To present the two stories here as separate but continuing chapters in a short novel is to act upon J. Middleton Murry's information. He said, in an introduction to her collected work, that the second story was written four years after the first, but that Miss Mansfield "conceived" At the Bay *"as a continuation of* Prelude.*" The student may determine whether or not Murry's advice is helpful. No judgment of any kind about the stories may rightfully be made, though, without awareness of Miss Mansfield's will-o'-the-wisp shifts in point of view. She quietly flits in and out of the minds of her people, lighting up atoms of consciousness here and there. What continuity, after all,* does *the life of the mind afford?*

Prelude

I

THERE WAS not an inch of room for Lottie and Kezia in the buggy. When Pat swung them on top of the luggage they wobbled; the grandmother's lap was full and Linda Burnell could not possibly have held a lump of a child on hers for any distance. Isabel, very superior, was perched beside the new handy-man on the driver's seat. Hold-alls, bags and boxes were piled upon the floor. "These are absolute necessities that I will not let out of my sight for one instant," said Linda Burnell, her voice trembling with fatigue and excitement.

Lottie and Kezia stood on the patch of lawn just inside the gate all ready for the fray in their coats with brass anchor buttons and little round caps with battleship ribbons. Hand in hand, they stared with round solemn eyes first at the absolute necessities and then at their mother.

"We shall simply have to leave them. That is all. We shall simply have to cast them off," said Linda Burnell. A strange little laugh flew from her lips; she leaned back against the buttoned leather cushions and shut her eyes, her lips trembling with laughter. Happily at that moment Mrs. Samuel Josephs, who had been watching the scene from behind her drawing-room blind, waddled down the garden path.

"Why nod leave the chudren with be for the afterdoon, Brs. Burnell? They could go on the dray with the storeban when he comes in the eveding. Those thigs on the path have to go, dod't they?"

"Yes, everything outside the house is supposed to go," said Linda Burnell, and she waved a white hand at the tables and chairs standing on their heads on the front lawn. How absurd they looked! Either they ought to be the other way up, or Lottie and Kezia ought to stand on their heads, too. And she longed to say: "Stand on your heads, children, and wait for the storeman." It seemed to her that would be so exquisitely funny that she could not attend to Mrs. Samuel Josephs.

The fat creaking body leaned across the gate, and the big jelly of a face smiled. "Dod't you worry, Brs. Burnell. Loddie and Kezia can have tea with by chudren in the dursery, and I'll see theb on the dray afterwards."

The grandmother considered. "Yes, it really is quite the best plan. We are very obliged to you, Mrs. Samuel Josephs. Children, say 'thank you' to Mrs. Samuel Josephs."

Two subdued chirrups: "Thank you, Mrs. Samuel Josephs."

"And be good little girls, and—come closer—" they advanced, "don't forget to tell Mrs. Samuel Josephs when you want to . . ."

"No, granma."

"Dod't worry, Brs. Burnell."

At the last moment Kezia let go Lottie's hand and darted towards the buggy.

"I want to kiss my granma good-bye again."

But she was too late. The buggy rolled off up the road, Isabel bursting with pride, her nose turned up at all the world, Linda Burnell prostrated, and the grandmother rummaging among the very curious oddments she had had put in her black silk reticule at the last moment, for something to give her daughter. The buggy twinkled away in the sunlight and fine golden dust up the hill and over. Kezia bit her lip, but Lottie, carefully finding her handkerchief first, set up a wail.

"Mother! Granma!"

Mrs. Samuel Josephs, like a huge warm black silk tea cosy, enveloped her.

"It's all right, by dear. Be a brave child. You come and blay in the dursery!"

She put her arm round weeping Lottie and led her away. Kezia followed, making a face at Mrs. Samuel Josephs' placket, which was undone as usual, with two long pink corset laces hanging out of it. . . .

Lottie's weeping died down as she mounted the stairs, but the sight of her at the nursery door with swollen eyes and a blob of a nose gave great satisfaction to the S. J.'s, who sat on two benches before a long table covered with American cloth and set out with immense plates of bread and dripping and two brown jugs that faintly steamed.

"Hullo! You've been crying!"

"Ooh! Your eyes have gone right in."

"Doesn't her nose look funny."

"You're all red-and-patchy."

Lottie was quite a success. She felt it and swelled, smiling timidly.

"Go and sid by Zaidee, ducky," said Mrs. Samuel Josephs, "and Kezia, you sid ad the end by Boses."

Moses grinned and gave her a nip as she sat down; but she pretended not to notice. She did hate boys.

"Which will you have?" asked Stanley, leaning across the table very politely, and smiling at her. "Which will you have to begin with—strawberries and cream or bread and dripping?"

"Strawberries and cream, please," said she.

"Ah-h-h-h." How they all laughed and beat the table with their teaspoons. Wasn't that a take in! Wasn't it now! Didn't he fox her! Good old Stan!

"Ma! She thought it was real."

Even Mrs. Samuel Josephs, pouring out the milk and water, could not help smiling. "You bustn't tease theb on their last day," she wheezed.

But Kezia bit a big piece out of her bread and dripping, and then stood the piece up on her plate. With the bite out it made a dear little sort of a gate. Pooh! She didn't care! A tear rolled down her cheek, but she wasn't crying. She couldn't have cried in front of those awful

Samuel Josephs. She sat with her head bent, and as the tear dripped slowly down, she caught it with a neat little whisk of her tongue and ate it before any of them had seen.

II

AFTER TEA Kezia wandered back to her own house. Slowly she walked up the back steps, and through the scullery into the kitchen. Nothing was left in it but a lump of gritty yellow soap in one corner of the kitchen window sill and a piece of flannel stained with a blue bag in another. The fireplace was choked up with rubbish. She poked among it but found nothing except a hair-tidy with a heart painted on it that had belonged to the servant girl. Even that she left lying, and she trailed through the narrow passage into the drawing-room. The Venetian blind was pulled down but not drawn close. Long pencil rays of sunlight shone through and the wavy shadow of a bush outside danced on the gold lines. Now it was still, now it began to flutter again, and now it came almost as far as her feet. Zoom! Zoom! a bluebottle knocked against the ceiling; the carpet-tacks had little bits of red fluff sticking to them.

The dining-room window had a square of coloured glass at each corner. One was blue and one was yellow. Kezia bent down to have one more look at a blue lawn with blue arum lilies growing at the gate, and then at a yellow lawn with yellow lilies and a yellow fence. As she looked a little Chinese Lottie came out on to the lawn and began to dust the tables and chairs with a corner of her pinafore. Was that really Lottie? Kezia was not quite sure until she had looked through the ordinary window.

Upstairs in her father's and mother's room she found a pill box black and shiny outside and red in, holding a blob of cotton wool.

"I could keep a bird's egg in that," she decided.

In the servant girl's room there was a stay-button stuck in a crack of the floor, and in another crack some beads and a long needle. She knew there was nothing in her grandmother's room; she had watched her pack. She went over to the window and leaned against it, pressing her hands against the pane.

Kezia liked to stand so before the window. She liked the feeling of the cold shining glass against her hot palms, and she liked to watch the funny white tops that came on her fingers when she pressed them hard against the pane. As she stood there, the day flickered out and dark came. With the dark crept the wind snuffling and howling. The windows of the empty house shook, a creaking came from the walls and floors, a piece of loose iron on the roof banged forlornly. Kezia was suddenly quite, quite still, with wide open eyes and knees pressed together. She was frightened. She wanted to call Lottie and to go on calling all the while she ran downstairs and out of the house. But IT was just behind her, waiting at the door, at the head of the stairs, at the bottom of the stairs, hiding in the passage, ready to dart out at the back door. But Lottie was at the back door, too.

"Kezia!" she called cheerfully. "The storeman's here. Everything is on the dray and three horses, Kezia. Mrs. Samuel Josephs has given us a big shawl to wear round us, and she says to button up your coat. She won't come out because of asthma."

Lottie was very important.

"Now then, you kids," called the storeman. He hooked his big thumbs under their arms and up they swung. Lottie arranged the shawl "most beautifully" and the storeman tucked up their feet in a piece of old blanket.

"Lift up. Easy does it."

They might have been a couple of young ponies. The storeman felt over the cords holding his load, unhooked the brakechain from the wheel, and whistling, he swung up beside them.

"Keep close to me," said Lottie, "because otherwise you pull the shawl away from my side, Kezia."

But Kezia edged up to the storeman. He towered beside her big as a giant and he smelled of nuts and new wooden boxes.

III

IT WAS the first time that Lottie and Kezia had ever been out so late. Everything looked different—the painted wooden houses far smaller than they did by day, the gardens far bigger and wilder. Bright stars speckled the sky and the moon hung over the harbour dabbling the waves with gold. They could see the lighthouse shining on Quarantine Island, and the green lights on the old coal hulks.

"There comes the Picton boat," said the storeman, pointing to a little steamer all hung with bright beads.

But when they reached the top of the hill and began to go down the other side the harbour disappeared, and although they were still in the town they were quite lost. Other carts rattled past. Everybody knew the storeman.

"Night, Fred."

"Night O," he shouted.

Kezia liked very much to hear him. Whenever a cart appeared in the distance she looked up and waited for his voice. He was an old friend; and she and her grandmother had often been to his place to buy grapes. The storeman lived alone in a cottage that had a glasshouse against one wall built by himself. All the glasshouse was spanned and arched over with one beautiful vine. He took her brown basket from her, lined it with three large leaves, and then he felt in his belt for a little horn knife, reached up and snapped off a big blue cluster and laid it on the leaves so tenderly that Kezia held her breath to watch. He was a very big man. He wore brown velvet trousers, and he had a long brown beard. But he never wore a collar, not even on Sunday. The back of his neck was burnt bright red.

"Where are we now?" Every few minutes one of the children asked him the question.

"Why, this is Hawk Street, or Charlotte Crescent."

"Of course it is," Lottie pricked up her ears at the last name; she always felt that Charlotte Crescent belonged specially to her. Very few people had streets with the same name as theirs.

"Look, Kezia, there is Charlotte Crescent. Doesn't it look different?" Now everything familiar was left behind. Now the big dray rattled into unknown country, along new roads with high clay banks on either side, up steep, steep hills, down into bushy valleys, through wide shallow rivers. Further and further. Lottie's head wagged; she drooped, she slipped half into Kezia's lap and lay there. But Kezia could not open her eyes wide enough. The wind blew and she shivered; but her checks and ears burned.

"Do stars ever blow about?" she asked.

"Not to notice," said the storeman.

"We've got a nuncle and a naunt living near our new house," said Kezia. "They have got two children, Pip, the eldest is called, and the youngest's name is Rags. He's got a ram. He has to feed it with a nenamuel teapot and a glove top over the spout. He's going to show us. What is the difference between a ram and a sheep?"

"Well, a ram has horns and runs for you."

Kezia considered. "I don't want to see it frightfully," she said. "I hate rushing animals like dogs and parrots. I often dream that animals rush at me—even camels—and while they are rushing, their heads swell e-enormous."

The storeman said nothing. Kezia peered up at him, screwing up her eyes. Then she put her finger out and stroked his sleeve; it felt hairy. "Are we near?" she asked.

"Not far off, now," answered the storeman. "Getting tired?"

"Well, I'm not an atom bit sleepy," said Kezia. "But my eyes keep curling up in such a funny sort of way." She gave a long sigh, and to stop her eyes from curling she shut them. . . . When she opened them again they were clanking through a drive that cut through the garden like a whip lash, looping suddenly an island of green, and behind the island, but out of sight until you came upon it, was the house. It was long and low built, with a pillared verandah and balcony all the way round. The soft white bulk of it lay stretched upon the green garden like a sleeping beast. And now one and now another of the windows leaped into light. Someone was walking through the empty rooms carrying a lamp. From a window downstairs the light of a fire flickered. A strange beautiful excitement seemed to stream from the house in quivering ripples.

"Where are we?" said Lottie, sitting up. Her reefer cap was all on one side and on her cheek there was the print of an anchor button she had pressed against while sleeping. Tenderly the storeman lifted her, set her cap straight, and pulled down her crumpled clothes. She stood blinking on the lowest verandah step watching Kezia who seemed to come flying through the air to her feet.

"Ooh!" cried Kezia, flinging up her arms. The grandmother came out of the dark hall carrying a little lamp. She was smiling.

"You found your way in the dark?" said she.

"Perfectly well."

But Lottie staggered on the lowest verandah step like a bird fallen out of the nest. If she stood still for a moment she fell asleep, if she leaned against anything her eyes closed. She could not walk another step.

"Kezia," said the grandmother, "can I trust you to carry the lamp?"

"Yes, my granma."

The old woman bent down and gave the bright breathing thing into her hands and then she caught up drunken Lottie. "This way."

Through a square hall filled with bales and hundreds of parrots (but the parrots were only on the wall-paper) down a narrow passage where the parrots persisted in flying past Kezia with her lamp.

"Be very quiet," warned the grandmother, putting down Lottie and opening the dining-room door. "Poor little mother has got such a headache."

Linda Burnell, in a long cane chair, with her feet on a hassock, and a plaid over her knees, lay before a crackling fire. Burnell and Beryl sat at the table in the middle of the room eating a dish of fried chops and drinking tea out of a brown china teapot. Over the back of her mother's chair leaned Isabel. She had a comb in her fingers and in a gentle absorbed fashion she was combing the curls from her mother's forehead. Outside the pool of lamp and firelight the room stretched dark and bare to the hollow windows.

"Are those the children?" But Linda did not really care; she did not even open her eyes to see.

"Put down the lamp, Kezia," said Aunt Beryl, "or we shall have the house on fire before we are out of packing cases. More tea, Stanley?"

"Well, you might just give me five-eighths of a cup," said Burnell, leaning across the table. "Have another chop, Beryl. Tip-top meat, isn't it? Not too lean and not too fat." He turned to his wife. "You're sure you won't change your mind, Linda darling?"

"The very thought of it is enough." She raised one eyebrow in the way she had. The grandmother brought the children bread and milk and they sat up to table, flushed and sleepy behind the wavy steam.

"I had meat for my supper," said Isabel, still combing gently.

"I had a whole chop for my supper, the bone and all and Worcester sauce. Didn't I, father?"

"Oh, don't boast, Isabel," said Aunt Beryl.

Isabel looked astounded. "I wasn't boasting, was I, Mummy? I never thought of boasting. I thought they would like to know. I only meant to tell them."

"Very well. That's enough," said Burnell. He pushed back his plate, took a tooth-pick out of his pocket and began picking his strong white teeth.

"You might see that Fred has a bite of something in the kitchen before he goes, will you, mother?"

"Yes, Stanley." The old woman turned to go.

"Oh, hold on half a jiffy. I suppose nobody knows where my slippers were put? I suppose I shall not be able to get at them for a month or two—what?"

"Yes," came from Linda. "In the top of the canvas hold-all marked 'urgent necessities.' "

"Well you might get them for me will you, mother?"

"Yes, Stanley."

Burnell got up, stretched himself, and going over to the fire he turned his back to it and lifted up his coat tails.

"By jove, this is a pretty pickle. Eh, Beryl?"

Beryl, sipping tea, her elbows on the table, smiled over the cup at him. She wore an unfamiliar pink pinafore; the sleeves of her blouse were rolled up to her shoulders showing her lovely freckled arms, and she had let her hair fall down her back in a long pig-tail.

"How long do you think it will take to get straight—couple of weeks—eh?" he chaffed.

"Good heavens, no," said Beryl airily. "The worst is over already. The servant girl and I have simply slaved all day, and ever since mother came she has worked like a horse, too. We have never sat down for a moment. We have had a day."

Stanley scented a rebuke.

"Well, I suppose you did not expect me to rush away from the office and nail carpets—did you?"

"Certainly not," laughed Beryl. She put down her cup and ran out of the dining-room.

"What the hell does she expect us to do?" asked Stanley. "Sit down and fan herself with a palm leaf fan while I have a gang of professionals to do the job? By Jove, if she can't do a hand's turn occasionally without shouting about it in return for . . ."

And he gloomed as the chops began to fight the tea in his sensitive stomach. But Linda put up a hand and dragged him down to the side of her long chair.

"This is a wretched time for you, old boy," she said. Her cheeks were very white but she smiled and curled her fingers into the big red hand she held. Burnell became quiet. Suddenly he began to whistle "Pure as a lily, joyous and free"—a good sign.

"Think you're going to like it?" he asked.

"I don't want to tell you, but I think I ought to, mother," said Isabel. "Kezia is drinking tea out of Aunt Beryl's cup."

IV

THEY WERE taken off to bed by the grandmother. She went first with a candle; the stairs rang to their climbing feet. Isabel and

Lottie lay in a room to themselves, Kezia curled in her grandmother's soft bed.

"Aren't there going to be any sheets, my granma?"

"No, not to-night."

"It's tickly," said Kezia, "but it's like Indians. She dragged her grandmother down to her and kissed her under the chin. "Come to bed soon and be my Indian brave."

"What a silly you are," said the old woman, tucking her in as she loved to be tucked.

"Aren't you going to leave me a candle?"

"No. Sh—h. Go to sleep."

"Well, can I have the door left open?"

She rolled herself up into a round but she did not go to sleep. From all over the house came the sounds of steps. The house itself creaked and popped. Loud whispering voices came from downstairs. Once she heard Aunt Beryl's rush of high laughter, and once she heard a loud trumpeting from Burnell blowing his nose. Outside the window hundreds of black cats with yellow eyes sat in the sky watching her—but she was not frightened. Lottie was saying to Isabel:

"I'm going to say my prayers in bed to-night."

"No you can't, Lottie." Isabel was very firm. "God only excuses you saying your prayers in bed if you've got a temperature." So Lottie yielded:

Gentle Jesus meek anmile,
Look pon a little chile.
Pity me, simple Lizzie
Suffer me to come to thee.

And then they lay down back to back, their little behinds just touching, and fell asleep.

Standing in a pool of moonlight Beryl Fairfield undressed herself. She was tired, but she pretended to be more tired than she really was—letting her clothes fall, pushing back with a languid gesture her warm, heavy hair.

"Oh, how tired I am—very tired."

She shut her eyes a moment, but her lips smiled. Her breath rose and fell in her breasts like two fanning wings. The window was wide open; it was warm, and somewhere out there in the garden a young man, dark and slender, with mocking eyes, tiptoed among the bushes, and gathered the flowers into a big bouquet, and slipped under her window and held it up to her. She saw herself bending forward. He thrust his head among the bright waxy flowers, sly and laughing. "No, no," said Beryl. She turned from the window and dropped her nightgown over her head.

"How frightfully unreasonable Stanley is sometimes," she thought, buttoning. And then, as she lay down, there came the old thought, the cruel thought—ah, if only she had money of her own.

A young man, immensely rich, has just arrived from England. He meets her quite by chance. . . . The new governor is unmarried. . . . There is a ball at Government House. . . . Who is that exquisite creature in *eau de nil* satin? Beryl Fairfield. . . .

"The thing that pleases me," said Stanley, leaning against the side of the bed and giving himself a good scratch on his shoulders and back before turning in, "is that I've got the place dirt cheap, Linda. I was talking about it to little Wally Bell to-day and he said he simply could not understand why they had accepted my figure. You see land about here is bound to become more and more valuable . . . in about ten years' time . . . of course we shall have to go very slow and cut down expenses as fine as possible. Not asleep—are you?"

"No, dear, I've heard every word," said Linda.

He sprang into bed, leaned over her and blew out the candle.

"Good night, Mr. Business Man," said she, and she took hold of his head by the ears and gave him a quick kiss. Her faint faraway voice seemed to come from a deep well.

"Good night, darling." He slipped his arm under her neck and drew her to him.

"Yes, clasp me," said the faint voice from the deep well.

Pat the handy-man sprawled in his little room behind the kitchen. His sponge-bag coat and trousers hung from the door-peg like a hanged man. From the edge of the blanket his twisted toes protruded, and on the floor beside him there was an empty cane bird-cage. He looked like a comic picture.

"Honk, honk," came from the servant girl. She had adenoids.

Last to go to bed was the grandmother.

"What! Not asleep yet?"

"No, I'm waiting for you," said Kezia. The old woman sighed and lay down beside her. Kezia thrust her head under the grandmother's arm and gave a little squeak. But the old woman only pressed her faintly, and sighed again, took out her teeth, and put them in a glass of water beside her on the floor.

In the garden some tiny owls, perched on the branches of a lace-bark tree, called: "More pork; more pork." And far away in the bush there sounded a harsh rapid chatter: "Ha-ha-ha . . . Ha-ha-ha."

V

DAWN CAME sharp and chill with red clouds on a faint green sky and drops of water on every leaf and blade. A breeze blew over the garden, dropping dew and dropping petals, shivered over the drenched paddocks, and was lost in the sombre bush. In the sky some tiny stars floated for a moment and then they were gone—they were dissolved like bubbles. And plain to be heard in the early quiet was the sound of the creek in the paddock running over the brown stones, running in and out of the sandy hollows, hiding under clumps of dark berry bushes, spilling into a swamp of yellow water flowers and cresses.

And then at the first beam of sun the birds began. Big cheeky birds, starlings and mynahs, whistled on the lawns, the little birds, the goldfinches and linnets and fan-tails flicked from bough to bough. A lovely kingfisher perched on the paddock fence preening his rich beauty, and a *tui* sang his three notes and laughed and sang them again.

"How loud the birds are," said Linda in her dream. She was walking with her father through a green paddock sprinkled with daisies. Suddenly he bent down and parted the grasses and showed her a tiny ball of fluff just at her feet. "Oh, Papa, the darling." She made a cup of her hands and caught the tiny bird and stroked its head with her finger. It was quite tame. But a funny thing happened. As she stroked it began to swell, it ruffled and pouched, it grew bigger and bigger and its round eyes seemed to smile knowingly at her. Now her arms were hardly wide enough to hold it and she dropped it into her apron. It had become a baby with a big naked head and a gaping bird-mouth, opening and shutting. Her father broke into a loud clattering laugh and she woke to see Burnell standing by the windows rattling the Venetian blind up to the very top.

"Hullo," he said. "Didn't wake you, did I? Nothing much wrong with the weather this morning."

He was enormously pleased. Weather like this set a final seal on his bargain. He felt, somehow, that he had bought the lovely day, too—got it chucked in dirt cheap with the house and ground. He dashed off to his bath and Linda turned over and raised herself on one elbow to see the room by daylight. All the furniture had found a place—all the old paraphernalia—as she expressed it. Even the photographs were on the mantlepiece and the medicine bottles on the shelf above the washstand. Her clothes lay across a chair—her outdoor things, a purple cape and a round hat with a plume in it. Looking at them she wished that she was going away from this house, too. And she saw herself driving away from them all in a little buggy, driving away from everybody and not even waving.

Back came Stanley girt with a towel, glowing and slapping his thighs. He pitched the wet towel on top of her hat and cape, and standing firm in the exact centre of a square of sunlight he began to do his exercises. Deep breathing, bending and squatting like a frog and shooting out his legs. He was so delighted with his firm, obedient body that he hit himself on the chest and gave a loud "Ah." But this amazing vigour seemed to set him worlds away from Linda. She lay on the white tumbled bed and watched him as if from the clouds.

"Oh, damn! Oh, blast!" said Stanley, who had butted into a crisp white shirt only to find that some idiot had fastened the neckband and he was caught. He stalked over to Linda waving his arms.

"You look like a big fat turkey," said she.

"Fat. I like that," said Stanley. "I haven't a square inch of fat on me. Feel that."

"It's rock—it's iron," mocked she.

"You'd be surprised," said Stanley, as though this were intensely interesting, "at the number of chaps at the club who have got a corporation. Young chaps, you know—men of my age." He began parting his bushy ginger hair, his blue eyes fixed and round in the glass, his knees bent, because the dressing table was always—confound it—a bit too low for him. "Little Wally Bell, for instance," and he straightened, describing upon himself an enormous curve with the hairbrush. "I must say I've a perfect horror . . ."

"My dear, don't worry. You'll never be fat. You are far too energetic."

"Yes, yes, I suppose that's true," said he, comforted for the hundredth time, and taking a pearl pen-knife out of his pocket he began to pare his nails.

"Breakfast, Stanley." Beryl was at the door. "Oh, Linda, mother says you are not to get up yet." She popped her head in at the door. She had a big piece of syringa stuck through her hair.

"Everything we left on the verandah last night is simply sopping this morning. You should see poor dear mother wringing out the tables and the chairs. However, there is no harm done—" this with the faintest glance at Stanley.

"Have you told Pat to have the buggy round in time? It's a good six and a half miles to the office."

"I can imagine what this early start for the office will be like," thought Linda. "It will be very high pressure indeed."

"Pat, Pat." She heard the servant girl calling. But Pat was evidently hard to find; the silly voice went baa-baaing through the garden.

Linda did not rest again until the final slam of the front door told her that Stanley was really gone.

Later she heard her children playing in the garden. Lottie's stolid, compact little voice cried: "Ke-zia. Isabel." She was always getting lost or losing people only to find them again, to her great surprise, round the next tree or the next corner. "Oh, there you are after all." They had been turned out after breakfast and told not to come back to the house until they were called. Isabel wheeled a neat pramload of prim dolls and Lottie was allowed for a great treat to walk beside her holding the doll's parasol over the face of the wax one.

"Where are you going to, Kezia?" asked Isabel, who longed to find some light and menial duty that Kezia might perform and so be roped in under her government.

"Oh, just away," said Kezia. . . .

Then she did not hear them any more. What a glare there was in the room. She hated blinds pulled up to the top at any time, but in the morning it was intolerable. She turned over to the wall and idly, with one finger, she traced a poppy on the wallpaper with a leaf and a stem and a fat bursting bud. In the quiet, and under her tracing finger, the poppy seemed to come alive. She could feel the sticky, silky petals, the stem, hairy like a gooseberry skin, the rough leaf and the tight glazed bud. Things had a habit of coming alive like that. Not only large substantial things like furniture, but curtains and the patterns of stuffs and fringes of quilts and cushions. How often she had seen the tassel fringe of her quilt change into a funny procession of dancers with priests attending. . . . For there were some tassels that did not dance at all but walked stately, bent forward as if praying or chanting. How often the medicine bottles had turned into a row of little men with brown top-hats on; and the washstand jug had a way of sitting in the basin like a fat bird in a round nest.

"I dreamed about birds last night," thought Linda. What was it? She had forgotten. But the strangest part of this coming alive of things was what they did. They listened, they seemed to swell out with some mysterious important content, and when they were full she felt that they smiled. But it was not for her, only, their sly secret smile; they were members of a secret society and they smiled among themselves. Sometimes, when she had fallen asleep in the daytime, she woke and could not lift a finger, could not even turn her eyes to left or right because THEY were there; sometimes when she went out of a room and left it empty, she knew as she clicked the door to that THEY were filling it. And there were times in the evenings when she was upstairs, perhaps, and everybody else was down, when she could hardly escape from them. Then she could not hurry, she could not hum a tune; if she tried to say ever so carelessly—"Bother that old thimble"—THEY were not deceived. THEY knew how frightened she was; THEY saw how she turned her head away as she passed the mirror. What Linda always felt was that THEY wanted something of her, and she knew that if she gave herself up and was quiet, more than quiet, silent, motionless, something would really happen.

"It's very quiet now," she thought. She opened her eyes wide, and she heard the silence spinning to its soft endless web. How lightly she breathed; she scarcely had to breathe at all.

Yes, everything had come alive down to the minutest, tiniest particle, and she did not feel her bed, she floated, held up in the air. Only she seemed to be listening with her wide open watchful eyes, waiting for someone to come who just did not come, watching for something to happen that just did not happen.

VI

IN THE kitchen at the long deal table under the two windows old Mrs. Fairfield was washing the breakfast dishes. The kitchen window looked out on to a big grass patch that led down to the vegetable garden and the rhubarb beds. On one side the grass patch was bordered by the scullery and washhouse and over this whitewashed lean-to there grew a knotted vine. She had noticed yesterday that a few tiny corkscrew tendrils had come right through some cracks in the scullery ceiling and all the windows of the lean-to had a thick frill of ruffled green.

"I am very fond of a grape vine," declared Mrs. Fairfield, "but I do not think that the grapes will ripen here. It takes Austrialian sun." And she remembered how Beryl when she was a baby had been picking some white grapes from the vine on the back verandah of their Tasmanian house and she had been stung on the leg by a huge red ant. She saw Beryl in a little plaid dress with red ribbon tie-ups on the shoulders screaming so dreadfully that half the street rushed in. And how the child's leg had swelled! "T-t-t-t!" Mrs. Fairfield caught her breath remembering. "Poor child, how terrifying it was." And she set her lips tight and went over to the stove for some more hot water. The water frothed up in the big soapy bowl with pink and blue bubbles on top of the foam. Old Mrs. Fairfield's arms were bare to the elbow and stained a bright pink. She wore a grey foulard dress patterned with large purple pansies, a white linen apron and a high cap shaped like a jelly mould of white muslin. At her throat there was a silver crescent moon with five little owls seated on it, and round her neck she wore a watchguard made of black beads.

It was hard to believe that she had not been in that kitchen for years; she was so much a part of it. She put the crocks away with a sure, precise touch, moving leisurely and amply from the stove to the dresser, looking into the pantry and the larder as though there were not an unfamiliar corner. When she had finished, everything in the kitchen had become part of a series of patterns. She stood in the middle of the room wiping her hands on a check cloth; a smile beamed on her lips; she thought it looked very nice, very satisfactory.

"Mother! Mother! Are you there?" called Beryl.

"Yes, dear. Do you want me?"

"No. I'm coming," and Beryl rushed in, very flushed, dragging with her two big pictures.

"Mother, whatever can I do with these awful hideous Chinese paintings that Chung Wah gave Stanley when he went bankrupt? It's absurd to say that they are valuable, because they were hanging in Chung Wah's fruit shop for months before. I can't make out why Stanley wants them kept. I'm sure he thinks them just as hideous as we do, but it's because of the frames," she said spitefully. "I suppose he thinks the frames might fetch something some day or other."

"Why don't you hang them in the passage?" suggested Mrs. Fairfield; "they would not be much seen there."

"I can't. There is no room. I've hung all the photographs of his office there before and after building, and the signed photos of his business friends, and that awful enlargement of Isabel lying on the mat in her singlet." Her angly glance swept the placid kitchen. "I know what I'll do. I'll hang them here. I will tell Stanley they got a little damp in the moving so I have put them in here for the time being."

She dragged a chair forward, jumped on it, took a hammer and a big nail out of her pinafore pocket and banged away.

"There! That is enough! Hand me the picture, mother."

"One moment, child." Her mother was wiping over the carved ebony frame.

"Oh, mother, really you need not dust them. It would take years to dust all those little holes." And she frowned at the top of her mother's head and bit her lip with impatience. Mother's deliberate way of doing things was simply maddening. It was old age, she supposed, loftily.

At last the two pictures were hung side by side. She jumped off the chair, stowing away the little hammer.

"They don't look so bad there, do they?" said she. "And at any rate nobody need gaze at them except Pat and the servant girl—have I got a spider's web on my face, mother? I've been poking into that cupboard under the stairs and now something keeps tickling my nose."

But before Mrs. Fairfield had time to look Beryl had turned away. Someone tapped on the window: Linda was there, nodding and smiling. They heard the latch of the scullery door lift and she came in. She had no hat on; her hair stood up on her head in curling rings and she was wrapped up in an old cashmere shawl.

"I'm so hungry," said Linda: "where can I get something to eat, mother? This is the first time I've been in the kitchen. It says 'mother' all over; everything is in pairs."

"I will make you some tea," said Mrs. Fairfield, spreading a clean napkin over a corner of the table, "and Beryl can have a cup with you."

"Beryl, do you want half my gingerbread?" Linda waved the knife at her. "Beryl, do you like the house now that we are here?"

"Oh yes, I like the house immensely and the garden is beautiful, but it feels very far away from everything to me. I can't imagine people coming out from town to see us in that dreadful jolting bus, and I am sure there is not anyone here to come and call. Of course it does not matter to you because—"

"But there's the buggy," said Linda. "Pat can drive you into town whenever you like."

That was a consolation, certainly, but there was something at the back of Beryl's mind, something she did not even put into words for herself.

"Oh, well, at any rate it won't kill us," she said dryly, putting down her empty cup and standing up and stretching. "I am going to hang curtains." And she ran away singing:

How many thousand birds I see
That sing aloud from every tree . . .

". . . birds I see That sing aloud from every tree. . . ." But when she reached the dining-room she stopped singing, her face changed; it became gloomy and sullen.

"One may as well rot here as anywhere else," she muttered savagely, digging the stiff brass safety-pins into the red serge curtains.

The two left in the kitchen were quiet for a little. Linda leaned her cheek on her fingers and watched her mother. She thought her mother looked wonderfully beautiful with her back to the leafy window. There was something comforting in the sight of her that Linda felt she could never do without. She needed the sweet smell of her flesh, and the soft feel of her cheeks and her arms and shoulders still softer. She loved the way her hair curled, silver at her forehead, lighter at her neck, and bright brown still in the big coil under the muslin cap. Exquisite were her mother's hands, and the two rings she wore seemed to melt into her creamy skin. And she was always so fresh, so delicious. The old woman could bear nothing but linen next to her body and she bathed in cold water winter and summer.

"Isn't there anything for me to do?" asked Linda.

"No, darling. I wish you would go into the garden and give an eye to your children; but that I know you will not do."

"Of course I will, but you know Isabel is much more grown up than any of us."

"Yes, but Kezia is not," said Mrs. Fairfield.

"Oh, Kezia has been tossed by a bull hours ago," said Linda, winding herself up in her shawl again.

But no, Kezia had seen a bull through a hole in a knot of wood in the paling that separated the tennis lawn from the paddock. But she had not liked the bull frightfully, so she had walked away back through the orchard, up the grassy slope, along the path by the lace-bark tree and so into the spread tangled garden. She did not believe that she would ever get lost in this garden. Twice she had found her way back to the big iron gates they had driven through the night before, and then had turned to walk up the drive that led to the house, but there were so many little paths on either side. On one side they all led into a tangle of tall dark trees and strange bushes with flat velvet leaves and feathery cream flowers that buzzed with flies when you shook them—this was the frightening side, and no garden at all. The little paths here were wet and clayey with tree roots spanned across them like the marks of big fowls' feet.

But on the other side of the drive there was a high box border and the paths had box edges and all of them led into a deeper and deeper tangle of flowers. The camellias were in bloom, white and crimson and pink and white striped with flashing leaves. You could not see a leaf on the syringa bushes for the white clusters. The roses were in flower—gentlemen's button-hole roses, little white ones, but far too full of insects to hold under anyone's nose, pink monthly roses with a ring of fallen petals round the bushes, cabbage roses on thick stalks, moss roses, always in bud, pink smooth beauties opening curl on curl, red ones so dark they seemed to turn black as they fell, and a certain exquisite cream kind with a slender red stem and bright scarlet leaves.

There were clumps of fairy bells, and all kinds of geraniums, and there were little trees of verbena and bluish lavender bushes and a bed of pelargoniums with velvet eyes and leaves like moths' wings. There was a bed of nothing but mignonette and another of nothing but pansies—borders of double and single daisies and all kinds of little tufty plants she had never seen before.

The red-hot pokers were taller than she; the Japanese sunflowers grew in a tiny jungle. She sat down on one of the box borders. By pressing hard at first it made a nice seat. But how dusty it was inside! Kezia bent down to look and sneezed and rubbed her nose.

And then she found herself at the top of the rolling grassy slope that led down to the orchard.... She looked down at the slope a moment; then she lay down on her back, gave a squeak and rolled over and over into the thick flowery orchard grass. As she lay waiting for things to stop spinning, she decided to go up to the house and ask the servant girl for an empty match-box. She wanted to make a surprise for the grandmother. . . . First she would put a leaf inside with a big violet lying on it, then she would put a very small white picotee, perhaps, on each side of the violet, and then she would sprinkle some lavender on the top, but not to cover their heads.

She often made these surprises for the grandmother, and they were always most successful.

"Do you want a match, my granny?"

"Why, yes, child, I believe a match is just what I'm looking for."

The grandmother slowly opened the box and came upon the picture inside.

"Good gracious, child! How you astonished me!"

"I can make her one every day here," she thought, scrambling up the grass on her slippery shoes.

But on her way back to the house she came to that island that lay in the middle of the drive, dividing the drive into two arms that met in front of the house. The island was made of grass banked up high. Nothing grew on the top except one huge plant with thick, grey-green, thorny leaves, and out of the middle there sprang up a tall stout stem. Some of the leaves of the plant were so old that they curled up in the air no longer; they turned back, they were split and broken; some of them lay flat and withered on the ground.

Whatever could it be? She had never seen anything like it before. She stood and stared. And then she saw her mother coming down the path.

"Mother, what is it?" asked Kezia.

Linda looked up at the fat swelling plant with its cruel leaves and fleshy stem. High above them, as though becalmed in the air, and yet holding so fast to the earth it grew from, it might have had claws instead of roots. The curving leaves seemed to be hiding something; the blind stem cut into the air as if no wind could ever shake it.

"That is an aloe, Kezia," said her mother.

"Does it ever have any flowers?"

"Yes, Kezia," and Linda smiled down at her, and half shut her eyes. "Once every hundred years."

VII

ON HIS way home from the office Stanley Burnell stopped the buggy at the Bodega, got out and bought a large bottle of oysters. At the Chinaman's shop next door he bought a pineapple in the pink of condition, and noticing a basket of fresh black cherries he told John to put him a pound of those as well. The oysters and the pine he stowed away in the box under the front seat, but the cherries he kept in his hand.

Pat, the handy-man, leapt off the box and tucked him up again in the brown rug.

"Lift yer feet, Mr. Burnell, while I give yer a fold under," said he.

"Right! Right! First-rate!" said Stanley. "You can make straight for home now."

Pat gave the grey mare a touch and the buggy sprang forward.

"I believe this man is a first-rate chap," thought Stanley. He liked the look of him sitting up there in his neat brown coat and brown bowler. He liked the way Pat had tucked him in, and he liked his eyes. There was nothing servile about him—and if there was one thing he hated more than another it was servility. And he looked as if he was pleased with his job—happy and contented already.

The grey mare went very well; Burnell was impatient to be out of the town. He wanted to be home. Ah, it was splendid to live in the country—to get right out of that hole of a town once the office was closed; and this drive in the fresh warm air, knowing all the while that his own house was at the other end, with its garden and paddocks, its three tip-top cows and enough fowls and ducks to keep them in poultry, was splendid too.

As they left the town finally and bowled away up the deserted road his heart beat hard for joy. He rooted in the bag and began to eat the cherries, three or four at a time, chucking the stones over the side of the buggy. They were delicious, so plump and cold, without a spot or a bruise on them.

Look at those two, now—black one side and white the other—perfect! A perfect little pair of Siamese twins. And he stuck them in his button-hole. . . . By Jove, he wouldn't mind giving that chap up there a handful—but no, better not. Better wait until he had been with him a bit longer.

He began to plan what he would do with his Saturday afternoons and his Sundays. He wouldn't go to the club for lunch on Saturday. No, cut away from the office as soon as possible and get them to give him a couple of slices of cold meat and half a lettuce when he got home. And then he'd get a few chaps out from town to play tennis in the afternoon. Not too many—three at most. Beryl was a good player, too. . . . He stretched out his right arm and slowly bent it, feeling the muscle. . . . A bath, a good rubdown, a cigar on the verandah after dinner. . . .

On Sunday morning they would go to church—children and all. Which reminded him that he must hire a pew, in the sun if possible and well forward so as to be out of the draught from the door. In fancy he heard himself intoning extremely well: "When thou didst overcome the *Sharp*ness of Death Thou didst open the *King*dom of Heaven to *all* Believers." And he saw the neat brass-edged card on the corner of the pew—Mr. Stanley Burnell and family. . . . The rest of the day he'd loaf about with Linda. . . . Now they were walking about the garden; she was on his arm, and he was explaining to her at length what he intended doing at the office the week following. He heard her saying: "My dear, I think that is most wise.". . . Talking things over with Linda was a wonderful help even though they were apt to drift away from the point.

Hang it all! They weren't getting along very fast. Pat had put the brake on again. Ugh! What a brute of a thing it was. He could feel it in the pit of his stomach.

A sort of panic overtook Burnell whenever he approached near home. Before he was well inside the gate he would shout to anyone within sight: "Is everything all right?" And then he did not believe it was until he heard Linda say: "Hullo! Are you home again?" That was the worst of living in the country—it took the deuce of a long time to get back. . . . But now they weren't far off. They were on the top of the last hill; it was a gentle slope all the way now and not more than half a mile.

Pat trailed the whip over the mare's back and he coaxed her: "Goop now. Goop now."

It wanted a few minutes to sunset. Everything stood motionless bathed in bright, metallic light and from the paddocks on either side there streamed the milky scent of ripe grass. The iron gates were open. They dashed through and up the drive and round the island, stopping at the exact middle of the verandah.

"Did she satisfy yer, Sir?" said Pat, getting off the box and grinning at his master.

"Very well indeed, Pat," said Stanley.

Linda came out of the glass door; her voice rang in the shadowy quiet. "Hullo! Are you home again?"

At the sound of her his heart beat so hard that he could hardly stop himself dashing up the steps and catching her in his arms.

"Yes, I'm home again. Is everything all right?"

Pat began to lead the buggy round to the side gate that opened into the courtyard.

"Here, half a moment," said Burnell. "Hand me those two parcels." And he said to Linda, "I've brought you back a bottle of oysters and a pineapple," as though he had brought her back all the harvest of the earth.

They went into the hall; Linda carried the oysters in one hand and the pineapple in the other. Burnell shut the glass door, threw his hat down, put his arms round her and strained her to him, kissing the top of her head, her ears, her lips, her eyes.

"Oh, dear! Oh, dear!" said she. "Wait a moment. Let me put down these silly things," and she put the bottle of oysters and the pine on a little carved chair. "What have you got in your button-hole—cherries?" She took them out and hung them over his ear.

"Don't do that, darling. They are for you."

So she took them off his ear again. "You don't mind if I save them. They'd spoil my appetite for dinner. Come and see your children. They are having tea."

The lamp was lighted on the nursery table. Mrs. Fairfield was cutting and spreading bread and butter. The three little girls sat up to table wearing large bibs embroidered with their names. They wiped their mouths as their father came in ready to be kissed. The windows were open; a jar of wild flowers stood on the mantelpiece, and the lamp made a big soft bubble of light on the ceiling.

"You seem pretty snug, mother," said Burnell, blinking at the light. Isabel and Lottie sat one on either side of the table, Kezia at the bottom—the place at the top was empty.

"That's where my boy ought to sit," thought Stanley. He tightened his arm round Linda's shoulder. By God, he was a perfect fool to feel as happy as this!

"We are, Stanley. We are very snug," said Mrs. Fairfield, cutting Kezia's bread into fingers.

"Like it better than town—eh, children?" asked Burnell.

"Oh, yes," said the three little girls, and Isabel added as an afterthought: "Thank you very much indeed, father dear."

"Come upstairs," said Linda. "I'll bring your slippers."

But the stairs were too narrow for them to go up arm in arm. It was quite dark in the room. He heard her ring tapping on the marble mantelpiece as she felt for the matches.

"I've got some, darling. I'll light the candles."

But instead he came up behind her and again he put his arms round her and pressed her head into his shoulder.

"I'm so confoundedly happy," he said.

"Are you?" She turned and put her hands on his breast and looked up at him.

"I don't know what has come over me," he protested.

It was quite dark outside now and heavy dew was falling. When Linda shut the window the cold dew touched her finger tips. Far away a dog barked. "I believe there is going to be a moon," she said.

At the words, and with the cold wet dew on her fingers, she felt as though the moon had risen—that she was being strangely discovered in a flood of cold light. She shivered; she came away from the window and sat down upon the box ottoman beside Stanley.

In the dining-room, by the flicker of a wood fire, Beryl sat on a hassock playing the guitar. She had bathed and changed all her clothes. Now she wore a white muslin dress with black spots on it and in her hair she had pinned a black silk rose.

Nature has gone to her rest, love,
See, we are alone.
Give me your hand to press, love,
Lightly within my own.

She played and sang half to herself, for she was watching herself playing and singing. The firelight gleamed on her shoes, on the ruddy belly of the guitar, and on her white fingers. . . .

"If I were outside the window and looked in and saw myself I really would be rather struck," thought she. Still more softly she played the accompaniment—not singing now but listening.

. . . "The first time that I ever saw you, little girl—oh, you had no idea that you were not alone—you were sitting with your little feet upon a hassock, playing the guitar. God, I can never forget. . . ." Beryl flung up her head and began to sing again:

Even the moon is aweary . . .

But there came a loud bang at the door. The servant girl's crimson face popped through.

"Please, Miss Beryl, I've got to come and lay."

"Certainly, Alice," said Beryl, in a voice of ice. She put the guitar in a corner. Alice lunged in with a heavy black iron tray.

"Well, I have had a job with that oving," said she. "I can't get nothing to brown."

"Really!" said Beryl.

But no, she could not stand that fool of a girl. She ran into the dark drawing-room and began walking up and down. . . . Oh, she was restless, restless. There was a mirror over the mantel. She leaned her arms along and looked at her pale shadow in it. How beautiful she looked, but there was nobody to see, nobody.

"Why must you suffer so?" said the face in the mirror. "You were not made for suffering. . . . Smile!"

Beryl smiled, and really her smile *was* so adorable that she smiled again—but this time because she could not help it.

VIII

"GOOD MORNING, Mrs. Jones."

"Oh, good morning, Mrs. Smith. I'm so glad to see you. Have you brought your children?"

"Yes, I've brought both my twins. I have had another baby since I saw you last, but she came so suddenly that I haven't had time to make her any clothes, yet. So I left her. . . . How is your husband?"

"Oh, he is very well, thank you. At least he had a nawful cold but Queen Victoria—she's my godmother, you know—sent him a case of pineapples and that cured it im-mediately. Is that your new servant?"

"Yes, her name's Gwen. I've only had her two days. Oh, Gwen, this is my friend, Mrs. Smith."

"Good morning, Mrs. Smith. Dinner won't be ready for about ten minutes."

"I don't think you ought to introduce me to the servant. I think I ought to just begin talking to her."

"Well, she's more of a lady-help than a servant and you do introduce lady-helps, I know, because Mrs. Samuel Josephs had one."

"Oh, well, it doesn't matter," said the servant, carelessly, beating up a chocolate custard with half a broken clothes peg. The dinner was baking beautifully on a concrete step. She began to lay the cloth on a pink garden seat. In front of each person she put two geranium leaf plates, a pine needle fork and a twig knife. There were three daisy heads on a laurel leaf for poached eggs, some slices of fuchsia petal cold beef, some lovely little rissoles made of earth and water and dandelion seeds, and the chocolate custard which she had decided to serve in the pawa shell she had cooked it in.

"You needn't trouble about my children," said Mrs. Smith graciously. "If you'll just take this bottle and fill it at the tap—I mean at the dairy."

"Oh, all right," said Gwen, and she whispered to Mrs. Jones: "Shall I go and ask Alice for a little bit of real milk?"

But someone called from the front of the house and the luncheon party melted away, leaving the charming table, leaving the rissoles and the poached eggs to the ants and to an old snail who pushed his quivering horns over the edge of the garden seat and began to nibble a geranium plate.

"Come round to the front, children. Pip and Rags have come."

The Trout boys were the cousins Kezia had mentioned to the storeman. They lived about a mile away in a house called Monkey Tree Cottage. Pip was tall for his age, with lank black hair and a white face, But Rags was very small and so thin that when he was undressed his shoulder blades stuck out like two little wings. They had a mongrel dog with pale blue eyes and a long tail turned up at the end who followed them everywhere; he was called Snooker. They spent half their time combing and brushing Snooker and dosing him with various awful mixtures concocted by Pip, and kept secretly by him in a broken jug covered with an old kettle lid. Even faithful little Rags was not allowed to know the full secret of these mixtures.... Take some carbolic tooth powder and a pinch of sulphur powdered up fine, and perhaps a bit of starch to stiffen up Snooker's coat.... But that was not all; Rags privately thought that the rest was gun-powder.... And he never was allowed to help with the mixing because of the danger.... "Why if a spot of this flew in your eye, you would be blinded for life," Pip would say, stirring the mixture with an iron spoon. "And there's always the chance—just the chance, mind you—of it exploding if you whack it hard enough. . . . Two spoons of this in a kerosene tin will be enough to kill thousands of fleas." But Snooker spent all his spare time biting and snuffling, and he stank abominably.

"It's because he is such a grand fighting dog," Pip would say. "All fighting dogs smell."

The Trout boys had often spent the day with the Burnells in town, but now that they lived in this fine house and boncer garden they were inclined to be very friendly. Besides, both of them liked playing with girls—Pip, because he could fox them so, and because Lottie was so easily frightened, and Rags for a shameful reason. He adored dolls. How he would look at a doll as it lay asleep, speaking in a whisper and smiling timidly, and what a treat it was to him to be allowed to hold one. . . .

"Curve your arms round her. Don't keep them stiff like that. You'll drop her," Isabel would say sternly.

Now they were standing on the verandah and holding back Snooker who wanted to go into the house but wasn't allowed to because Aunt Linda hated decent dogs.

"We came over in the bus with Mum," they said, "and we're going to spend the afternoon with you. We brought over a batch of our gingerbread for Aunt Linda. Our Minnie made it. It's all over nuts."

"I skinned the almonds," said Pip. "I just stuck my hand into a saucepan of boiling water and grabbed them out and gave them a kind of pinch and the nuts flew out of the skins, some of them as high as the ceiling. Didn't they, Rags?"

Rags nodded. "When they make cakes at our place," said Pip, "we always stay in the kitchen, Rags and me, and I get the bowl and he gets the spoon and the egg beater. Sponge cake's best. It's all frothy stuff, then.

He ran down the verandah steps to the lawn, planted his hands on the grass, bent forward, and just did not stand on his head.

"That lawn's all bumpy," he said. "You have to have a flat place for standing on your head. I can walk round the monkey tree on my head at our place. Can't I, Rags?"

"Nearly," said Rags faintly.

"Stand on your head on the verandah. That's quite flat," said Kezia.

"No, smarty," said Pip. "You have to do it on something soft. Because if you give a jerk and fall over, something in your neck goes click, and it breaks off. Dad told me."

"Oh, do let's play something," said Kezia.

"Very well," said Isabel quickly, "we'll play hospitals. I will be the nurse and Pip can be the doctor and you and Lottie and Rags can be the sick people."

Lottie didn't want to play that, because last time Pip had squeezed something down her throat and it hurt awfully.

"Pooh," scoffed Pip. "It was only the juice out of a bit of mandarin peel."

"Well, let's play ladies," said Isabel. "Pip can be the father and you can be all our dear little children."

"I hate playing ladies," said Kezia. "You always make us go to church hand in hand and come home and go to bed."

Suddenly Pip took a filthy handkerchief out of his pocket. "Snooker! Here, sir," he called. But Snooker, as usual, tried to sneak away, his tail between his legs. Pip leapt on top of him, and pressed him between his knees.

"Keep his head firm, Rags," he said, and he tied the handkerchief round Snooker's head with a funny knot sticking up at the top.

"Whatever is that for?" asked Lottie.

"It's to train his ears to grow more close to his head—see?" said Pip. "All fighting dogs have ears that lie back. But Snooker's ears are a bit too soft."

"I know," said Kezia. "They are always turning inside out. I hate that."

Snooker lay down, made one feeble effort with his paw to get the handkerchief off, but finding he could not, trailed after the children, shivering with misery.

IX

PAT CAME swinging along; in his hand he held a little tomahawk that winked in the sun.

"Come with me," he said to the children, "and I'll show you how the kings of Ireland chop the head off a duck."

They drew back—they didn't believe him, and besides, the Trout boys had never seen Pat before.

"Come on now," he coaxed, smiling and holding out his hand to Kezia.

"Is it a real duck's head? One from the paddock?"

"It is," said Pat. She put her hand in his hard dry one, and he stuck the tomahawk in his belt and held out the other to Rags. He loved little children.

"I'd better keep hold of Snooker's head if there's going to be any blood about," said Pip, "because the sight of blood makes him awfully wild." He ran ahead dragging Snooker by the hankerchief.

"Do you think we ought to go?" whispered Isabel. "We haven't asked or anything. Have we?"

At the bottom of the orchard a gate was set in the paling fence. On the other side a steep bank led down to a bridge that spanned the creek, and once up the bank on the other side you were on the fringe of the paddocks. A little old stable in the first paddock had been turned into a fowl-house. The fowls had strayed far away across the paddock down to a dumping ground in a hollow, but the ducks kept close to that part of the creek that flowed under the bridge.

Tall bushes overhung the stream with red leaves and yellow flowers and clusters of blackberries. At some places the stream was wide and shallow, but at others it tumbled into deep little pools with foam at the edges and quivering bubbles. It was in these pools that the big white ducks had made themselves at home, swimming and guzzling along the weedy banks.

Up and down they swam, preening their dazzling breasts, and other ducks with the same dazzling breasts and yellow bills swam upside down with them.

"There is the little Irish navy," said Pat, "and look at the old admiral there with the green neck and the grand little flagstaff on his tail."

He pulled a handful of grain from his pocket and began to walk towards the fowl-house, lazy, his straw hat with the broken crown pulled over his eyes.

"Lid. Lid—lid—lid—lid—" he called.

"Qua. Qua—qua—qua—qua—" answered the ducks, making for land, and flapping and scrambling up the bank they streamed after him in a long waddling line. He coaxed them, pretending to throw the grain, shaking it in his hands and calling to them until they swept round him in a white ring.

From far away the fowls heard the clamour and they too came running across the paddock, their heads thrust forward, their wings spread, turning in their feet in the silly way fowls run and scolding as they came.

Then Pat scattered the grain and the greedy ducks began to gobble. Quickly he stooped, seized two, one under each arm, and strode across to the children. Their darting heads and round eyes frightened the children—all except Pip.

"Come on, sillies," he cried, "they can't bite. They haven't any teeth. They've only got those two little holes in their beaks for breathing through."

"Will you hold one while I finish with the other?" asked Pat. Pip let go of Snooker. "Won't I? Won't I? Give us one. I don't mind how much he kicks."

He nearly sobbed with delight when Pat gave the white lump into his arms.

There was an old stump beside the door of the fowl-house. Pat grabbed the duck by its legs, laid it flat across the stump, and almost at the same moment down came the little tomahawk and the duck's head flew off the stump. Up the blood spurted over the white feathers and over his hand.

When the children saw the blood they were frightened no longer. They crowded round him and began to scream. Even Isabel leaped about crying: "The blood! The blood!" Pip forgot all about his duck. He simply threw it away from him and shouted, "I saw it. I saw it," and jumped round the wood block.

Rags, with cheeks as white as paper, ran up to the little head, put out a finger as if he wanted to touch it, shrank back again and then again put out a finger. He was shivering all over.

Even Lottie, frightened little Lottie, began to laugh and point at the duck and shrieked: "Look, Kezia, look."

"Watch it!" shouted Pat. He put down the body and it began to waddle—with only a long spurt of blood where the head had been; it began to pad away without a sound towards the steep bank that led to the stream. . . . That was the crowning wonder.

"Do you see that? Do you see that?" yelled Pip. He ran among the little girls tugging at their pinafores.

"It's like a little engine. It's like a funny little railway engine," squealed Isabel.

But Kezia suddenly rushed at Pat and flung her arms round his legs and butted her head as hard as she could against his knees.

"Put head back! Put head back!" she screamed.

When he stooped to move her she would not let go or take her head away. She held on as hard as she could and sobbed: "Head back! Head back!" until it sounded like a loud strange hiccup.

"It's stopped. It's tumbled over. It's dead," said Pip.

Pat dragged Kezia up into his arms. Her sunbonnet had fallen back, but she would not let him look at her face. No, she pressed her face into a bone in his shoulder and clasped her arms round his neck.

The children stopped screaming as suddenly as they had begun. They stood round the dead duck. Rags was not frightened of the head any more. He knelt down and stroked it, now.

"I don't think the head is quite dead yet," he said. "Do you think it would keep alive if I gave it something to drink?"

But Pip got very cross: "Bah! You baby." He whistled to Snooker and went off.

When Isabel went up to Lottie, Lottie snatched away.

"What are you always touching me for, Isabel?"

"There now," said Pat to Kezia. "There's the grand little girl."

She put up her hands and touched his ears. She felt something. Slowly she raised her quivering face and looked. Pat wore little round gold ear-rings. She never knew that men wore ear-rings. She was very much surprised.

"Do they come on and off?" she asked huskily.

X

UP IN the house, in the warm tidy kitchen, Alice, the servant girl, was getting the afternoon tea. She was "dressed." She had on a black stuff dress that smelt under the arms, a white apron like a large sheet of paper, and a lace bow pinned on to her hair with two jetty pins. Also her comfortable carpet slippers were changed for a pair of black leather ones that pinched her corn on her little toe something dreadful. . . .

It was warm in the kitchen. A blow-fly buzzed, a fan of whity steam came out of the kettle, and the lid kept up a rattling jig as the water bubbled. The clock ticked in the warm air, slow and deliberate, like the click of an old woman's knitting needle, and sometimes—for no reason at all, for there wasn't any breeze—the blind swung out and back, tapping the window.

Alice was making water-cress sandwiches. She had a lump of butter on the table, a barracouta loaf, and the cresses tumbled in a white cloth.

But propped against the butter dish there was a dirty, greasy little book, half unstitched, with curled edges, and while she mashed the butter she read:

"To dream of black-beetles drawing a hearse is bad. Signifies death of one you hold near or dear, either father, husband, brother, son, or intended. If beetles crawl backwards as you watch them it means death from fire or from great height such as flight of stairs, scaffolding, etc.

"Spiders. To dream of spiders creeping over you is good. Signifies large sum of money in near future. Should party be in family way an easy confinement may be expected. But care should be taken in sixth month to avoid eating of probable present of shell fish. . . ."

How many thousand birds I see.

Oh, life. There was Miss Beryl. Alice dropped the knife and slipped the *Dream Book* under the butter dish. But she hadn't time to hide it quite, for Beryl ran into the kitchen and up to the table, and the first thing her eye lighted on were those greasy edges. Alice saw Miss Beryl's meaning little smile and the way she raised her eyebrows and screwed up here eyes as though she were not quite sure what that could be. She decided to answer if Miss Beryl should ask her: "Nothing as belongs to you, Miss." But she knew Miss Beryl would not ask her.

Alice was a mild creature in reality, but she had the most marvellous retorts ready for questions that she knew would never be put to her. The composing of them and the turning of them over and over in her mind comforted her just as much as if they'd been expressed. Really, they kept her alive in places where she'd been that chivvied she'd been afraid to go to bed at night with a box of matches on the chair in case she bit the tops off in her sleep, as you might say.

"Oh, Alice," said Miss Beryl. "There's one extra to tea, so heat a plate of yesterday's scones, please. And put on the Victoria sandwich as well as the coffee cake. And don't forget to put little doyleys under the plates—will you? You did yesterday, you know, and the tea looked so ugly and common. And, Alice, don't put that dreadful old pink and green cosy on the afternoon teapot again. That is only for the mornings.

Really, I think it ought to be kept for the kitchen—it's so shabby, and quite smelly. Put on the Japanese one. You quite understand, don't you?"

Miss Beryl had finished.

That sing aloud from every tree . . .

she sang as she left the kitchen, very pleased with her firm handling of Alice.

Oh, Alice was wild. She wasn't one to mind being told, but there was someting in the way Miss Beryl had of speaking to her that she couldn't stand. Oh, that she couldn't. It made her curl up inside, as you might say, and she fair trembled. But what Alice really hated Miss Beryl for was that she made her feel low. She talked to Alice in a special voice as though she wasn't quite all there; and she never lost her temper with her—never. Even when Alice dropped anything or forgot anything important Miss Beryl seemed to have expected it to happen.

"If you please, Mrs. Burnell," said an imaginary Alice, as she buttered the scones, "I'd rather not take my orders from Miss Beryl. I may be only a common servant girl as doesn't know how to play the guitar, but . . ."

This last thrust pleased her so much that she quite recovered her temper.

"The only thing to do," she heard, as she opened the dining-room door, "is to cut the sleeves out entirely and just have a broad band of black velvet over the shoulders instead. . . ."

XI

THE WHITE duck did not look as if it had ever had a head when Alice placed it in front of Stanley Burnell that night. It lay, in beautifully basted resignation, on a blue dish—its legs tied together with a piece of string and a wreath of little balls of stuffing round it.

It was hard to say which of the two, Alice or the duck, looked the better basted; they were both such a rich colour and they both had the same air of gloss and strain. But Alice was fiery red and the duck a Spanish mahogany.

Burnell ran his eye along the edge of the carving knife. He prided himself very much upon his carving, upon making a first-class job of it.

He hated seeing a woman carve; they were always too slow and they never seemed to care what the meat looked like afterwards. Now he did; he took a real pride in cutting delicate shaves of cold beef, little wads of mutton, just the right thickness, and in dividing a chicken or a duck with nice precision. . . .

"Is this the first of the home products?" he asked, knowing perfectly well that it was.

"Yes, the butcher did not come. We have found out that he only calls twice a week."

But there was no need to apologise. It was a superb bird. It wasn't meat at all, but a kind of very superior jelly. "My father would say," said Burnell, "this must have been one of those birds whose mother played to it in infancy upon the German flute. And the sweet strains of the dulcet instrument acted with such effect upon the infant mind. . . . Have some more, Beryl? You and I are the only ones in this house with a real feeling for food. I'm perfectly willing to state, in a court of law, if necessary, that I love good food."

Tea was served in the drawing-room, and Beryl, who for some reason had been very charming to Stanley ever since he came home, suggested a game of crib. They sat at a little table near one of the open windows. Mrs. Fairfield disappeared, and Linda lay in a rocking-chair, her arms above her head, rocking to and fro.

"You don't want the light—do you, Linda?" said Beryl. She moved the lamp so that she sat under its soft light.

How remote they looked, those two, from where Linda sat and rocked. The green table, the polished cards, Stanley's big hands and Beryl's tiny ones, all seemed to be part of one mysterious movement. Stanley himself, big and solid, in his dark suit, took his ease, and Beryl tossed her bright head and pouted. Round her throat she wore an unfamiliar velvet ribbon. It changed her, somehow—altered the shape of her face—but it was charming, Linda decided. The room smelled of lilies; there were two big jars of arums in the fire-place.

"Fifteen two—fifteen four—and a pair is six and a run of three is nine," said Stanley, so deliberately, he might have been counting sheep.

"I've nothing but two pairs," said Beryl, exaggerating her woe because she knew how he loved winning.

The cribbage pegs were like two little people going up the road together, turning round the sharp corner, and coming down the road again. They were pursuing each other. They did not so much want to get ahead as to keep near enough to talk—to keep near, perhaps that was all.

But no, there was always one who was impatient and hopped away as the other came up, and would not listen. Perhaps the white peg was frightened of the red one, or perhaps he was cruel and would not give the red one a chance to speak. . . .

In the front of her dress Beryl wore a bunch of pansies, and once when the little pegs were side by side, she bent over and the pansies dropped out and covered them.

"What a shame," said she, picking up the pansies. "Just as they had a chance to fly into each other's arms."

"Farewell, my girl," laughed Stanley, and away the red peg hopped.

The drawing-room was long and narrow with glass doors that gave on to the verandah. It had a cream paper with a pattern of gilt roses, and the furniture, which had belonged to old Mrs. Fairfield, was dark and plain. A little piano stood against the wall with yellow pleated silk let into the carved front. Above it hung an oil painting by Beryl of a large cluster of surprised looking clematis. Each flower was the size of a small saucer, with a centre like an astonished eye fringed in black. But the room was not finished yet. Stanley had set his heart on a Chesterfield and two decent chairs. Linda liked it best as it was. . . .

Two big moths flew in through the window and round and round the circle of lamplight.

"Fly away before it is too late. Fly out again."

Round and round they flew; they seemed to bring the silence and the moonlight in with them on their silent wings. . . .

"I've two kings," said Stanley. "Any good?"

"Quite good," said Beryl.

Linda stopped rocking and got up. Stanley looked across. "Anything the matter, darling?"

"No, nothing. I'm going to find mother."

She went out of the room and standing at the foot of the stairs she called, but her mother's voice answered her from the verandah.

The moon that Lottie and Kezia had seen from the storeman's wagon was full, and the house, the garden, the old woman and Linda—all were bathed in dazzling light.

"I have been looking at the aloe," said Mrs. Fairfield. "I believe it is going to flower this year. Look at the top there. Are those buds, or is it only an effect of light?"

As they stood on the steps, the high grassy bank on which the aloe rested rose up like a wave, and the aloe seemed to ride upon it like a ship with the oars lifted. Bright moonlight hung upon the lifted oars like water, and on the green wave glittered the dew.

"Do you feel it, too," said Linda, and she spoke to her mother with the special voice that women use at night to each other as though they spoke in their sleep or from some hollow cave— "Don't you feel that it is coming towards us?"

She dreamed that she was caught up out of the cold water into the ship with the lifted oars and the budding mast. Now the oars fell striking quickly, quickly. They rowed far away over the top of the garden trees, the paddocks and the dark bush beyond. Ah, she heard herself cry: "Faster! Faster!" to those who were rowing.

How much more real this dream was than that they should go back to the house where the sleeping children lay and where Stanley and Beryl played cribbage.

"I believe those are buds," said she. "Let us go down into the garden, mother. I like that aloe. I like it more than anything here. And I am sure I shall remember it long after I've forgotten all the other things."

She put her hand on her mother's arm and they walked down the steps, round the island and on to the main drive that led to the front gates.

Looking at it from below she could see the long sharp thorns that edged the aloe leaves, and at the sight of them her heart grew hard.... She particularly liked the long sharp thorns.... Nobody would dare to come near the ship or to follow after.

"Not even my Newfoundland dog," thought she, "that I'm so fond of in the daytime."

For she really was fond of him; she loved and admired and respected him tremendously. Oh, better than anyone else in the world. She knew him through and through. He was the soul of truth and decency, and for all his practical experience he was awfully simple, easily pleased and easily hurt....

If only he wouldn't jump at her so, and bark so loudly, and watch her with such eager, loving eyes. He was too strong for her; she had always hated things that rush at her, from a child. There were times when he was frightening—really frightening. When she just had not screamed at the top of her voice: "You are killing me." And at those times she had longed to say the most coarse, hateful things....

"You know I'm very delicate. You know as well as I do that my heart is affected, and the doctor has told you I may die any moment. I have had three great lumps of children already. . . ."

Yes, yes, it was true. Linda snatched her hand from her mother's arm. For all her love and respect and admiration she hated him. And how tender he always was after times like those, how submissive, how thoughtful. He would do anything for her; he longed to serve her. . . . Linda heard herself saying in a weak voice:

"Stanley, would you light a candle?"

And she heard his joyful voice answer: "Of course I will my darling." And he leapt out of bed as though he were going to leap at the moon for her.

It had never been so plain to her as it was at this moment. There were all her feelings for him, sharp and defined, one as true as the other. And there was this other, this hatred, just as real as the rest. She could have done her feelings up in little packets and given them to Stanley. She longed to hand him that last one, for a surprise. She could see his eyes as he opened that . . .

She hugged her folded arms and began to laugh silently. How absurd life was—it was laughable, simply laughable. And why this mania of hers to keep alive at all? For it really was a mania, she thought, mocking and laughing.

"What am I guarding myself for so preciously? I shall go on having children and Stanley will go on making money and the children and the gardens will grow bigger and bigger, with the whole fleets of aloes in them for me to choose from."

She had been walking with her head bent, looking at nothing. Now she looked up and about her. They were standing by the red and white camellia trees. Beautiful were the rich dark leaves spangled with light and the round flowers that perch among them like red and white birds. Linda pulled a piece of verbena and crumpled it, and held her hands to her mother.

"Delicious," said the old woman. "Are you cold, child? Are you trembling? Yes, your hands are cold. We had better go back to the house."

"What have you been thinking about?" said Linda. "Tell me."

"I haven't really been thinking of anything. I wondered as we passed the orchard what the fruit trees were like and whether we should be able to make much jam this autumn. There are splendid healthy currant bushes in the vegetable garden. I noticed them to-day. I should like to see those pantry shelves thoroughly well stocked with our own jam. . . ."

XII

"MY DARLING NAN,

Don't think me a piggy wig because I haven't written before. I haven't had a moment, dear, and even now I feel so exhausted that I can hardly hold a pen.

Well, the dreadful deed is done. We have actually left the giddy whirl of town, and I can't see how we shall ever go back again, for my brother-in-law has bought this house 'lock, stock and barrel,' to use his own words.

In a way, of course, it is an awful relief, for he has been threatening to take a place in the country ever since I've lived with them—and I must say the house and garden are awfully nice—a million times better than that awful cubby-hole in town.

But, buried, my dear. Buried isn't the word.

We have got neighbours, but they are only farmers—big louts of boys who seem to be milking all day, and two dreadful females with rabbit teeth who brought us some scones when we were moving and said they would be pleased to help. But my sister who lives a mile away doesn't know a soul here, so I am sure we never shall. It's pretty certain nobody will ever come out from town to see us, because though there is a bus it's an awful old rattling thing with black leather sides that any decent person would rather die than ride in for six miles.

Such is life. It's a sad ending for poor little B. I'll get to be a most awful frump in a year or two and come and see you in a mackintosh and a sailor hat tied on with a white china silk motor veil. So pretty.

Stanley says that now we are settled—for after the most awful week of my life we really are settled—he is going to bring out a couple of men from the club on Saturday afternoons for tennis. In fact, two are promised as a great treat to-day. But, my dear, if you could see Stanley's men from the club . . . rather fattish, the type who look frightfully indecent without waistcoats—always with toes that turn in rather—so conspicuous when you are walking about a court in white shoes. And they are pulling up their trousers every minute—don't you know—and whacking at imaginary things with their rackets.

I used to play with them at the club last summer, and I am sure you will know the type when I tell you that after I'd been there about three times they all called me Miss Beryl. It's a weary world. Of course mother simply loves the place, but then I suppose when I am mother's age I shall be content to sit in the sun and shell peas into a basin. But I'm not—not—not.

What Linda thinks about the whole affair, per usual, I haven't the slightest idea. Mysterious as ever....

My dear, you know that white satin dress of mine. I have taken the sleeves out entirely, put bands of black velvet across the shoulders and two big red poppies off my dear sister's *chapeau*. It is a great success, though when I shall wear it I do not know."

Beryl sat writing this letter at a little table in her room. In a way, of course, it was all perfectly true, but in another way it was all the greatest rubbish and she didn't believe a word of it. No, that wasn't true. She felt all those things, but she didn't really feel them like that.

It was her other self who had written that letter. It not only bored, it rather disgusted her real self.

"Flippant and silly," said her real self. Yet she knew that she'd send it and she'd always write that kind of twaddle to Nan Pym. In fact, it was a very mild example of the kind of letter she generally wrote.

Beryl leaned her elbows on the table and read it through again. The voice of the letter seemed to come up to her from the page. It was faint already, like a voice heard over the telephone, high, gushing, with something bitter in the sound. Oh, she detested it to-day.

"You've always got so much animation," said Nan Pym. "That's why men are so keen on you." And she added, rather mournfully, for men were not at all keen on Nan, who was a solid kind of girl, with fat hips and a high colour—"I can't understand how you can keep it up. But it is your nature, I suppose."

What rot. What nonsense. It wasn't her nature at all. Good heavens, if she had ever been her real self with Nan Pym, Nannie would have jumped out of the window with surprise.... My dear, you know that white satin of mine.... Beryl slammed the letter-case to.

She jumped up and half unconsciously, half consciously she drifted over to the looking-glass.

There stood a slim girl in white—a white serge skirt, a white silk blouse, and a leather belt drawn in very tightly at her tiny waist.

Her face was heart-shaped, wide at the brows and with a pointed chin—but not too pointed. Her eyes, her eyes were perhaps her best feature; they were such a strange uncommon colour—greeny blue with little gold points in them.

She had fine black eyebrows and long lashes—so long, that when they lay on her cheeks you positively caught the light in them, someone or other had told her.

Her mouth was rather large. Too large? No, not really. Her underlip protruded a little; she had a way of sucking it in that somebody else had told her was awfully fascinating.

Her nose was her least satisfactory feature. Not that it was really ugly. But it was not half as fine as Linda's. Linda really had a perfect little nose. Hers spread rather—not badly. And in all probability she exaggerated the spreadiness of it just because it was her nose, and she was so awfully critical of herself. She pinched it with a thumb and first finger and made a little face. . . .

Lovely, lovely hair. And such a mass of it. It had the colour of fresh fallen leaves, brown and red with a glint of yellow. When she did it in a long plait she felt it on her backbone like a long snake. She loved to feel the weight of it dragging her head back, and she loved to feel it loose, covering her bare arms. "Yes, my dear, there is no doubt about it, you really are a lovely little thing."

At the words her bosom lifted; she took a long breath of delight, half closing her eyes.

But even as she looked the smile faded from her lips and eyes. Oh God, there she was, back again, playing the same old game. False—false as ever. False as when she'd written to Nan Pym. False even when she was alone with herself, now.

What had that creature in the glass to do with her, and why was she staring? She dropped down to one side of her bed and buried her face in her arms.

"Oh," she cried, "I am so miserable—so frightfully miserable. I know that I'm silly and spiteful and vain; I'm always acting a part. I'm never my real self for a moment." And plainly, plainly, she saw her false self running up and down the stairs, laughing a special trilling laugh if they had visitors, standing under the lamp if a man came to dinner, so that he should see the light on her hair, pouting and pretending to be a little girl when she was asked to play the guitar. Why? She even kept it up for Stanley's benefit. Only last night when he was reading the paper her false self had stood beside him and leaned against his shoulder on purpose. Hadn't she put her hand over his, pointing out something so that he should see how white her hand was beside his brown one.

How despicable! Despicable! Her heart was cold with rage. "It's marvellous how you keep it up," said she to the false self. But then it was only because she was so miserable—so miserable. If she had been happy and leading her own life, her false life would cease to be. She saw the real Beryl—a shadow . . . a shadow. Faint and unsubstantial she shone. What was there of her except the radiance? And for what tiny moments she was really she. Beryl could almost remember every one of them. At those times she had felt: "Life is rich and mysterious and good, and I am rich and mysterious and good, too." Shall I ever be that Beryl for ever? Shall I? How can I? And was there ever a time when I did not have a false self? . . . But just as she had got that far she heard the sound of little steps running along the passage; the door handle rattled. Kezia came in.

"Aunt Beryl, mother says will you please come down? Father is home with a man and lunch is ready."

Botheration! How she had crumpled her skirt, kneeling in that idiotic way.

"Very well, Kezia." She went over to the dressing table and powdered her nose.

Kezia crossed too, and unscrewed a little pot of cream and sniffed it. Under her arm she carried a very dirty calico cat.

When Aunt Beryl ran out of the room she sat the cat up on the dressing table and stuck the top of the cream jar over its ear.

"Now look at yourself," said she sternly.

The calico cat was so overcome by the sight that it toppled over backwards and bumped and bumped on to the floor. And the top of the cream jar flew through the air and rolled like a penny in a round on the linoleum—and did not break.

But for Kezia it had broken the moment it flew through the air, and she picked it up, hot all over, and put it back on the dressing table.

Then she tip-toed away, far too quickly and airily. . . .

At the Bay

I

VERY EARLY morning. The sun was not yet risen, and the whole of Crescent Bay was hidden under a white sea-mist. The big bush-covered hills at the back were smothered. You could not see where they ended and the paddocks and bungalows began. The sandy road was gone and the paddocks and bungalows the other side of it; there were no white dunes covered with reddish grass beyond them; there was nothing to mark which was beach and where was the sea. A heavy dew had fallen. The grass was blue. Big drops hung on the bushes and just did not fall; the silvery, fluffy toi-toi was limp on its long stalks, and all the marigolds and the pinks in the bungalow gardens were bowed to the earth with wetness. Drenched were the cold fuchsias, round pearls of dew lay on the flat nasturtium leaves. It looked as though the sea had beaten up softly in the darkness, as though one immense wave had come rippling, rippling—how far? Perhaps if you had waked up in the middle of the night you might have seen a big fish flicking in at the window and gone again. . . .

Ah-aah! sounded the sleepy sea. And from the bush there came the sound of little streams flowing, quickly, lightly, slipping between the smooth stones, gushing into ferny basins and out again; and there was the splashing of big drops on large leaves, and something else—what was it?—a faint stirring and shaking, the snapping of a twig and then such silence that it seemed some one was listening.

Round the corner of Crescent Bay, between the piled-up masses of broken rock, a flock of sheep came pattering. They were huddled together, a small, tossing, wooly mass, and their thin, stick-like legs trotted along quickly as if the cold and the quiet had frightened them. Behind them an old sheep-dog, his soaking paws covered with sand, ran along with his nose to the ground, but carelessly, as if thinking,

---

of something else. And then in the rocky gateway the shepherd himself appeared. He was a lean, upright old man, in a frieze coat that was covered with a web of tiny drops, velvet trousers tied under the knee, and a wideawake with a folded blue handkerchief round the brim. One hand was crammed into his belt, the other grasped a beautifully smooth yellow stick. And as he walked, taking his time, he kept up a very soft light whistling, an airy, far-away fluting that sounded mournful and tender. The old dog cut an ancient caper or two and then drew up sharp, ashamed of his levity, and walked a few dignified paces by his master's side. The sheep ran forward in little pattering rushes; they began to bleat, and ghostly flocks and herds answered them from under the sea. "Baa! Baaa!" For a time they seemed to be always on the same piece of ground. There ahead was stretched the sandy road with shallow puddles; the same soaking bushes showed on either side and the same shadowy palings. Then something immense came into view; an enormous shock-haired giant with his arms stretched out. It was the big gum-tree outside Mrs. Stubb's shop, and as they passed by there was a strong whiff of eucalyptus. And now big spots of light gleamed in the mist. The shepherd stopped whistling; he rubbed his red nose and wet beard on his wet sleeve and, screwing up his eyes, glanced in the direction of the sea. The sun was rising. It was marvellous how quickly the mist thinned, sped away, dissolved from the shallow plain, rolled up from the bush and was gone as if in a hurry to escape; big twists and curls jostled and shouldered each other as the silvery beams broadened. The far-away sky—a bright, pure blue—was reflected in the puddles, and the drops, swimming along the telegraph poles, flashed into points of light. Now the leaping, glittering sea was so bright it made one's eyes ache to look at it. The shepherd drew a pipe, the bowl as small as an acorn, out of his breast pocket, fumbled for a chunk of speckled tobacco, pared off a few shavings and stuffed the bowl. He was a grave, fine-looking old man. As he lit up and the blue smoke wreathed his head, the dog, watching, looked proud of him.

"Baa! Baaa!" The sheep spread out onto a fan. They were just clear of the summer colony before the first sleeper turned over and lifted a drowsy head; their cry sounded in the dreams of little children . . . who lifted their arms to drag down, to cuddle the darling little woolly lambs of sleep. Then the first inhabitant appeared; it was the Burnells' cat Florrie, sitting on the gatepost, far too early as usual, looking for their milk-girl. When she saw the old sheep-dog she sprang up quickly, arched her back, drew in her tabby head, and seemed to give a little fastidious shiver. "Ugh! What a coarse, revolting creature!" said Florrie. But the old sheep-dog, not looking up, waggled past, flinging out his legs from side to side. Only one of his ears twitched to prove that he saw, and thought her a silly young female.

The breeze of the morning lifted in the bush and the smell of leaves and wet black earth mingled with the sharp smell of the sea. Myriads of birds were singing. A goldfinch flew over the shepherd's head and, perching on the tiptop of a spray, it turned to the sun, ruffling its small breast feathers. And now they had passed the fisherman's hut, passed the charred-looking little whare where Leila the milk-girl lived with her old Gran. The sheep strayed over a yellow swamp and Wag, the sheep-dog, padded after, rounded them up and headed them for the steeper, narrower rocky pass that led out of Crescent Bay towards Daylight Cove. "Baa! Baa!" Faint the cry came as they rocked along the fast-drying road. The shepherd put away his pipe, dropping it into his breast pocket so that the little bowl hung over. And straightway the soft airy whistling began again. Wag ran out along a ledge of rock after something that smelled, and ran back again disgusted. Then pushing, nudging, hurrying, the sheep rounded the bend and the shepherd followed after out of sight.

II

A FEW moments later the back door of one of the bungalows opened, and a figure in a broad-striped bathing suit flung down the paddock, cleared the stile, rushed through the tussock grass into the hollow, staggered up the sandy hillock, and raced for dear life over the big porous stones, over the cold, wet pebbles, on to the hard sand that gleamed like oil. Splish-splosh! Splish-splosh! The water bubbled round his legs as Stanley Burnell waded out exulting. First man in as usual! He'd beaten them all again. And he swooped down to souse his head and neck.

"Hail, brother! All hail, Thou Mighty One!" A velvety bass voice came booming over the water.

Great Scott! Damnation take it! Stanley lifted up to see a dark head bobbing far out and an arm lifted. It was Jonathan Trout—there before him! "Glorious morning!" sang the voice.

"Yes, very fine!" said Stanley briefly. Why the dickens didn't the fellow stick to his part of the sea? Why should he come barging over to this exact spot? Stanley gave a kick, a lunge and struck out, swimming overarm. But Jonathan was a match for him. Up he came, his black hair sleek on his forehead, his short beard sleek.

"I had an extraordinary dream last night!" he shouted.

What was the matter with the man? This mania for conversation irritated Stanley beyond words. And it was always the same—always some piffle about a dream he'd had, or some cranky idea he'd got hold of, or some rot he'd been reading. Stanley turned over on his back and kicked with his legs til he was a living waterspout. But even then ... "I dreamed I was hanging over a terrifically high cliff, shouting to some one below." You would be! thought Stanley. He could stick no more of it. He stopped splashing. "Look here, Trout," he said, "I'm in rather a hurry this morning."

"You're WHAT?" Jonathan was so surprised—or pretended to be—that he sank under the water, then appeared again blowing.

"All I mean is," said Stanley, "I've no time to—to—to fool about. I want to get this over. I'm in a hurry. I've work to do this morning—see?"

Jonathan was gone before Stanley had finished. "Pass, friend!" said the bass voice gently, and he slid away through the water with scarcely a ripple.... But curse the fellow! He'd ruined Stanley's bathe. What an unpractical idiot the man was! Stanley struck out to sea again, and then as quickly swam in again, and away he rushed up the beach. He felt cheated.

Jonathan stayed a little longer in the water. He floated, gently moving his hands like fins, and letting the sea rock his long, skinny body. It was curious, but in spite of everything he was fond of Stanley Burnell. True, he had a fiendish desire to tease him sometimes, to poke fun at him, but at bottom he was sorry for the fellow. There was something pathetic in his determination to make a job of everything. You couldn't help feeling he'd be caught out one day, and then what an almighty cropper he'd come! At that moment an immense wave lifted Jonathan, rode past him, and broke along the beach with a joyful sound. What a beauty! And now there came another. That was the way to live—carelessly, recklessly, spending oneself. He got on to his feet and began to wade towards the shore, pressing his toes into the firm, wrinkled sand. To take things easy, not to fight against the ebb and flow of life, but to give way to it—that was what was needed.

It was this tension that was all wrong. To live—to live! And the perfect morning, so fresh and fair, basking in the light, as though laughing at its own beauty, seemed to whisper, "Why not?"

But now he was out of the water Jonathan turned blue with cold. He ached all over; it was as though some one was wringing the blood out of him. And stalking up the beach, shivering, all his muscles tight, he too felt his bathe was spoilt. He'd stayed in too long.

III

BERYL was alone in the living-room when Stanley appeared, wearing a blue serge suit, a stiff collar and a spotted tie. He looked almost uncannily clean and brushed; he was going to town for the day. Dropping into his chair, he pulled out his watch and put it beside his plate.

"I've just got twenty-five minutes," he said. "You might go and see if the porridge is ready, Beryl?"

"Mother's just gone for it," said Beryl. She sat down at the table and poured out his tea.

"Thanks!" Stanley took a sip. "Hallo!" he said in an astonished voice, "you've forgotten the sugar."

"Oh, sorry!" But even then Beryl didn't help him; she pushed the basin across. What did this mean? As Stanley helped himself his blue eyes widened; they seemed to quiver. He shot a quick glance at his sister-in-law and leaned back.

"Nothing wrong, is there?" he asked carelessly, fingering his collar.

Beryl's head was bent; she turned her plate in her fingers.

"Nothing," said her light voice. Then she too looked up, and smiled at Stanley. "Why should there be?"

"O-oh! No reason at all as far as I know. I thought you seemed rather—"

At that moment the door opened and the three little girls appeared, each carrying a porridge plate. They were dressed alike in blue jerseys and knickers; their brown legs were bare, and each had her hair plaited and pinned up in what was called a horse's tail. Behind them came Mrs. Fairfield with the tray.

"Carefully, children," she warned. But they were taking the very greatest care. They loved being allowed to carry things. "Have you said good morning to your father?"

"Yes, grandma." They settled themselves on the bench opposite Stanley and Beryl.

"Good morning, Stanley!" Old Mrs. Fairfield gave him his plate.

"Morning, mother! How's the boy?"

"Splendid! He only woke up once last night. What a perfect morning!" The old woman paused, her hand on thc loaf of bread, to gaze out of the open door into the graden. The sea sounded. Through the wide-open window streamed the sun on to the yellow varnished walls and bare floor. Everything on the table flashed and glittered. In the middle there was an old salad bowl filled with yellow and red nasturtiums. She smiled, and a look of deep content shone in her eyes.

"You might *cut* me a slice of that bread, mother," said Stanley. "I've only twelve and a half minutes before the coach passes. Has any one given my shoes to the servant girl?"

"Yes, they're ready for you." Mrs. Fairfield was quite unruffled.

"Oh, Kezia! Why are you such a messy child!" cried Beryl despairingly.

"Me, Aunt Beryl?" Kezia stared at her. What had she done now? She had only dug a river down the middle of her porridge, filled it, and was eating the banks away. But she did that every single morning, and no one had said a word up till now.

"Why can't you eat your food properly like Isabel and Lottie?" How unfair grown-ups are!

"But Lottie always makes a floating island, don't you, Lottie?"

"I don't," said Isabel smartly. "I just sprinkle mine with sugar and put on the milk and finish it. Only babies play with their food."

Stanley pushed back his chair and got up.

"Would you get me those shoes, mother? And, Beryl, if you've finished, I wish you'd cut down to the gate and stop the coach. Run in to your mother, Isabel, and ask her where my bowler hat's been put. Wait a minute—have you children been playing with my stick?"

"No, father!"

"But I put it here." Stanley began to bluster. "I remember distinctly putting it in this corner. Now, who's had it? There's no time to lose. Look sharp! The stick's got to be found."

Even Alice, the servant-girl, was drawn into the chase. "You haven't been using it to poke the kitchen fire with by any chance?"

Stanley dashed into the bedroom where Linda was lying. "Most extraordinary thing. I can't keep a single possession to myself. They've made away with my stick, now!"

"Stick, dear? What stick?" Linda's vagueness on these occasions could not be real, Stanley decided. Would nobody sympathize with him?

"Coach! Coach, Stanley!" Beryl's voice cried from the gate.

Stanley waved his arm to Linda. "No time to say good-bye!" he cried. And he meant that as a punishment to her.

He snatched his bowler hat, dashed out of the house, and swung down the garden path. Yes, the coach was there waiting, and Beryl, leaning over the open gate, was laughing up at somebody or other just as if nothing had happened. The heartlessness of women! The way they took it for granted it was your job to slave away for them while they didn't even take the trouble to see that your walking-stick wasn't lost. Kelly trailed his whip across the horses.

"Good-bye, Stanley," called Beryl, sweetly and gaily. It was easy enough to say good-bye! And there she stood, idle, shading her eyes with her hand. The worst of it was Stanley had to shout good-bye too, for the sake of appearances. Then he saw her turn, give a little skip and run back to the house. She was glad to be rid of him!

Yes, she was thankful. Into the living-room she ran and called "He's gone!" Linda cried from her room: "Beryl! Has Stanley gone?" Old Mrs. Fairfield appeared, carrying the boy in his little flannel coatee.

"Gone?"

"Gone!"

Oh, the relief, the difference it made to have the man out of the house. Their very voices were changed as they called to one another; they sounded warm and loving and as if they shared a secret. Beryl went over to the table. "Have another cup of tea, mother. It's still hot." She wanted, somehow, to celebrate the fact that they could do what they liked now. There was no man to disturb them; the whole perfect day was theirs.

"No, thank you, child," said old Mrs. Fairfield, but the way at that moment she tossed the boy up and said "a-goos-a-goos-a-ga!" to him meant that she felt the same. The little girls ran into the paddock like chickens let out of a coop.

Even Alice, the servant-girl, washing up the dishes in the kitchen, caught the infection and used the precious tank water in a perfectly reckless fashion.

"Oh, these men!" said she, and she plunged the teapot into the bowl and held it under the water even after it had stopped bubbling, as if it too was a man and drowning was too good for them.

IV

"WAIT FOR me, Isa-bel! Kezia, wait for me!"

There was poor little Lottie, left behind again, because she found it so fearfully hard to get over the stile by herself. When she stood on the first step her knees began to wobble; she grasped the post. Then you had to put one leg over. But which leg? She never could decide. And when she did finally put one leg over with a sort of stamp of despair—then the feeling was awful. She was half in the paddock still and half in the tussock grass. She clutched the post desperately and lifted up her voice. "Wait for me!"

"No, don't you wait for her, Kezia!" said Isabel. "She's such a little silly. She's always making a fuss. Come on!" And she tugged Kezia's jersey. "You can use my bucket if you come with me," she said kindly. "It's bigger than yours." But Kezia couldn't leave Lottie all by herself. She ran back to her. By this time Lottie was very red in the face and breathing heavily.

"Here, put your other foot over," said Kezia.

"Where?"

Lottie looked down at Kezia as if from a mountain height.

"Here where my hand is." Kezia patted the place.

"Oh, *there* do you mean!" Lottie gave a deep sigh and put the second foot over.

"Now—sort of turn round and sit down and slide," said Kezia.

"But there's nothing to sit down *on*, Kezia," said Lottie.

She managed it at last, and once it was over she shook herself and began to beam.

"I'm getting better at climbing over stiles, aren't I, Kezia?"

Lottie's was a very hopeful nature.

The pink and the blue sunbonnet followed Isabel's bright red sunbonnet up that sliding, slipping hill. At the top they paused to decide where to go and to have a good stare at who was there already. Seen from behind, standing against the skyline, gesticulating largely with their spades, they looked like minute puzzled explorers.

The whole family of Samuel Josephs was there already with their lady-help, who sat on a camp-stool and kept order with a whistle that she wore tied round her neck, and a small cane with which she directed operations. The Samuel Josephs never played by themselves or managed their own game. If they did, it ended in the boys pouring water down the girls' necks or the girls trying to put little black crabs into the boys' pockets. So Mrs. S. J. and the poor lady-help drew up what she called a "brogramme" every morning to keep them "abused and out of bischief." It was all competitions or races or round games. Everything began with a piercing blast of the lady-help's whistle and ended with another. There were even prizes—large, rather dirty paper parcels which the lady-help with a sour little smile drew out of a bulging string kit. The Samuel Josephs fought fearfully for the prizes and cheated and pinched one another's arms—they were all expert pinchers. The only time the Burnell children ever played with them Kezia had got a prize, and when she undid three bits of paper she found a very small rusty button-hook. She couldn't understand why they made such a fuss. . . .

But they never played with the Samuel Josephs now or even went to their parties. The Samuel Josephs were always giving children's parties at the Bay and there was always the same food. A big washhand basin of very brown fruit-salad, buns cut into four and a washhand jug full of something the lady-help called "Limonadear." And you went away in the evening with half the frill torn off your frock or something spilled all down the front of your open-work pinafore, leaving the Samuel Josephs leaping like savages on their lawn. No! They were too awful.

On the other side of the beach, close down to the water, two little boys, their knickers rolled up, twinkled like spiders. One was digging, the other pattered in and out of the water, filling a small bucket. They were the Trout boys, Pip and Rags. But Pip was so busy digging and Rags was so busy helping that they didn't see their little cousins until they were quite close.

"Look!" said Pip. " Look what I've discovered." And he showed them an old, wet, squashed-looking boot. The three little girls stared.

"Whatever are you going to do with it?" asked Kezia.

"Keep it, of course!" Pip was very scornful. "It's a find—see?"

Yes, Kezia saw that. All the same . . .

"There's lots of things buried in the sand," explained Pip. "They get chucked up from wrecks. Treasure. Why—you might find—"

"But why does Rags have to keep on pouring water in?" asked Lottie.

"Oh, that's to moisten it," said Pip, "to make the work a bit easier. Keep it up, Rags."

And good little Rags ran up and down, pouring in the water that turned brown like cocoa.

"Here, shall I show you what I found yesterday?" said Pip mysteriously, and he stuck his spade into the sand. "Promise not to tell."

They promised.

"Say, cross my heart straight dinkum."

The little girls said it.

Pip took something out of his pocket, rubbed it a long time on the front of his jersey, then breathed on it and rubbed it again.

"Now turn round!" he ordered.

They turned round.

"All look the same way! Keep still! Now!"

And his hand opened; he held up to the light something that flashed, that winked, that was a most lovely green.

"It's a nemeral," said Pip solemnly.

"Is it really, Pip?" Even Isabel was impressed.

The lovely green thing seemed to dance in Pip's fingers. Aunt Beryl had a nemeral in a ring, but it was a very small one. This one was as big as a star and far more beautiful.

V

AS THE morning lengthened whole parties appeared over the sand-hills and came down on the beach to bathe. It was understood that at eleven o'clock the women and children of the summer colony had the sea to themselves. First the women undressed, pulled on their bathing dresses and covered their heads in hideous caps like sponge bags; then the children were unbuttoned. The beach was strewn with little heaps of clothes and shoes; the big summer hats, with stones on them to keep them from blowing away, looked like immense shells. It was strange that even the sea seemed to sound differently when all those leaping, laughing figures ran into the waves. Old Mrs. Fairfield, in a lilac cotton dress and a black hat tied under the chin, gathered her little brood and got them ready. The little Trout boys whipped their shirts over their heads, and away the five sped, while their grandma sat with one hand in her knitting-bag ready to draw out the ball of wool when she was satisfied they were safely in.

The firm compact little girls were not half so brave as the tender, delicate-looking little boys. Pip and Rags, shivering, crouching down, slapping the water, never hesitated. But Isabel, who could swim twelve strokes, and Kezia, who could nearly swim eight, only followed on the strict understanding they were not to be splashed. As for Lottie, she didn't follow at all. She liked to be left to go in her own way, please. And that way was to sit down at the edge of the water, her legs straight, her knees pressed together, and to make vague motions with her arms as if she expected to be wafted out to sea. But when a bigger wave than usual, an old whiskery one, came lolloping along in her direction, she scrambled to her feet with a face of horror and flew up the beach again.

"Here, mother, keep those for me, will you?"

Two rings and a thin gold chain were dropped into Mrs. Fairfield's lap.

"Yes, dear. But aren't you going to bathe here?"

"No-o," Beryl drawled. She sounded vague. "I'm undressing farther along. I'm going to bathe with Mrs. Henry Kember."

"Very well." But Mrs. Fairfield's lips set. She disapproved of Mrs. Harry Kember. Beryl knew it.

Poor old mother, she smiled, as she skimmed over the stones. Poor old mother! Old! Oh, what joy, what bliss it was to be young. . . .

"You look very pleased," said Mrs. Harry Kember. She sat hunched up on the stones, her arms round her knees, smoking .

"It's such a lovely day," said Beryl, smiling down at her.

"Oh, my *dear!*" Mrs Harry Kember's voice sounded as though she knew better than that. But then her voice always sounded as though she knew something better about you than you did yourself. She was a long, strange-looking woman with narrow hands and feet. Her face, too, was long and narrow and exhausted-looking; even her fair curled fringe looked burnt out and withered. She was the only woman at the Bay who smoked, and she smoked incessantly, keeping the cigarette between her lips while she talked, and only taking it out when the ash was so long you could not understand why it did not fall. When she was not playing bridge—she played bridge every day of her life—she spent her time lying in the full glare of the sun. She could stand any amount of it; she never had enough. All the same, it did not seem to warm her. Parched, withered, cold, she lay stretched on the stones like a piece of tossed-up driftwood. The women at the Bay thought she was very, very fast. Her lack of vanity, her slang, the way she treated men as though she was one of them, and the fact that she didn't care twopence about her house and called the servant Gladys "Glad-eyes," was disgraceful. Standing on the verandah steps Mrs. Kember would call in her indifferent, tired voice, "I say, Glad-eyes, you might heave me a handkerchief if I've got one, will you?" And Glad-eyes, a red bow in her hair instead of a cap, and white shoes, came running with an impudent smile. It was an absolute scandal! True, she had no children, and her husband. . . . Here the voices were always raised; they became fervent. How can he have married her? How can he, how can he? It must have been money, of course, but even then!

Mrs. Kember's husband was at least ten years younger than she was, and so incredibly handsome that he looked like a mask or a most perfect illustration in an American novel rather than a man. Black hair, dark blue eyes, red lips, a slow sleepy smile, a fine tennis player, a perfect dancer, and with it all a mystery. Harry Kember was like a man walking in his sleep. Men couldn't stand him, they couldn't get a word out of the chap; he ignored his wife just as she ignored him. How did he live? Of course there were stories, but such stories! They simply couldn't be told. The women he'd been seen with, the places he'd been seen in . . . but nothing was ever certain, nothing definite. Some of the women at the Bay privately thought he'd commit a murder one day. Yes, even while they talked to Mrs. Kember and took in the awful concoction she was wearing, they saw her, stretched as she lay on the beach; but cold, bloody, and still with a cigarette stuck in the corner of her mouth.

Mrs. Kember rose, yawned, unsnapped her belt buckle, and tugged at the tape of her blouse. And Beryl stepped out of her skirt and shed her jersey, and stood up in her short white petticoat, and her camisole with ribbon bows on the shoulders.

"Mercy on us," said Mrs. Harry Kember, "what a little beauty you are!"

"Don't!" said Beryl softly; but, drawing off one stocking and then the other, she felt a little beauty.

"My dear—why not?" said Mrs. Harry Kember, stamping on her own petticoat. Really—her underclothes! A pair of blue cotton knickers and a linen bodice that reminded one somehow of a pillow-case. . . . "And you don't wear stays, do you?" She touched Beryl's waist, and Beryl sprang away with a small affected cry. Then "Never!" she said firmly.

"Lucky little creature," sighed Mrs. Kember, unfastening her own.

Beryl turned her back and began the complicated movements of some one who is trying to take off her clothes and to pull on her bathing-dress all at one and the same time.

"Oh, my dear—don't mind me," said Mrs. Harry Kember. "Why be shy? I shan't eat you. I shan't be shocked like those other ninnies." And she gave her strange neighing laugh and grimaced at the other women.

But Beryl was shy. She never undressed in front of anybody. Was that silly? Mrs. Harry Kember made her feel it was silly, even something to be ashamed of. Why be shy indeed! She glanced quickly at her friend standing so boldly in her torn chemis and lighting a fresh cigarette; and a quick, bold, evil feeling started up in her breast. Laughing recklessly, she drew on the limp, sandy-feeling bathing-dress that was not quite dry and fastened the twisted buttons.

"That's better," said Mrs. Harry Kember. They began to go down the beach together. "Really, it's a sin for you to wear clothes, my dear. Somebody's got to tell you some day."

The water was quite warm. It was that marvellous transparent blue, flecked with silver, but the sand at the bottom looked gold; when you kicked with your toes there rose a little puff of gold-dust. Now the waves just reached her breast. Beryl stood, her arms outstretched, gazing out, and as each wave came she gave the slightest little jump, so that it seemed it was the wave which lifted her so gently.

"I believe in pretty girls having a good time," said Mrs. Harry Kember. "Why not? Don't you make a mistake, my dear. Enjoy yourself." And suddenly she turned turtle, disappeared, and swam away quickly, quickly, like a rat. Then she flicked round and began swimming back. She was going to say something else. Beryl felt that she was being poisoned by this cold woman, but she longed to hear. But oh, how strange, how horrible! As Mrs. Harry Kember came up close she looked, in her black water-proof bathing cap, with her sleepy face lifted above the water, just her chin touching, like a horrible caricature of her husband.

VI

IN A steamer chair, under a manuka tree that grew in the middle of the front grass patch, Linda Burnell dreamed the morning away. She did nothing. She looked up at the dark, close, dry leaves of the manuka, at the chinks of blue between, and now and again a tiny yellowish flower dropped on her. Pretty—yes, if you held one of those flowers on the palm of your hand and looked at it closely, it was an exquisite small thing. Each pale yellow petal shone as if each was the careful work of a loving hand. The tiny tongue in the centre gave it the shape of a bell. And when you turned it over the outside was a deep bronze colour. But as soon as they flowered, they fell and were scattered. You brushed them off your frock as you talked; the horrid little things got caught in one's hair. Why, then, flower at all? Who takes the trouble—or the joy—to make all these things that are wasted, wasted. . . . It was uncanny.

On the grass beside her, lying between two pillows, was the boy. Sound asleep he lay, his head turned away from his mother. His fine dark hair looked more like a shadow than like real hair, but his ear was a bright, deep coral. Linda clasped her hands above her head and crossed her feet. It was very pleasant to know that all these bungalows were empty, that everybody was down on the beach, out of sight, out of hearing. She had the garden to herself; she was alone.

Dazzling white the picotees shone; the golden-eyed marigolds glittered; the nasturtiums wreathed the verandah poles in green and gold flame. If one only had time to look at these flowers long enough, time to get over the sense of novelty and strangeness, time to know them! But as soon as one paused to part the petals, to discover the underside of the leaf, along came Life and one was swept away. And, lying in her cane chair, Linda felt so light; she felt like a leaf. Along came Life like a wind and she was seized and shaken; she had to go. Oh dear, would it always be so? Was there no escape?

. . . Now she sat on the verandah of their Tasmanian home, leaning against her father's knee. And he promised, "As soon as you and I are old enough, Linny, we'll cut off somewhere, we'll escape. Two boys together. I have a fancy I'd like to sail up a river in China." Linda saw that river, very wide, covered with little rafts and boats. She saw the yellow hats of the boatmen and she heard their high, thin voices as they called. . . .

"Yes, papa."

But just then a very broad young man with bright ginger hair walked slowly past their house, and slowly, solemnly even, uncovered. Linda's father pulled her ear teasingly, in the way he had.

"Linny's beau," he whispered.

"Oh, papa, fancy being married to Stanley Burnell!"

Well, she was married to him. And what was more she loved him. Not the Stanley whom every one saw, not the everyday one; but a timid, sensitive, innocent Stanley who knelt down every night to say his prayers, and who longed to be good. Stanley was simple. If he believed in people—as he believed in her, for instance—it was with his whole heart. He could not be disloyal; he could not tell a lie. And how terribly he suffered if he thought any one—she—was not being dead straight, dead sincere with him! "This is too subtle for me!" He flung out the words, but his open, quivering, distraught look was like the look of a trapped beast.

But the trouble was—here Linda felt almost inclined to laugh, though Heaven knows it was no laughing matter—she saw *her* Stanley so seldom. There were glimpses, moments, breathing spaces of calm, but all the rest of the time it was like living in a house that couldn't be cured of the habit of catching on fire, on a ship that got wrecked every day. And it was always Stanley who was in the thick of the danger. Her whole time was spent in rescuing him, and restoring him, and calming him down, and listening to his story. And what was left of her time was spent in the dread of having children.

Linda frowned; she sat up quickly in her steamer chair and clasped her ankles. Yes, that was her real grudge against life; that was what she could not understand. That was the question she asked and asked, and listened in vain for the answer. It was all very well to say it was the common lot of women to bear children. It wasn't true. She, for one, could prove that wrong. She was broken, made weak, her courage was gone, through childbearing. And what made it doubly hard to bear was, she did not love her children. It was useless pretending. Even if she had had the strength she never would have nursed and played with the little girls. No, it was as though a cold breath had chilled her through and through on each of those awful journeys; she had no warmth left to give them. As to the boy—well, thank Heaven, mother had taken him; he was mother's, or Beryl's, or anybody's who wanted him. She had hardly held him in her arms. She was so indifferent about him that as he lay there . . . Linda glanced down.

The boy had turned over. He lay facing her, and he was no longer asleep. His dark-blue, baby eyes were open; he looked as though he was peeping at his mother. And suddenly his face dimpled; it broke into a wide, toothless smile, a perfect beam, no less.

"I'm here!" that happy smile seemed to say. "Why don't you like me?"

There was something so quaint, so unexpected about that smile that Linda smiled herself. But she checked herself and said to the boy coldly, "I don't like babies."

"Don't like babies?" The boy couldn't believe her. Don't like *me*?" He waved his arms foolishly at his mother.

Linda dropped off her chair on to the grass.

"Why do you keep on smiling?" she said severely. "If you knew what I was thinking about, you wouldn't."

But he only squeezed up his eyes, slyly, and rolled his head on the pillow. He didn't believe a word she said.

"We know all about that!" smiled the boy.

Linda was so astonished at the confidence of this little creature.... Ah no, be sincere. That was not what she felt; it was something far different, it was something so new, so . . . The tears danced in her eyes; she breathed in a small whisper to the boy, "Hallo, my funny!"

But by now the boy had forgotten his mother. He was serious again. Something pink, something soft waved in front of him. He made a grab at it and it immediately disappeared. But when he lay back, another, like the first, appeared. This time he determined to catch it. He made a tremendous effort and rolled right over.

VII

THE TIDE was out; the beach was deserted; lazily flopped the warm sea. The sun beat down, beat down hot and fiery on the fine sand, baking the grey and blue and black and white-veined pebbles. It sucked up the little drop of water that lay in the hollow of the curved shells; it bleached the pink convolvulus that threaded through and through the sand-hills. Nothing seemed to move but the small sand-hoppers. Pit-pit-pit! They were never still.

Over there on the weed-hung rocks that looked at low tide like shaggy beasts come down to the water to drink, the sunlight seemed to spin like a silver coin dropped into each of the small rock pools. They danced, they quivered, and minute ripples laved the porous shores. Looking down, bending over, each pool was like a lake with pink and blue houses clustered on the shores; and oh! the vast mountainous country behind those houses—the ravines, the passes, the dangerous creeks and fearful tracks that led to the water's edge. Underneath waved the sea-forest—pink thread-like trees, velvet anemones, and orange berry-spotted weeds. Now a stone on the bottom moved, rocked, and there was a glimpse of a black feeler; now a thread-like creature wavered by and was lost. Something was happening to the pink, waving trees; they were changing to a cold moonlight blue. And now there sounded the faintest "plop." Who made that sound? What was going on down there? And how strong, how damp the seaweed smelt in the hot sun. . . .

The green blinds were drawn in the bungalows of the summer colony. Over the verandahs, prone on the paddock, flung over the fences, there were exhausted-looking bathing-dresses and rough striped towels. Each back window semed to have a pair of sand-shoes on the sill and some lumps of rock or a bucket or a collection of pawa shells. The bush quivered in a haze of heat; the sandy road was empty except for the Trouts' dog Snooker, who lay stretched in the very middle of it. His blue eye was turned up, his legs stuck out stiffly, and he gave an occasional desperate-sounding puff, as much as to say he had decided to make an end of it and was only waiting for some kind of cart to come along.

"What are you looking at, my grandma? Why do you keep stopping and sort of staring at the wall?"

Kezia and her grandmother were taking their siesta together. The little girl, wearing only her short drawers and her underbodice, her arms and legs bare, lay on one of the puffed-up pillows of her grandma's bed, and the old woman, in a white ruffled dressing-gown, sat in a rocker at the window, with a long piece of pink knitting in her lap. This room that they shared, like the other rooms of the bungalow, was of light varnished wood and the floor was bare. The furniture was of the shabbiest, the simplest. The dressing table, for instance, was a packing-case in a sprigged muslin petticoat, and the mirror above was very strange; it was as though a little piece of forked lightning was imprisoned in it. On the table there stood a jar of sea-pinks, pressed so tightly together they looked more like a velvet pincushion, and a special shell which Kezia had given her grandma for a pin-tray, and another even more special which she had thought would make a very nice place for a watch to curl up in.

"Tell me, grandma," said Kezia.

The old woman sighed, whipped the wool twice round her thumb, and drew the bone needle through. She was casting on.

"I was thinking of your Uncle William, darling," she said quietly.

"My Australian Uncle William?" said Kezia. She had another.

"Yes, of course."

"The one I never saw?"

"That was the one."

"Well, what happened to him?" Kezia knew perfectly well, but she wanted to be told again.

"He went to the mines, and he got a sunstroke there and died," said old Mrs. Fairfield.

Kezia blinked and considered the picture again . . . a little man fallen over like a tin soldier by the side of a big black hole.

"Does it make you sad to think about him, grandma?" She hated her grandma to be sad.

It was the old woman's turn to consider. Did it make her sad? To look back, back. To stare down the years, as Kezia had seen her doing. To look after *them* as a woman does, long after *they* were out of sight. Did it make her sad? No, life was like that.

"No, Kezia."

"But why?" asked Kezia. She lifted one bare arm and began to draw things in the air. "Why did Uncle William have to die? He wasn't old."

Mrs. Fairfield began counting the stitches in threes. "It just happened," she said in an absorbed voice.

"Does everybody have to die?" asked Kezia.

"Everybody!"

"*Me?*" Kezia sounded fearfully incredulous.

"Some day, my darling."

"But, grandma." Kezia waved her left leg and waggled the toes. They felt sandy. "What if I just won't?"

The old woman sighed again and drew a long thread from the ball.

"We're not asked, Kezia," she said sadly. "It happens to all of us sooner or later."

Kezia lay still thinking this over. She didn't want to die. It meant she would have to leave here, leave everywhere, for ever, leave—leave her grandma. She rolled over quickly.

"Grandma," she said in a startled voice.

"What, my pet!"

"*You're* not to die." Kezia was very decided.

"Ah, Kezia"—her grandma looked up and smiled and shook her head—"don't let's talk about it."

"But you're not to. You couldn't leave me. You couldn't not be there." This was awful. "Promise me you won't ever do it, grandma," pleaded Kezia.

The old woman went on knitting.

"Promise me! Say never!"

But still her grandma was silent.

Kezia rolled off the bed; she couldn't bear it any longer, and lightly she leapt on to her grandma's knees, clasped her hands round the old woman's throat and began kissing her, under the chin, behind the ear, and blowing down her neck.

"Say never . . . say never . . . say never—" she gasped between the kisses, And then she began, very softly and lightly, to tickle her grandma.

"Kezia!" The old woman dropped her knitting. She swung back in the rocker. She began to tickle Kezia. "Say never, say never, say never," gurgled Kezia, while they lay there laughing in each other's arms. "Come, that's enough, my squirrel! That's enough, my wild pony!" said old Mrs. Fairfield, setting her cap straight. "Pick up my knitting."

Both of them had forgotten what the "never" was about.

VIII

THE SUN was still full on the garden when the back door of the Burnells' shut with a bang, and a very gay figure walked down the the path to the gate. It was Alice, the servant-girl, dressed for her afternoon out. She wore a white cotton dress with such large red spots on it and so many that they made you shudder, white shoes and a leghorn turned up under the brim with poppies. Of course she wore gloves, white ones, stained at the fastenings with iron-mould, and in one hand she carried a very dashed-looked sunshade which she referred to as her *perishall*.

Beryl, sitting in the window, fanning her freshly-washed hair, thought she had never seen such a guy. If Alice had only blacked her face with a piece of cork before she started out, the picture would have been complete. And where did a girl like that go to in a place like this? The heart-shaped Fijian fan beat scornfully at that lovely bright mane. She supposed Alice had picked up some horrible common larrikin and they'd go off into the bush together. Pity to make herself so conspicuous; they'd have hard work to hide with Alice in that rig-out.

But no, Beryl was unfair. Alice was going to tea with Mrs. Stubbs, who'd sent her an "invite" by the little boy who called for orders. She had taken ever such a liking to Mrs. Stubbs ever since the first time she went to the shop to get something for her mosquitoes.

"Dear heart!" Mrs. Stubbs had clapped her hand to her side. "I never seen any one so eaten. You might have been attacked by canningbals."

Alice did wish there'd been a bit of life on the road though. Made her feel so queer, having nobody behind her. Made her feel all weak in the spine. She couldn't believe that some one wasn't watching her. And yet it was silly to turn round; it gave you away. She pulled up her gloves, hummed to herself and said to the distant gum-tree, "Shan't be long now." But that was hardly company.

Mrs. Stubbs's shop was perched on a little hillock just off the road. It had two big windows for eyes, a broad verandah for a hat, and the sign on the roof, scrawled MRS. STUBBS'S, was like a little card stuck rakishly in the hat crown.

On the verandah there hung a long string of bathing-dresses, clinging together as though they'd just been rescued from the sea rather than waiting to go in, and beside them there hung a cluster of sand-shoes so extraordinarily mixed that to get at one pair you had to tear apart and forcibly separate at least fifty. Even then it was the rarest thing to find the left that belonged to the right. So many people had lost patience and gone off with one shoe that fitted and one that was a little too big. . . . Mrs. Stubbs prided herself on keeping something of everything. The two windows, arranged in the form of precarious pyramids, were crammed so tight, piled so high, that it seemed only a conjuror could prevent them from toppling over. In the left-hand corner of one window, glued to the pane by four gelatine lozenges, there was—and there had been from time immemorial—a notice.

LOST! HANDSOME GOLD BROOCH
SOLID GOLD
ON OR NEAR BEACH
REWARD OFFERED

Alice pressed open the door. The bell jangled, the red serge curtains parted, and Mrs. Stubbs appeared. With her broad smile and the long bacon knife in her hand, she looked like a friendly brigand. Alice was welcomed so warmly that she found it quite difficult to keep up her "manners." They consisted of persistent little coughs and hems, pulls at her gloves, tweaks at her skirt, and a curious difficulty in seeing what was set before her or understanding what was said.

Tea was laid on the parlour table—ham, sardines, a whole pound of butter, and such a large johnny cake that it looked like an advertisement for somebody's baking-powder. But the Primus stove roared so loudly that it was useless to try to talk above it. Alice sat down on the edge of a basket-chair while Mrs. Stubbs pumped the stove still higher. Suddenly Mrs. Stubbs whipped the cushion off a chair and disclosed a large brown-paper parcel.

"I've just had some new photers taken, my dear," she shouted cheerfully to Alice. "Tell me what you think of them."

In a very dainty, refined way Alice wet her finger and put the tissue back from the first one. Life! How many there were! There were three dozzing at least. And she held it up to the light.

Mrs. Stubbs sat in an arm-chair, leaning very much to one side. There was a look of mild astonishment on her large face, and well there might be. For though the arm-chair stood on a carpet, to the left of it, miraculously skirting the carpet-border, there was a dashing water-fall. On her right stood a Grecian pillar with a giant fern-tree on either side of it, and in the background towered a gaunt mountain, pale with snow.

"It is a nice style, isn't it?" shouted Mrs. Stubbs; and Alice had just screamed "Sweetly" when the roaring of the Primus stove died down, fizzled out, ceased, and she said "Pretty" in a silence that was frightening.

"Draw up your chair, my dear," said Mrs. Stubbs, beginning to pour out. "Yes," she said thoughtfully, as she handed the tea, "but I don't care about the size. I'm having an enlargemint. All very well for Christmas cards, but I never was the one for small photers myself. You get no comfort out of them. To say the truth, I find them dis'eartening."

Alice quite saw what she meant.

"Size," said Mrs. Stubbs. "Give me size. That was what my poor dear husband was always saying. He couldn't stand anything small. Gave him the creeps. And, strange as it may seem, my dear"—here Mrs. Stubbs creaked and seemed to expand herself at the memory—"it was dropsy that carried him off at the larst. Many's the time they drawn one and a half pints from 'im at the 'ospital. . . . It seemed like a judgmint."

Alice burned to know exactly what it was that was drawn from him. She ventured, "I suppose it was water."

But Mrs. Stubbs fixed Alice with her eyes and replied meaningly, "It was *liquid*, my dear."

Liquid! Alice jumped away from the word like a cat and came back to it, nosing and wary.

"That's 'im!" said Mrs. Stubbs, and she pointed dramatically to the life-size head and shoulders of a burly man with a dead white rose in the button-hole of his coat that made you think of a curl of cold mutting fat. Just below, in silver letters on a red cardboard ground, were the words, "Be not afraid, it is I."

"It's ever such a fine face," said Alice faintly.

The pale-blue bow on the top of Mrs. Stubbs's fair frizzy hair quivered. She arched her plump neck. What a neck she had! It was bright pink where it began and then it changed to warm apricot, and that faded to the colour of a brown egg and then to a deep creamy.

"All the same, my dear," she said surprisingly, "freedom's best!" Her soft, fat chuckle sounded like a purr. "Freedom's best," said Mrs. Stubbs again.

Freedom! Alice gave a loud, silly little titter. She felt awkward. Her mind flew back to her own kitching. Ever so queer! She wanted to be back in it again.

IX

A STRANGE company assembled in the Burnells' washhouse after tea. Round the table there sat a bull, a rooster, a donkey that kept forgetting it was a donkey, a sheep and a bee. The washhouse was the perfect place for such a meeting because they could make as much noise as they liked, and nobody ever interrupted. It was a small tin shed standing apart from the bungalow. Against the wall there was a deep trough and in the corner a copper with a basket of clothes-pegs on top of it. The little window, spun over with cobwebs, had a piece of candle and a mouse-trap on the dusty sill. There were clotheslines crisscrossed overhead and, hanging from a peg on the wall, a very big, a huge, rusty horseshoe. The table was in the middle with a form at either side.

"You can't be a bee, Kezia. A bee's not an animal. It's a ninseck."

"Oh, but I do want to be a bee frightfully," wailed Kezia. . . . A tiny bee, all yellow-furry, with striped legs. She drew her legs up under her and leaned over the table. She felt she was a bee.

"A ninseck must be an animal," she said stoutly. "It makes a noise. It's not like a fish."

"I'm a bull, I'm a bull!" cried Pip. And he gave such a tremendous bellow—how did he make that noise?—that Lottie looked quite alarmed.

"I'll be a sheep," said little Rags. "A whole lot of sheep went past this morning."

"How do you know?"

"Dad heard them. Baa!" He sounded like the little lamb that trots behind and seems to wait to be carried.

"Cock-a-doodle-do!" shrilled Isabel. With her red cheeks and bright eyes she looked like a rooster.

"What'll I be?" Lottie asked everybody, and she sat there smiling, waiting for them to decide for her. It had to be an easy one.

"Be a donkey, Lottie." It was Kezia's suggestion. "Hee-haw! You can't forget that."

"Hee-haw!" said Lottie solemnly. "When do I have to say it?"

"I'll explain, I'll explain," said the bull. It was he who had the cards. He waved them round his head. "All be quiet! All listen!" And he waited for them. "Look here, Lottie." He turned up a card. "It's got two spots on it—see? Now, if you put that card in the middle and somebody else has one with two spots as well, you say 'Hee-haw,' and the card's yours."

"Mine?" Lottie was round-eyed. "To keep?"

"No, silly. Just for the game, see? Just while we're playing." The bull was very cross with her.

"Oh, Lottie, you *are* a little silly," said the proud rooster.

Lottie looked at both of them. Then she hung her head; her lip quivered. "I don't not want to play," she whispered. The others glanced at one another like conspirators. All of them knew what that meant. She would go away and be discovered somewhere standing with her pinny thrown over her head, in a corner, or against a wall, or even behind a chair.

"Yes, you *do*, Lottie. It's quite easy," said Kezia.

And Isabel, repentant, said exactly like a grown-up, "Watch *me*, Lottie, and you'll soon learn."

"Cheer up, Lot," said Pip. "There, I know what I'll do. I'll give you the first one. It's mine, really, but I'll give it to you. Here you are." And he slammed the card down in front of Lottie.

Lottie revived at that. But now she was in another difficulty. "I haven't got a hanky," she said; "I want one badly, too."

"Here, Lottie, you can use mine." Rags dipped into his sailor blouse and brought up a very wet-looking one, knotted together. "Bc very careful," he warned her. "Only use that corner. Don't undo it. I've got a little starfish inside I'm going to try and tame."

"Oh, come on, you girls," said the bull. "And mind—you're not to look at your cards. You've got to keep your hands under the table till I say 'Go.' "

Smack went the cards round the table. They tried with all their might to see, but Pip was too quick for them. It was very exciting, sitting there in the washhouse; it was all they could do not to burst into a little chorus of animals before Pip had finished dealing.

"Now, Lottie, you begin."

Timidly Lottie stretched out a hand, took the top card off her pack, had a good look at it—it was plain she was counting the spots—and put it down.

"No, Lottie, you can't do that. You mustn't look first. You must turn it the other way over."

"But then everybody will see it the same time as me," said Lottie.

The game proceeded. Mooe-ooo-er! The bull was terrible. He charged over the table and seemed to eat the cards up.

Bss-ss! said the bee.

Cock-a-doodle-do! Isabel stood up in her excitement and moved her elbows like wings.

Baa! Little Rags put down the King of Diamonds and Lottie put down the one they called the King of Spain. She had hardly any cards left.

"Why don't you call out, Lottie?"

"I've forgotten what I am," said the donkey woefully.

"Well, change! Be a dog instead! Bow-wow!"

"Oh yes. That's *much* easier." Lottie smiled again. But when she and Kezia both had a one Kezia waited on purpose. The others made signs to Lottie and pointed. Lottie turned very red; she looked bewildered, and at last she said, "Hee-haw! Ke-zia."

"Ss! Wait a minute!" They were in the very thick of it when the bull stopped them, holding up his hand. "What's that? What's that noise?"

"What noise? What do you mean?" asked the rooster.

"Ss! Shut up! Listen!" They were mouse-still. "I thought I heard a—a sort of knocking," said the bull.

"What was it like?" asked the sheep faintly.

No answer.

The bee gave a shudder. "Whatever did we shut the door for?" she said softly. Oh, why, why had they shut the door?

While they were playing, the day had faded; the gorgeous sunset had blazed and died. And now the quick dark came racing over the sea, over the sand-hills, up the paddock. You were frightened to look in the corners of the washhouse, and yet you had to look with all your might. And somewhere, far away, grandma was lighting a lamp. The blinds were being pulled down; the kitchen fire leapt in the tins on the mantelpiece.

"It would be awful now," said the bull, "if a spider was to fall from the ceiling on to the table, wouldn't it?"

"Spiders don't fall from ceilings."

"Yes, they do. Our Min told us she'd seen a spider as big as a saucer, with long hairs on it like a gooseberry."

Quickly all the little heads were jerked up; all the little bodies drew together, pressed together.

"Why doesn't somebody come and call us?" cried the rooster.

Oh, those grown-ups, laughing and snug, sitting in the lamplight, drinking out of cups! They'd forgotten about them. No, not really forgotten. That was what their smile meant. They had decided to leave them there all by themselves.

Suddenly Lottie gave such a piercing scream that all of them jumped off the forms, all of them screamed too. "A face—a face looking!" shrieked Lottie.

It was true, it was real. Pressed against the window was a pale face, black eyes, a black beard.

"Grandma! Mother! Somebody!"

But they had not got to the door, tumbling over one another, before it opened for Uncle Jonathan. He had come to take the little boys home.

X

HE HAD meant to be there before, but in the front garden he had come upon Linda walking up and down the grass, stopping to pick off a dead pink or give a top-heavy carnation something to lean against, or to take a deep breath of something, and then walking on again, with her little air of remoteness. Over her white frock she wore a yellow, pink-fringed shawl from the Chinaman's shop.

"Hallo, Jonathan!" called Linda. And Jonathan whipped off his shabby panama, pressed it against his breast, dropped on one knee, and kissed Linda's hand.

"Greeting, my Fair One! Greeting, my Celestial Peach Blossom!" boomed the bass voice gently. "Where are the other noble dames?"

"Beryl's out playing bridge and mother's giving the boy his bath. . . . Have you come to borrow something?"

The Trouts were for ever running out of things and sending across to the Burnells' at the last moment.

But Jonathan only answered, "A little love, a little kindness"; and he walked by his sister-in-law's side.

Linda dropped into Beryl's hammock under the manuka tree, and Jonathan stretched himself on the grass beside her, pulled a long stalk and began chewing it. They knew each other well. The voices of children cried from the other gardens. A fisherman's light cart shook along the sandy road, and from far away they heard a dog barking; it was muffled as though the dog had its head in a sack. If you listened you could just hear the soft swish of the sea at full tide sweeping the pebbles. The sun was sinking.

"And so you go back to the office on Monday, do you, Jonathan?" asked Linda.

"On Monday the cage door opens and clangs to upon the victim for another eleven months and a week," answered Jonathan.

Linda swung a little. "It must be awful," she said slowly.

"Would ye have me laugh, my fair sister? Would ye have me weep?"

Linda was so accustomed to Jonathan's way of talking that she paid no attention to it.

"I suppose," she said vaguely, "one gets used to it. One gets used to anything."

"Does one? Hum!" The "Hum" was so deep it seemed to boom from underneath the ground. "I wonder how it's done," brooded Jonathan; "I've never managed it."

Looking at him as he lay there, Linda thought again how attractive he was. It was strange to think that he was only an ordinary clerk, that Stanley earned twice as much money as he. What was the matter with Jonathan? He had no ambition; she supposed that was it. And yet one felt he was gifted, exceptional. He was passionately fond of music; every spare penny he had went on books. He was always full of new ideas, schemes, plans. But nothing came of it all. The new fire blazed in Jonathan; you almost heard it roaring softly as he explained, described and dilated on the new thing; but a moment later it had fallen in and there was nothing but ashes, and Jonathan went about with a look like hunger in his black eyes. At these times he exaggerated his absurd manner of speaking, and he sang in church—he was the leader of the choir—with such fearful dramatic intensity that the meanest hymn put on an unholy splendour.

"It seems to me just as imbecile, just as infernal, to have to go to the office on Monday," said Jonathan, "as it always has done and always will do. To spend all the best years of one's life sitting on a stool from nine to five, scratching in somebody's ledger! It's a queer use to make of one's . . . one and only life, isn't it? Or do I fondly dream?" He rolled over on the grass and looked up at Linda. "Tell me, what is the difference between my life and that of an ordinary prisoner? The only difference I can see is that I put myself in jail and nobody's ever going to let me out. That's a more intolerable situation than the other. For if I'd been—pushed in, against my will—kicking, even—once the door was locked, or at any rate in five years or so, I might have accepted the fact and begun to take an interest in the flight of flies or counting the warder's steps along the passage with particular attention to variations of tread and so on. But as it is, I'm like an insect that's flown into a room of its own accord. I dash against the walls, dash against the windows, flop against the ceiling, do everything on God's earth, in fact, except fly out again. And all the while I'm thinking, like that moth, or that butterfly, or whatever it is, 'The shortness of life! The shortness of life!' I've only one night or one day, and there's this vast dangerous garden, waiting out there, undiscovered, unexplored."

"But, if you feel like that, why—" began Linda quickly.

"*Ah!*" cried Jonathan. And that "ah!" was somehow almost exultant. "There you have me. Why? Why indeed? There's the maddening, mysterious question. Why don't I fly out again? There's the window or the door or whatever it was I came in by. It's not hopelessly shut—is it? Why don't I find it and be off? Answer me that, little sister." But he gave her no time to answer.

"I'm exactly like that insect again. For some reason"—Jonathan paused between the words—"it's not allowed, it's forbidden, it's against the insect law, to stop banging and flopping and crawling up the pane even for an instant. Why don't I leave the office? Why don't I seriously consider, this moment, for instance, what it is that prevents me leaving? It's not as though I'm tremendously tied. I've two boys to provide for, but, after all, they're boys. I could cut off to sea, or get a job up-country, or—" Suddenly he smiled at Linda and said in a changed voice, as if he were confiding a secret, "Weak . . . weak. No stamina. No anchor. No guiding principle, let us call it." But then the dark velvety voice rolled out:

Would ye hear the story
How it unfolds itself . . .

and they were silent.

The sun had set. In the western sky there were great masses of crushed-up rose-coloured clouds. Broad beams of light shone through the clouds and beyond them as if they would cover the whole sky. Overhead the blue faded; it turned a pale gold, and the bush outlined against it gleamed dark and brilliant like metal. Sometimes when those beams of light show in the sky they are very awful. They remind you that up there sits Jehovah, the jealous God, the Almighty, Whose eye is upon you, ever watchful, never weary. You remember that at his coming the whole earth will shake into one ruined graveyard; the cold, bright angels will drive you this way and that, and there will be no time to explain what could be explained so simply. . . . But to-night it seemed to Linda there was something infinitely joyful and loving in those silver beams. And now no sound came from the sea. It breathed softly as if it would draw that tender, joyful beauty into its own bosom.

"It's all wrong, it's all wrong," came the shadowy voice of Jonathan. "It's not the scene, it's not the setting for . . . three stools, three desks, three inkpots and a wire blind."

Linda knew that he would never change, but she said, "Is it too late, even now?"

"I'm old—I'm old," intoned Jonathan. He bent towards her, he passed his hand over his head. "Look!" His black hair was speckled all over with silver, like the breast plumage of a black fowl.

Linda was surprised. She had no idea that he was grey. And yet, as he stood up beside her and sighed and stretched, she saw him, for the first time, not resolute, not gallant, not careless, but touched already with age. He looked very tall on the darkening grass, and the thought crossed her mind, "He is like a weed."

Jonathan stooped again and kissed her fingers.

"Heaven reward thy sweet patience, lady mine," he murmured. "I must go seek those heirs to my fame and fortune. . . ." He was gone.

XI

LIGHT shone in the windows of the bungalow. Two square patches of gold fell upon the pinks and the peaked marigolds. Florrie, the cat, came out on to the verandah, and sat on the top step, her white paws close together, her tail curled round. She looked content, as though she had been waiting for this moment all day.

"Thank goodness, it's getting late," said Florrie. "Thank goodness, the long day is over." Her greengage eyes opened.

Presently there sounded the rumble of the coach, the crack of Kelly's whip. It came near enough for one to hear the voices of the men from town, talking loudly together. It stopped at the Burnells' gate.

Stanley was half-way up the path before he saw Linda. "Is that you, darling?"

"Yes, Stanley."

He leapt across the flower-bed and seized her in his arms. She was enfolded in that familiar, eager, strong embrace.

"Forgive me, darling, forgive me," stammered Stanley, and he put his hand under her chin and lifted her face to him.

"Forgive you?" smiled Linda. "But whatever for?"

"Good God! You can't have forgotten," cried Stanley Burnell. "I've thought of nothing else all day. I've had the hell of a day. I made up my mind to dash out and telegraph, and then I thought the wire mightn't reach you before I did. I've been in tortures, Linda."

"But Stanley," said Linda, "what must I forgive you for?"

"Linda!"—Stanley was very hurt—"didn't you realize—you must have realized—I went away without saying good-bye to you this morning? I can't imagine how I can have done such a thing. My confounded temper, of course. But—well"—and he sighed and took her in his arms again—"I've suffered for it enough to-day."

"What's that you've got in your hand?" asked Linda. "New gloves? Let me see."

"Oh, just a cheap pair of wash-leather ones," said Stanley. "I noticed Bell was wearing some in the coach this morning, so, as I was passing the shop, I dashed in and got myself a pair. What are you smiling at? You don't think it was wrong of me, do you?"

"On the *con*-trary, darling," said Linda, "I think it was most sensible."

She pulled one of the large, pale gloves on her own fingers and looked at her hand, turning it this way and that. She was still smiling.

Stanley wanted to say, "I was thinking of you the whole time I bought them." It was true, but for some reason he couldn't say it. "Let's go in," said he.

XII

WHY DOES one feel so different at night? Why is it so exciting to be awake when everybody else is asleep? Late—it is very late! And yet every moment you feel more and more wakeful, as though you were slowly, almost with every breath, waking up into a new, wonderful, far more thrilling and exciting world than the daylight one. And what is this queer sensation that you're a conspirator? Lightly, stealthily you move about your room. You take something off the dressing table and put it down again without a sound. And everything, even the bed-post, knows you, responds, shares your secret. . . .

You're not very fond of your room by day. You never think about it. You're in and out, the door opens and slams, the cupboard creaks. You sit down on the side of your bed, change your shoes and dash out again. A dive down to the glass, two pins in your hair, powder your nose and off again. But now—it's suddenly dear to you. It's a darling little funny room. It's yours. Oh, what a joy it is to own things! Mine—my own!

"My very own for ever?"

"Yes." Their lips met.

No, of course, that had nothing to do with it. That was all nonsense and rubbish. But, in spite of herself, Beryl saw so plainly two people standing in the middle of her room. Her arms were round his neck; he held her. And now he whispered, "My beauty, my little beauty!"

She jumped off her bed, ran over to the window and kneeled on the window-seat, with her elbows on the sill. But the beautiful night, the garden, every bush, every leaf, even the white palings, even the stars, were conspirators too. So bright was the moon that the flowers were bright as by day; the shadow of the nasturtiums, exquisite lily-like leaves and wide-open flowers, lay across the silvery verandah. The manuka tree, bent by the southerly winds, was like a bird on one leg stretching out a wing.

But when Beryl looked at the bush, it seemed to her the bush was sad.

"We are dumb trees, reaching up in the night, imploring we know not what," said the sorrowful bush.

It is true when you are by yourself and you think about life, it is always sad. All that excitement and so on has a way of suddenly leaving you, and it's as though, in the silence, somebody called your name, and you heard your name for the first time. "Beryl!"

"Yes, I'm here. I'm Beryl. Who wants me?"

"Beryl!"

"Let me come."

It is lonely living by oneself. Of course, there are relations, friends, heaps of them; but that's not what she means. She wants some one who will find the Beryl they none of them know, who will expect her to be that Beryl always. She wants a lover.

"Take me away from all these other people, my love. Let us go far away. Let us live our life, all new, all ours, from the very beginning. Let us make our fire. Let us sit down to eat together. Let us have long talks at night."

And the thought was almost, "Save me, my love. Save me!"

. . . "Oh, go on! Don't be a prude, my dear. You enjoy yourself while you're young. That's my advice." And a high rush of silly laughter joined Mrs. Harry Kember's loud, indifferent neigh.

You see, it's so frightfully difficult when you've nobody. You're so at the mercy of things. You can't just be rude. And you've always this horror of seeming inexperienced and stuffy like the other ninnies at the Bay. And—and it's fascinating to know you've power over people. Yes, that is fascinating. . . .

Oh why, oh why doesn't "he" come soon?

If I go on living here, thought Beryl, anything may happen to me.

"But how do you know he is coming at all?" mocked a small voice within her.

But Beryl dismissed it. She couldn't be left. Other people, perhaps, but not she. It wasn't possible to think that Beryl Fairfield never married, that lovely fascinating girl.

"Do you remember Beryl Fairfield?"

"Remember her! As if I could forget her! It was one summer at the Bay that I saw her. She was standing on the beach in a blue"—no, pink—"muslin frock, holding on a big cream"—no, black—"straw hat. But it's years ago now."

"She's as lovely as ever, more so if anything."

Beryl smiled, bit her lip, and gazed over the garden. As she gazed, she saw somebody, a man, leave the road, step along the paddock beside their palings as if he was coming straight towards her. Her heart beat. Who was it? Who could it be? It couldn't be a burglar, certainly not a burglar, for he was smoking and he strolled lightly. Beryl's heart leapt; it seemed to turn right over, and then to stop. She recognized him.

"Good evening, Miss Beryl," said the voice softly.

"Good evening."

"Won't you come for a little walk?" it drawled.

Come for a walk—at that time of night! "I couldn't. Everybody's in bed. Everybody's asleep."

"Oh," said the voice lightly, and a whiff of sweet smoke reached her. "What does everybody matter? Do come! It's such a fine night. There's not a soul about."

Beryl shook her head. But already something stirred in her, something reared its head.

The voice said, "Frightened?" It mocked, "Poor little girl!"

"Not in the least," said she. As she spoke that weak thing within her seemed to uncoil, to grow suddenly tremendously strong; she longed to go!

And just as if this was quite understood by the other, the voice said, gently and softly, but finally, "Come along!"

Beryl stepped over her low window, crossed the verandah, ran down the grass to the gate. He was there before her.

"That's right," breathed the voice, and it teased, "You're not frightened, are you? You're not frightened?"

She was; now she was here she was terrified, and it seemed to her everything was different. The moonlight stared and glittered; the shadows were like bars of iron. Her hand was taken.

"Not in the least," she said lightly. "Why should I be?"

Her hand was pulled gently, tugged. She held back.

"No, I'm not coming any farther," said Beryl.

"Oh, rot!" Harry Kember didn't believe her. "Come along! We'll just go as far as that fuchsia bush. Come along!"

The fuchsia bush was tall. It fell over the fence in a shower. There was a little pit of darkness beneath.

"No, really, I don't want to," said Beryl.

For a moment Harry Kember didn't answer. Then he came close to her, turned to her, smiled and said quickly, "Don't be silly! Don't be silly!"

His smile was something she'd never seen before. Was he drunk? That bright, blind, terrifying smile froze her with horror. What was she doing? How had she got here? the stern garden asked her as the gate pushed open, and quick as a cat Harry Kember came through and snatched her to him.

"Cold little devil! Cold little devil!" said the hateful voice.

But Beryl was strong. She slipped, ducked, wrenched free.

"You are vile, vile," said she.

"Then why in God's name did you come?" stammered Harry Kember.

Nobody answered him.

XIII

A CLOUD, small, serene, floated across the moon. In that moment of darkness the sea sounded deep, troubled. Then the cloud sailed away, and the sound of the sea was a vague murmur, as though it waked out of a dark dream. All was still.

KATHERINE ANNE PORTER

❡ *Miss Porter's reputation, in a world that worships quantity rather than quality, is a heartening sign. The very short list of her published fiction emphasizes rather than denigrates her dedication to literature. Only the small portion of writing that meets her own fastidious standards is allowed to reach the press. Her incandescent style prompts admiration; her permanent literary values serve as example.*

In Pale Horse, Pale Rider, *the past, present, and future coalesce and merge. The novella has an integrity, an intactness that bespeaks an organically whole product. Form and content are artistically so well united that violence to the union may result from probing any of the individual parts.*

The heroine of Pale Horse *arrives at life by passing through death. Miranda (at the close of the short novel) calls for the trappings of her new existence, items symbolic of death. Now she will embrace life with that inner freedom possible for one possessing a true knowledge of life's final outcome. "Now there would be time for everything." The woman has, by an ordeal of fire, made a break with her past; and she will suffer despair as she tries to replace the lost old love with a new. But her life will be lived with a compassionate understanding deriving from an apprehension of death.*

Adam is innocence and, through Miranda, he is infected with the seeds of physical death. Symbolically, the man is sacrificed and the woman is released from the world of her childhood. This personal drama is acted out as mankind grinds itself out in World War I. "There was no more hating then,/ And no more love: Gone is the heart of Man," wrote the poet Edith Sitwell, following World War II. Similarly, Pale Horse *articulates this experience of our times. Miranda dies, and gains her soul. Her destiny, like ours, is to live with guilt and adumbrated hope. Revelation vi:8 says, ". . . and behold a pale horse: and his name that sat on him was Death, and Hell followed with him."*

The reader is exposed to these meanings through a feverish admixture of Miranda's dreams and hallucinations, yet no appreciable confusion occurs between the actual and the imagined, the real and the dreamed. Through Miranda's cerebrations, the two facets of her reality—what she dreams and what she has experienced in the objective world—are projected onto the narrative screen. Miss Porter writes from the omniscient point of view, turning with unusual fluidity from the inward to the outward world. Macrocosm and microcosm meet.

As minor classics of modern American fiction, Katherine Anne Porter's works are secure. Recognition of the seriousness, indeed, the anguish underlying her theme may perhaps best be gained by looking to the delicate symbolic treatment of the people, the events, and the objects. The reader must be as perceptive as the author, however, to appreciate the way in which the writer explores the psyche of sensitive people confronting their destinies, seeking alleviation.

Pale Horse, Pale Rider

IN SLEEP she knew she was in her bed, but not the bed she had lain down in a few hours since, and the room was not the same but it was a room she had known somewhere. Her heart was a stone lying upon her breast outside of her; her pulses lagged and paused, and she knew that something strange was going to happen, even as the early morning winds were cool through the lattice, the streaks of light were dark blue and the whole house was snoring in its sleep.

Now I must get up and go while they are all quiet. Where are my things? Things have a will of their own in this place and hide where they like. Daylight will strike a sudden blow on the roof startling them all up to their feet; faces will beam asking, Where are you going, What are you doing, What are you thinking, How do you feel, Why do you say such things, What do you mean? No more sleep. Where are my boots and what horse shall I ride? Fiddler or Graylie or Miss Lucy with the long nose and the wicked eye? How I have loved this house in the morning before we are all awake and tangled together like badly cast fishing lines. Too many people have been born here, and have wept too much here, and have laughed too much, and have been too angry and outrageous with each other here. Too many have died in this bed already, there are far too many ancestral bones propped up on the mantelpieces, there have been too damned many antimacassars in this house, she said loudly, and oh, what accumulation of storied dust never allowed to settle in peace for one moment.

And the stranger? Where is that lank greenish stranger I remember hanging about the place, welcomed by my grandfather, my great-aunt, my five times removed cousin, my decrepit hound and my silver kitten? Why did they take to him, I wonder? And where are they now? Yet I saw him pass the window in the evening. What else besides them did I have in the world? Nothing. Nothing is mine, I have only nothing but it is enough, it is beautiful and it is all mine. Do I even walk about in my own skin or is it something I have borrowed to spare my modesty? Now what horse shall I borrow for this journey I do not mean to take, Graylie or Miss Lucy or Fiddler who can jump ditches in the dark and knows how to get the bit between his teeth? Early morning is best for me because trees are trees in one stroke, stones are stones set in shades known to be grass, there are no false shapes or surmises, the road is still asleep with the crust of dew unbroken. I'll take Graylie because he is not afraid of bridges.

Come now, Graylie, she said, taking his bridle, we must outrun Death and the Devil. You are no good for it, she told the other horses standing saddled before the stable gate, among them the horse of the stranger, gray also, with tarnished nose and ears. The stranger swung into his saddle beside her, leaned far towards her and regarded her without meaning, the blank still stare of mindless malice that makes no threats and can bide its time. She drew Graylie around sharply, urged him to run. He leaped the low rose hedge and the narrow ditch beyond, and the dust of the lane flew heavily under his beating hoofs. The stranger rode beside her, easily, lightly, his reins loose in his half-closed hand, straight and elegant in dark shabby garments that flapped upon his bones; his pale face smiled in an evil trance, he did not glance at her.

Ah, I have seen this fellow before, I know this man if I could place him. He is no stranger to me.

She pulled Graylie up, rose in her stirrups and shouted, I'm not going with you this time—ride on! Without pausing or turning his head the stranger rode on. Graylie's ribs heaved under her, her own ribs rose and fell, Oh, why am I so tired, I must wake up. "But let me get a fine yawn first," she said, opening her eyes and stretching, "a slap of cold water in my face, for I've been talking in my sleep again, I heard myself but what was I saying?"

Slowly, unwillingly, Miranda drew herself up inch by inch out of the pit of sleep, waited in a daze for life to begin again. A single word struck in her mind, a gong of warning, reminding her for the daylong what she forgot happily in sleep, and only in sleep. The war, said the gong, and she shook her head. Dangling her feet idly with their slippers hanging, she was reminded of the way all sorts of persons sat upon her desk at the newspaper office. Every day she found someone there, sitting upon her desk instead of the chair provided, dangling his legs, eyes roving, full of his important affairs, waiting to pounce about something or other. "*Why* won't they sit in the chair? Should I put a sign on it, saying, 'For God's sake, sit here'?"

Far from putting up a sign, she did not even frown at her visitors. Usually she did not notice them at all until their determination to be seen was greater than her determination not to see them. Saturday, she thought, lying comfortably in her tub of hot water, will be payday, as always. Or I hope always. Her thoughts roved hazily in a continual effort to bring together and unite firmly the disturbing oppositions in her day-to-day existence, where survival, she could see clearly, had become a series of feats of sleight of hand. I owe—let me see, I wish I had pencil and paper—well, suppose I *did* pay five dollars now on a Liberty Bond, I couldn't possibly keep it up. Or maybe. Eighteen dollars a week. So much for rent, so much for food, and I mean to have a few things besides. About five dollars' worth. Will leave me twenty-seven cents. I suppose I can make it. I suppose I should be worried. I am worried. Very well, now I am worried and what next? Twenty-seven cents. That's not so bad. Pure profit, really. Imagine if they should suddenly raise me to twenty I should then have two dollars and twenty-seven cents left over. But they aren't going to raise me to twenty. They are in fact going to throw me out if I don't buy a Liberty Bond. I hardly believe that. I'll ask Bill. (Bill was the city editor.) I wonder if a threat like that isn't a kind of blackmail. I don't believe even a Lusk Committeeman can get away with that.

Yesterday there had been two pairs of legs dangling, on either side of her typewriter, both pairs stuffed thickly into funnels of dark expensive-looking material. She noticed at a distance that one of them was oldish and one was youngish, and they both of them had a stale air of borrowed importance which apparently they had got from the same source. They were both much too well nourished and the younger one wore a square little mustache. Being what they were, no matter what their business was it would be something unpleasant. Miranda had nodded at them, pulled out her chair and without removing her cap or gloves had reached into a pile of letters and sheets from the copydesk as if she had not a moment to spare. They did not move, or take off their hats. At last she had said "Good morning" to them, and asked if they were, perhaps, waiting for her?

The two men slid off the desk, leaving some of her papers rumpled, and the oldish man had inquired why she had not bought a Liberty Bond. Miranda had looked at him then, and got a poor impression. He was a pursy-faced man, gross-mouthed, with little lightless eyes, and Miranda wondered why nearly all of those selected to do the war work at home were of his sort. He might be anything at all, she thought; advance agent for a road show, promoter of a wildcat oil company, a former saloon keeper announcing the opening of a new cabaret, an automobile salesman—any follower of any one of the crafty, haphazard callings. But he was now all Patriot, working for the government. "Look here," he asked her, "do you know there's a war, or don't you?"

Did he expect an answer to that? Be quiet, Miranda told herself, this was bound to happen. Sooner or later it happens. Keep your head. The man wagged his finger at her. "Do you?" he persisted, as if he were prompting an obstinate child.

"Oh, the war," Miranda had echoed on a rising note and she almost smiled at him. It was habitual, automatic, to give that solemn, mystically uplifted grin when you spoke the words or heard them spoken. "*C'est la guerre*," whether you could pronounce it or not, was even better, and always, always, you shrugged.

"Yeah," said the younger man in a nasty way, "the war." Miranda, startled by the tone, met his eye; his stare was really stony, really viciously cold, the kind of thing you might expect to meet behind a pistol on a deserted corner. This expression gave temporary meaning to a set of features otherwise nondescript, the face of those men who have no business of their own. "We're having a war, and some people are buying Liberty Bonds and others just don't seem to get around to it," he said. "That's what we mean."

Miranda frowned with nervousness, the sharp beginnings of fear. "Are you selling them?" she asked, taking the cover off her typewriter and putting it back again.

"No, we're not selling them," said the older man. "We're just asking you why you haven't bought one." The voice was persuasive and ominous.

Miranda began to explain that she had no money, and did not know where to find any, when the older man interrupted: "That's no excuse, no excuse at all, and you know it, with the Huns overrunning martyred Belgium."

"With our American boys fighting and dying in Belleau Wood," said the younger man, "anybody can raise fifty dollars to help beat the Boche."

Miranda said hastily, "I have eighteen dollars a week and not another cent in the world. I simply cannot buy anything."

"You can pay for it five dollars a week," said the older man (they had stood there cawing back and forth over her head), "like a lot of other people in this office, and a lot of other offices besides are doing."

Miranda, desperately silent, had thought, "Suppose I were not a coward, but said what I really thought? Suppose I said to hell with this filthy war? Suppose I asked that little thug, What's the matter with you, why aren't you rotting in Belleau Wood? I wish you were. . . ."

She began to arrange her letters and notes, her fingers refusing to pick up things properly. The older man went on making his little set speech. It was hard, of course. Everybody was suffering, naturally. Everybody had to do his share. But as to that, a Liberty Bond was the safest investment you could make. It was just like having the money in the bank. Of course. The government was back of it and where better could you invest?

"I agree with you about that," said Miranda, "but I haven't any money to invest."

And of course, the man had gone on, it wasn't so much her fifty dollars that was going to make any difference. It was just a pledge of good faith on her part. A pledge of good faith that she was a loyal American doing her duty. And the thing was safe as a church. Why, if he had a million dollars he'd be glad to put every last cent of it in these Bonds. . . . "You can't lose by it," he said, almost benevolently, "and you can lose a lot if you don't. Think it over. You're the only one in this whole newspaper office that hasn't come in. And every firm in this city has come in one hundred per cent. Over at the *Daily Clarion* nobody had to be asked twice."

"They pay better over there," said Miranda. "But next week, if I can. Not now, next week."

"See that you do," said the younger man. "This ain't any laughing matter."

They lolled away, past the Society Editor's desk, past Bill the City Editor's desk, past the long copydesk where old man Gibbons sat all night shouting at intervals, "Jarge! Jarge!" and the copyboy would come flying. "Never say *people* when you mean *persons*," old man Gibbons had instructed Miranda, "and never say *practically*, say *virtually*, and don't for God's sake ever so long as I am at this desk use the barbarism *inasmuch* under any circumstances whatsoever. Now you're educated, you may go." At the head of the stairs her inquisitors had stopped in their fussy pride and vainglory, lighting cigars and wedging their hats more firmly over their eyes.

Miranda turned over in the soothing water, and wished she might fall asleep there, to wake up only when it was time to sleep again. She had a burning slow headache, and noticed it now, remembering she had waked up with it and it had in fact begun the evening before. While she dressed she tried to trace the insidious career of her headache, and it seemed reasonable to suppose it had started with the war. "It's been a headache, all right, but not quite like this." After the Committeemen had left, yesterday, she had gone to the cloakroom and had found Mary Townsend, the Society Editor, quietly hysterical about something. She was perched on the edge of the shabby wicker couch with ridges down the center, knitting on something rose-colored. Now and then she would put down her knitting, seize her head with both hands and rock, saying, "My *God*," in a surprised, inquiring voice. Her column was called Ye Towne Gossyp, so of course everybody called her Towney. Miranda and Towney had a great deal in common, and liked each other. They had both been real reporters once, and had been sent together to "cover" a scandalous elopement in which no marriage had taken place, after all, and the recaptured girl, her face swollen, had sat with her mother who was moaning steadily under a mound of blankets. They had both wept painfully and implored the young reporter to suppress the worst of the story. They had suppressed it, and the rival newspaper printed it all the next day. Miranda and Towney had then taken their punishment together, and had been degraded publicly to routine female jobs, one to the theaters, the other to society. They had this in common, that neither of them could see what else they could possibly have done, and they knew they were considered fools by the rest of the staff—nice girls, but fools. At sight of Miranda, Towney had broken out in a rage, "I can't do it, I'll never be able to raise the money, I told them, I can't, I can't, but they wouldn't listen."

Miranda said, "I knew I wasn't the only person in this office who couldn't raise five dollars. I told them I couldn't, too, and I can't."

"My *God*," said Towney, in the same voice, "they told me I'd lose my job—"

"I'm going to ask Bill," Miranda said; "I don't believe Bill would do that."

"It's not up to Bill," said Towney. "He'd have to if they got after him. Do you suppose they could put us in jail?"

"I don't know," said Miranda. "If they do, we won't be lonesome." She sat down beside Towney and held her own head. "What kind of soldier are you knitting that for? It's a sprightly color, it ought to cheer him up."

"Like hell," said Towney, her needles going again. "I'm making this for myself. That's that."

"Well," said Miranda, "we won't be lonesome and we'll catch up on our sleep." She washed her face and put on fresh makeup. Taking clean gray gloves out of her pocket she went out to join a group of young women fresh from the country club dances, the morning bridge, the charity bazaar, the Red Cross workrooms, who were wallowing in good works. They gave tea dances and raised money, and with the money they bought quantities of sweets, fruit, cigarettes, and magazines for the men in the cantonment hospitals. With this loot they were now setting out, a gay procession of high-powered cars and brightly tinted faces to cheer the brave boys who already, you might very well say, had fallen in defense of their country. It must be frightfully hard on them, the dears, to be floored like this when they're all crazy to get overseas and into the trenches as quickly as possible. Yes, and some of them are the cutest things you ever saw, I didn't know there were so many good-looking men in this country, good heavens, I said, where do they come from? Well, my dear, you may ask yourself that question, who knows where they did come from? You're quite right, the way I feel about it is this, we must do everything we can to make them contented, but I draw the line at talking to them. I told the chaperons at those dances for enlisted men, I'll dance with them, every dumbbell who asks me, but I will NOT talk to them, I said, even if there is a war. So I danced hundreds of miles without opening my mouth except to say, Please keep your knees to yourself. I'm glad we gave those dances up. Yes, and the men stopped coming, anyway. But listen, I've heard that a great many of the enlisted men come from very good families; I'm not good at catching names, and those I did catch I'd never heard before, so I don't know . . . but it seems to me if they were from good families, you'd know it, wouldn't you? I mean, if a man is well bred he doesn't step on your feet, does he? At least not that. I used to have a pair of sandals ruined at every one of those dances. Well, I think any kind of social life is in very poor taste just now, I think we should all put on our Red Cross headdresses and wear them for the duration of the war—

Miranda, carrying her basket and her flowers, moved in among the young women, who scattered out and rushed upon the ward uttering girlish laughter meant to be refreshingly gay, but there was a grim determined clang in it calculated to freeze the blood. Miserably embarrassed at the idiocy of her errand, she walked rapidly between the long rows of high beds, set foot to foot with a narrow aisle between. The men, a selected presentable lot, sheets drawn up to their chins, not seriously ill, were bored and restless, most of them willing to be amused at anything. They were for the most part picturesquely bandaged as to arm or head, and those who were not visibly wounded invariably replied "Rheumatism" if some tactless girl, who had been solemnly warned never to ask this question, still forgot and asked a man what his illness was. The good-natured, eager ones, laughing and calling out from their hard narrow beds, were soon surrounded. Miranda, with her wilting bouquet and her basket of sweets and cigarettes, looking about, caught the unfriendly bitter eye of a young fellow lying on his back, his right leg in a cast and pulley. She stopped at the foot of his bed and continued to look at him, and he looked back with an unchanged, hostile face. Not having any, thank you and be damned to the whole business, his eyes said plainly to her, and will you be so good as to take your trash off my bed? For Miranda had set it down, leaning over to place it where he might be able to reach it if he would. Having set it down, she was incapable of taking it up again, but hurried away, her face burning, down the long aisle and out into the cool October sunshine, where the dreary raw barracks swarmed and worked with an aimless life of scurrying, dun-colored insects, and going around to a window near where he lay, she looked in, spying upon her soldier. He was lying with his eyes closed, his eyebrows in a sad bitter frown. She could not place him at all, she could not imagine where he came from nor what sort of being he might have been "in life," she said to herself. His face was young and the features sharp and plain, the hands were not laborer's hands but not well-cared-for hands either. They were good useful properly shaped hands, lying there on the coverlet. It occurred to her that it would be her luck to find him, instead of a jolly hungry puppy glad of a bite to eat and a little chatter. It is like turning a corner absorbed in your painful thoughts and meeting your state of mind embodied, face to face, she said. "My own feelings about this whole thing, made flesh. Never again will I come here, this is no sort of thing to be doing. This is disgusting," she told herself plainly. "Of course I would pick him out," she thought, getting into the back seat of the car she came in, "serves me right, I know better."

Another girl came out looking very tired and climbed in beside her. After a short silence, the girl said in a puzzled way, "I don't know what good it does, really. Some of them wouldn't take anything at all. I don't like this, do you?"

"I hate it," said Miranda.

"I suppose it's all right, though," said the girl, cautiously.

"Perhaps," said Miranda, turning cautious also.

That was for yesterday. At this point Miranda decided there was no good in thinking of yesterday, except for the hour after midnight she had spent dancing with Adam. He was in her mind so much, she hardly knew when she was thinking about him directly. His image was simply always present in more or less degree, he was sometimes nearer the surface of her thoughts, the pleasantest, the only really pleasant thought she had. She examined her face in the mirror between the windows and decided that her uneasiness was not all imagination. For three days at least she had felt odd and her expression was unfamiliar. She would have to raise that fifty dollars somehow, she supposed, or who knows what can happen? She was hardened to stories of personal disaster, of outrageous accusations and extraordinarily bitter penalties that had grown monstrously out of incidents very little more important than her failure—her refusal—to buy a Bond. No, she did not find herself a pleasing sight, flushed and shiny, and even her hair felt as if it had decided to grow in the other direction. I must do something about this, I can't let Adam see me like this, she told herself, knowing that even now at that moment he was listening for the turn of her doorknob, and he would be in the hallway, or on the porch when she came out, as if by the sheerest coincidence. The noon sunlight cast cold slanting shadows in the room where, she said, I suppose I live, and this day is beginning badly, but they all do now, for one reason or another. In a drowse, she sprayed perfume on her hair, put on her moleskin cap and jacket, now in their second winter, but still good, still nice to wear, again being glad she had paid a frightening price for them. She had enjoyed them all this time, and in no case would she have had the money now. Maybe she could manage for that Bond. She could not find the lock without leaning to search for it, then stood undecided a moment possessed by the notion that she had forgotten something she would miss seriously later on.

Adam was in the hallway, a step outside his own door; he swung about as if quite startled to see her, and said, "Hello. I don't have to go back to camp today after all—isn't that luck?"

Miranda smiled at him gaily because she was always delighted at the sight of him. He was wearing his new uniform, and he was all olive and tan and tawny, hay colored and sand colored from hair to boots. She half noticed again that he always began by smiling at her; that his smile faded gradually; that his eyes became fixed and thoughtful as if he were reading in a poor light.

They walked out together into the fine fall day, scuffling bright ragged leaves under their feet, turning their faces up to a generous sky really blue and spotless. At the first corner they waited for a funeral to pass, the mourners seated straight and firm as if proud in their sorrow.

"I imagine I'm late," said Miranda, "as usual. What time is it?"

"Nearly half past one," he said, slipping back his sleeve with an exaggerated thrust of his arm upward. The young soldiers were still self-conscious about their wristwatches. Such of them as Miranda knew were boys from southern and southwestern towns, far off the Atlantic seaboard, and they had always believed that only sissies wore wristwatches. "I'll slap you on the wristwatch," one vaudeville comedian would simper to another, and it was always a good joke, never stale.

"I think it's a most sensible way to carry a watch," said Miranda. "You needn't blush."

"I'm nearly used to it," said Adam, who was from Texas. "We've been told time and again how all the he-manly regular army men wear them. It's the horrors of war," he said; "are we downhearted? I'll say we are."

It was the kind of patter going the rounds. "You look it," said Miranda.

He was tall and heavily muscled in the shoulders, narrow in the waist and flanks, and he was infinitely buttoned, strapped, harnessed into a uniform as tough and unyielding in cut as a straitjacket, though the cloth was fine and supple. He had his uniforms made by the best tailor he could find, he confided to Miranda one day when she told him how squish he was looking in his new soldier suit. "Hard enough to make anything out of the outfit, anyhow," he told her. "It's the least I can do for my beloved country, not to go around looking like a tramp." He was twenty-four years old and a Second Lieutenant in an Engineers Corps, on leave because his outfit expected to be sent over shortly. "Came in to make my will," he told Miranda, "and get a supply of toothbrushes and razor blades. By what gorgeous luck do you suppose," he asked her, "I happened to pick on your rooming house? How did I know you were there?"

Strolling, keeping step, his stout polished well-made boots setting themselves down firmly beside her thin-soled black suede, they put off as long as they could the end of their moment together, and kept up as well as they could their small talk that flew back and forth over little grooves worn in the thin upper surface of the brain, things you could say and hear clink reassuringly at once without disturbing the radiance which played and darted about the simple and lovely miracle of being two persons named Adam and Miranda, twenty-four years old each, alive and on the earth at the same moment: "Are you in the mood for dancing, Miranda?" and "I'm always in the mood for dancing, Adam!" but there were things in the way, the day that ended with dancing was a long way to go.

He really did look, Miranda thought, like a fine healthy apple this morning. One time or another in their talking, he had boasted that he had never had a pain in his life that he could remember. Instead of being horrified at this monster, she approved his monstrous uniqueness. As for herself, she had had too many pains to mention, so she did not mention them. After working for three years on a morning newspaper she had an illusion of maturity and experience; but it was fatigue merely, she decided, from keeping what she had been brought up to believe were unnatural hours, eating casually at dirty little restaurants, drinking bad coffee all night, and smoking too much. When she said something of her way of living to Adam, he studied her face a few seconds as if he had never seen it before, and said in a forthright way, "Why, it hasn't hurt you a bit, I think you're beautiful," and left her dangling there, wondering if he had thought she wished to be praised. She did wish to be praised, but not at that moment. Adam kept unwholesome hours too, or had in the ten days they had known each other, staying awake until one o'clock to take her out for supper; he smoked also continually, though if she did not stop him he was apt to explain to her exactly what smoking did to the lungs. "But," he said, "does it matter so much if you're going to war, anyway?"

"No," said Miranda, "and it matters even less if you're staying at home knitting socks. Give me a cigarette, will you?" They paused at another corner, under a half-foliaged maple, and hardly glanced at a funeral procession approaching. His eyes were pale tan with orange flecks in them, and his hair was the color of a haystack when you turn the weathered top back to the clear straw beneath. He fished out his cigarette case and snapped his silver lighter at her, snapped it several times in his own face, and they moved on, smoking.

"I can see you knitting socks," he said. "That would be just your speed. You know perfectly well you can't knit."

"I do worse," she said, soberly; "I write pieces advising other young women to knit and roll bandages and do without sugar and help win the war."

"Oh, well," said Adam, with the easy masculine morals in such questions, "that's merely your job, that doesn't count."

"I wonder," said Miranda. "How did you manage to get an extension of leave?"

"They just gave it," said Adam, "for no reason. The men are dying like flies out there, anyway. This funny new disease. Simply knocks you into a cocked hat."

"It seems to be a plague," said Miranda, "something out of the Middle Ages. Did you ever see so many funerals, ever?"

"Never did. Well, let's be strong-minded and not have any of it. I've got four days more straight from the blue and not a blade of grass must grow under our feet. What about tonight?"

"Same thing," she told him, "but make it about half past one. I've got a special job beside my usual run of the mill."

"What a job you've got," said Adam, "nothing to do but run from one dizzy amusement to another and then write a piece about it."

"Yes, it's too dizzy for words," said Miranda. They stood while a funeral passed, and this time they watched it in silence. Miranda pulled her cap to an angle and winked in the sunlight, her head swimming slowly "like goldfish," she told Adam, "my head swims. I'm only half awake, I must have some coffee.

They lounged on their elbows over the counter of a drugstore. "No more cream for the stay-at-homes," she said, "and only one lump of sugar. I'll have two or none; that's the kind of martyr I'm being. I mean to live on boiled cabbage and wear shoddy from now on and get in good shape for the next round. No war is going to sneak up on me again."

"Oh, there won't be any more wars, don't you read the newspapers?" asked Adam. "We're going to mop 'em up this time, and they're going to stay mopped, and this is going to be all."

"So they told me," said Miranda, tasting her bitter lukewarm brew and making a rueful face. Their smiles approved of each other, they felt they had got the right tone, they were taking the war properly. Above all, thought Miranda, no tooth-gnashing, no hair-tearing, it's noisy and unbecoming and it doesn't get you anywhere.

"Swill," said Adam rudely, pushing back his cup. "Is that all you're having for breakfast?"

"It's more than I want," said Miranda.

"I had buckwheat cakes, with sausage and maple syrup, and two bananas, and two cups of coffee, at eight o'clock, and right now, again, I feel like a famished orphan left in the ashcan. I'm all set," said Adam, "for broiled steak and fried potatoes and—"

"Don't go on with it," said Miranda, "it sounds delirious to me. Do all that after I'm gone." She slipped from the high seat, leaned against it slightly, glanced at her face in her round mirror, rubbed rouge on her lips and decided that she was past praying for.

"There's something terribly wrong," she told Adam. "I feel too rotten. It can't just be the weather, and the war."

"The weather is perfect," said Adam, "and the war is simply too good to be true. But since when? You were all right yesterday."

"I don't know," she said slowly, her voice sounding small and thin. They stopped as always at the open door before the flight of littered steps leading up to the newspaper loft. Miranda listened for a moment to the rattle of typewriters above, the steady rumble of presses below. "I wish we were going to spend the whole afternoon on a park bench," she said, "or drive to the mountains."

"I do too," he said; "let's do that tomorrow."

"Yes, tomorrow, unless something else happens. I'd like to run away," she told him; "let's both."

"Me?" said Adam. "Where I'm going there's no running to speak of. You mostly crawl about on your stomach here and there among the debris. You know, barbed wire and such stuff. It's going to be the kind of thing that happens once in a lifetime." He reflected a moment, and went on, "I don't know a darned thing about it, really, but they make it sound awfully messy. I've heard so much about it I feel as if I had been there and back. It's going to be an anticlimax," he said, "like seeing the pictures of a place so often you can't see it at all when you actually get there. Seems to me I've been in the army all my life."

Six months, he meant. Eternity. He looked so clear and fresh, and he had never had a pain in his life. She had seen them when they had been there and back and they never looked like this again. "Already the returned hero," she said, "and don't I wish you were."

"When I learned the use of the bayonet in my first training camp," said Adam, "I gouged the vitals out of more sandbags and sacks of hay than I could keep track of. They kept bawling at us, 'Get him, get that Boche, stick him before he sticks you'—and we'd go for those sandbags like wildfire, and honestly, sometimes I felt a perfect fool for getting so worked up when I saw the stand trickling out. I used to wake up in the night sometimes feeling silly about it."

"I can imagine," said Miranda. "It's perfect nonsense." They lingered, unwilling to say good-by. After a little pause, Adam, as if keeping up the conversation, asked, "Do you know what the average life expectation of a sapping party is after it hits the job?"

"Something speedy, I suppose."

"Just nine minutes," said Adam; "I read that in your own newspaper not a week ago."

"Make it ten and I'll come along," said Miranda.

"Not another second," said Adam, "exactly nine minutes, take it or leave it."

"Stop bragging," said Miranda. "Who figured that out?"

"A noncombatant," said Adam, "a fellow with rickets."

This seemed very comic, they laughed and leaned towards each other and Miranda heard herself being a little shrill. She wiped the tears from her eyes. "My, it's a funny war," she said; "isn't it? I laugh every time I think about it."

Adam took her hand in both of his and pulled a little at the tips of her gloves and sniffed them. "What nice perfume you have," he said, "and such a lot of it, too. I like a lot of perfume on gloves and hair," he said, sniffing again.

"I've got probably too much," she said. "I can't smell or see or hear today. I must have a fearful cold."

"Don't catch cold," said Adam; "my leave is nearly up and it will be the last, the very last." She moved her fingers in her gloves as he pulled at the fingers and turned her hands as if they were something new and curious and of great value, and she turned shy and quiet. She liked him, she liked him, and there was more than this but it was no good even imagining, because he was not for her nor for any woman, being beyond experience already, committed without any knowledge or act of his own to death. She took back her hands. "Good-by," she said finally, "until tonight."

She ran upstairs and looked back from the top. He was still watching her, and raised his hand without smiling. Miranda hardly ever saw anyone look back after he had said good-by. She could not help turning sometimes for one glimpse more of the person she had been talking with, as if that would save too rude and too sudden a snapping of even the lightest bond. But people hurried away, their faces already changed, fixed, in their straining towards their next stopping place, already absorbed in planning their next act or encounter. Adam was waiting as if he expected her to turn, and under his brows fixed in a strained frown, his eyes were very black.

At her desk she sat without taking off jacket or cap, slitting envelopes and pretending to read the letters. Only Chuck Rouncivale, the sports reporter, and Ye Towne Gossyp were sitting on her desk today, and them she liked having there. She sat on theirs when she pleased. Towney and Chuck were talking and they went on with it.

"They say," said Towney, "that it is really caused by germs brought by a German ship to Boston, a camouflaged ship, naturally, it didn't come in under its own colors. Isn't that ridiculous?"

"Maybe it was a submarine," said Chuck, "sneaking in from the bottom of the sea in the dead of night. Now that sounds better."

"Yes, it does," said Towney; "they always slip up somewhere in these details . . . and they think the germs were sprayed over the city—it started in Boston, you know—and somebody reported seeing a strange, thick, greasy-looking cloud float up out of Boston Harbor and spread slowly all over that end of town. I think it was an old woman who saw it."

"Should have been," said Chuck.

"I read it in a New York newspaper," said Towney; "so it's bound to be true."

Chuck and Miranda laughed so loudly at this that Bill stood up and glared at them. "Towney still reads the newspapers," explained Chuck.

"Well, what's funny about that?" asked Bill, sitting down again and frowning into the clutter before him.

"It was a noncombatant saw that cloud," said Miranda.

"Naturally," said Towney.

"Member of the Lusk Committee, maybe," said Miranda.

"The Angel of Mons," said Chuck, "or a dollar-a-year man."

Miranda wished to stop hearing and talking, she wished to think for just five minutes of her own about Adam, really to think about him, but there was no time. She had seen him first ten days ago, and since then they had been crossing streets together, darting between trucks and limousines and pushcarts and farm wagons; he had waited for her in doorways and in little restaurants that smelled of stale frying fat; they had eaten and danced to the urgent whine and bray of jazz orchestras, they had sat in dull theaters because Miranda was there to write a piece about the play. Once they had gone to the mountains and, leaving the car, had climbed a stony trail, and had come out on a ledge upon a flat stone, where they sat and watched the lights change on a valley landscape that was, no doubt, Miranda said, quite apocryphal—"We need not believe it, but it is fine poetry," she told him; they had leaned their shoulders together there, and had sat quite still, watching. On two Sundays they had gone to the geological museum, and had pored in shared fascination over bits of meteors, rock formations, fossilized tusks and trees, Indian arrows, grottoes from the silver and gold lodes. "Think of those old miners washing out their fortunes in little pans besides the streams," said Adam, "and inside the earth there was this—" and he had told her he liked better those things that took long to make; he loved airplanes too, all sorts of machinery, things carved out of wood or stone. He knew nothing much about them, but he recognized them when he saw them. He had confessed that he simply could not get through a book, any kind of book, except textbooks on engineering; reading bored him to crumbs; he regretted now he hadn't brought his roadster, but he hadn't thought he would need a car; he loved driving, he wouldn't expect her to believe how many hundreds of miles he could get over in a day . . . he had showed her snapshots of himself at the wheel of his roadster; of himself sailing a boat, looking very free and windblown, all angles, hauling on the ropes; he would have joined the air force but his mother had hysterics every time he mentioned it. She didn't seem to realize that dogfighting in the air was a good deal safer than sapping parties on the ground at night. But he hadn't argued, because of course she did not realize about sapping parties. And here he was, stuck, on a plateau a mile high with no water for a boat and his car at home, otherwise they could really have had a good time. Miranda knew he was trying to tell her what kind of person he was when he had his machinery with him. She felt she knew pretty well what kind of person he was, and would have liked to tell him that if he thought he had left himself at home in a boat or an automobile, he was much mistaken. The telephones were ringing, Bill was shouting at somebody who kept saying, "Well, but listen, well, but listen—" but nobody was going to listen, of course, nobody. Old man Gibbons bellowed in despair, "Jarge, Jarge—"

"Just the same," Towney was saying in her most complacent patriotic voice, "Hut Service is a fine idea, and we should all volunteer even if they don't want us." Towney does well at this, thought Miranda, look at her; remembering the rose-colored sweater and the tight rebellious face in the cloakroom. Towney was now all open-faced glory and goodness, willing to sacrifice herself for her country. "After all," said Towney, "I *can* sing and dance well enough for the Little Theater, and I could write their letters for them, and at a pinch I might drive an ambulance. I have driven a Ford for years."

Miranda joined in: "Well, I can sing and dance too, but who's going to do the bed-making and the scrubbing up? Those huts are hard to keep, and it would be a dirty job and we'd be perfectly miserable;

and as I've got a hard dirty job and am perfectly miserable, I'm going to stay at home."

"I think the women should keep out of it," said Chuck Rouncivale. "They just add skirts to the horrors of war." Chuck had bad lungs and fretted a good deal about missing the show. "I could have been there and back with a leg off by now; it would have served the old man right. Then he'd either have to buy his own hooch or sober up."

Miranda had seen Chuck on payday giving the old man money for hooch. He was a good-humored ingratiating old scoundrel, too, that was the worst of him. He slapped his son on the back and beamed upon him with the bleared eye of paternal affection while he took his last nickel.

"It was Florence Nightingale ruined wars," Chuck went on. "What's the idea of petting soldiers and binding up their wounds and soothing their fevered brows? That's not war. Let 'em perish where they fall. That's what they're there for."

"You can talk," said Towney, with a slantwise glint at him.

"What's the idea?" asked Chuck, flushing and hunching his shoulders. "You know I've got this lung, or maybe half of it anyway by now."

"You're much too sensitive," said Towney. "I didn't mean a thing."

Bill had been raging about, chewing his half-smoked cigar, his hair standing up in a brush, his eyes soft and lambent but wild, like a stag's. He would never, thought Miranda, be more than fourteen years old if he lived for a century, which he would not at the rate he was going. He behaved exactly like city editors in the moving pictures, even to the chewed cigar. Had he formed his style on the films, or had scenario writers seized once for all on the type Bill in its inarguable purity? Bill was shouting to Chuck: "*And* if he comes back here take him up the alley and saw his head off *by hand!*"

Chuck said, "He'll be back, don't worry." Bill said mildly, already off on another track, "Well, saw him off." Towney went to her own desk, but Chuck sat waiting amiably to be taken to the new vaudeville show. Miranda, with two tickets, always invited one of the reporters to go with her on Monday. Chuck was lavishly hardboiled and professional in his sports writing, but he had told Miranda that he didn't give a damn about sports, really; the job kept him out in the open, and paid him enough to buy the old man's hooch. He preferred shows and didn't see why women always had the job.

"Who does Bill want sawed today?" asked Miranda.

"That hoofer you panned in this morning's," said Chuck. "He was up here bright and early asking for the guy that writes up the show business. He said he was going to take the goof who wrote that piece up the alley and bop him in the nose. He said . . ."

"I hope he's gone," said Miranda; "I do hope he had to catch a train."

Chuck stood up and arranged his maroon-colored turtle-necked sweater, glanced down at the peasoup tweed plus fours and the hob-nailed tan boots which he hoped would help to disguise the fact that he had a bad lung and didn't care for sports, and said, "He's long gone by now, don't worry. Let's get going; you're late as usual."

Miranda, facing about, almost stepped on the toes of a little drab man in a derby hat. He might have been a pretty fellow once, but now his mouth drooped where he had lost his side teeth, and his sad red-rimmed eyes had given up coquetry. A thin brown wave of hair was combed out with brilliantine and curled against the rim of the derby. He didn't move his feet, but stood planted with a kind of inert resistance, and asked Miranda: "Are you the so-called dramatic critic on this hick newspaper?"

"I'm afraid I am," said Miranda.

"Well," said the little man, "I'm just asking for one minute of your valuable time." His underlip shot out, he began with shaking hands to fish about in his waistcoat pocket. "I just hate to let you get away with it, that's all." He riffled through a collection of shabby newspaper clippings. "Just give these the once-over, will you? And then let me ask you if you think I'm gonna stand for being knocked by a tanktown critic," he said, in a toneless voice; "look here, here's Buffalo, Chicago, Saint Looey, Philadelphia, Frisco, besides New York. Here's the best publications in the business, *Variety* the *Billboard*, they all broke down and admitted that Danny Dickerson knows his stuff. So you don't think so, hey? That's all I wanta ask you."

"No, I don't," said Miranda, as bluntly as she could, "and I can't stop to talk about it."

The little man leaned nearer, his voice shook as if he had been nervous for a long time. "Look here, what was there you didn't like about me? Tell me that."

Miranda said, "You shouldn't pay any attention at all. What does it matter what I think?"

"I don't care what you think, it ain't that," said the little man, "but these things get round and booking agencies back East don't know how it is out here. We get panned in the sticks and they think it's the same as getting panned in Chicago, see? They don't know the difference. They don't know that the more high class an act is the more the hick critics pan it. But I've been called the best in the business by the best in the business and I wanta know what you think is wrong with me."

Chuck said, "Come on, Miranda, curtain's going up." Miranda handed the little man his clippings, they were mostly ten years old, and tried to edge past him. He stepped before her again and said without much conviction. "If you was a man I'd knock your block off." Chuck got up at that and lounged over, taking his hands out of his pockets, and said, "Now you've done your song and dance you'd better get out. Get the hell out now before I throw you downstairs."

The little man pulled at the top of his tie, a small blue tie with red polka dots, slightly frayed at the knot. He pulled it straight and repeated as if he had rehearsed it, "Come out in the alley." The tears filled his thickened red lids. Chuck said, "Ah, shut up," and followed Miranda, who was running towards the stairs. He overtook her on the sidewalk. "I left him sniveling and shuffling his publicity trying to find the joker," said Chuck, "the poor old heel."

Miranda said, "There's too much of everything in this world just now. I'd like to sit down here on the curb, Chuck, and die, and never again see—I wish I could lose my memory and forget my own name. . . . I wish—"

Chuck said, "Toughen up, Miranda. This is no time to cave in. Forget that fellow. For every hundred people in show business, there are ninety-nine like him. But you don't manage right, anyway. You bring it on yourself. All you have to do is play up the headliners, and you needn't even mention the also-rans. Try to keep in mind that Rypinsky has got show business cornered in this town; please Rypinsky and you'll please the advertising department, please them and you'll get a raise. Hand-in-glove, my poor dumb child, will you never learn?"

"I seem to keep learning all the wrong things," said Miranda hopelessly.

"You do for a fact," Chuck told her cheerfully. "You are as good at it as I ever saw. Now do you feel better?"

"This is a rotten show you've invited me to," said Chuck. "Now what are you going to do about it? If I were writing it up. I'd—"

"Do write it up," said Miranda. "You write it up this time. I'm getting ready to leave, anyway, but don't tell anybody yet."

"You mean it? All my life," said Chuck, "I've yearned to be a so-called dramatic critic on a hick newspaper, and this is positively my first chance."

"Better take it," Miranda told him. "It may be your last." She thought, This is the beginning of the end of something. Something terrible is going to happen to me. I shan't need bread and butter where I'm going. I'll will it to Chuck, he has a venerable father to buy hooch for. I hope they let him have it. Oh, Adam, I hope I see you once more before I go under with whatever is the matter with me. "I wish the war were over," she said to Chuck, as if they had been talking about that. "I wish it were over and I wish it had never begun."

Chuck had got out his pad and pencil and was already writing his review. What she had said seemed safe enough but how would he take it? "I don't care how it started or where it ends," said Chuck, scribbling away. "I'm not going to be there."

All the rejected men talked like that, thought Miranda. War was the only thing they wanted, now they couldn't have it. Maybe they had wanted badly to go, some of them. All of them had a sidelong eye for the women they talked with about it, a guarded resentment which said, "Don't pin a white feather on me, you bloodthirsty female. I've offered my meat to the crows and they won't have it." The worst thing about war for the stay-at-homes is there isn't anyone to talk to any more. The Lusk Committee will get you if you don't watch out. Bread will win the war. Work will win, sugar will win, peach pits will win the war. Nonsense. *Not* nonsense, I tell you, there's some kind of valuable high explosive to be got out of peach pits. So all the happy housewives hurry during the canning season to lay their baskets of peach pits on the altar of their country. It keeps them busy and makes them feel useful, and all these women running wild with the men away are dangerous, if they aren't given something to keep their little minds out of mischief. So rows of young girls, the intact cradles of the future, with their pure serious faces framed becomingly in Red Cross wimples, roll cock-eyed bandages that will never reach a base hospital, and knit sweaters that will never warm a manly chest, their minds dwelling lovingly on all the blood and mud and the next dance at the Acanthus Club for the officers of the flying corps. Keeping still and quiet will win the war.

"I'm simply not going to be there," said Chuck, absorbed in his review. No, Adam will be there, thought Miranda. She slipped down in the chair and leaned her head against the dusty plush, closed her eyes and faced for one instant that was a lifetime the certain, the overwhelming and awful knowledge that there was nothing at all ahead for Adam and for her. Nothing. She opened her eyes and held her hands together palms up, gazing at them and trying to understand oblivion.

"Now look at this," said Chuck, for the lights had come on and the audience was rustling and talking again. "I've got it all done, even before the headliner comes on. It's old Stella Mayhew, and she's always good, she's been good for forty years, and she's going to sing, 'O the blues ain't nothin' but the easy-going heart disease.' That's all you need to know about her. Now just glance over this. Would you be willing to sign it?"

Miranda took the pages and stared at them conscientiously, turning them over, she hoped, at the right moment, and gave them back. "Yes, Chuck, yes, I'd sign that. But I won't. We must tell Bill you wrote it, because it's your start, maybe."

"You don't half appreciate it," said Chuck. "You read it too fast. Here, listen to this—" and he began to mutter excitedly. While he was reading she watched his face. It was a pleasant face with some kind of spark of life in it, and a good severity in the modeling of the brow above the nose. For the first time since she had known him she wondered what Chuck was thinking about. He looked preoccupied and unhappy, he wasn't so frivolous as he sounded. The people were crowding into the aisle, bringing out their cigarette cases ready to strike a match the instant they reached the lobby; women with waved hair clutched at their wraps, men stretched their chins to ease them of their stiff collars, and Chuck said, "We might as well go now." Miranda, buttoning her jacket, stepping into the moving crowd, thinking, What did I ever know about them? There must be a great many of them here who think as I do, and we dare not say a word to each other of our desperation, we are speechless animals letting ourselves be destroyed, and why? Does anybody here believe the things we say to each other?

Stretched in unease on the ridge of the wicker couch in the cloakroom, Miranda waited for time to pass and leave Adam with her. Time seemed to proceed with more than usual eccentricity, leaving twilight gaps in her mind for thirty minutes which seemed like a second, and then hard flashes of light that shone clearly on her watch proving that three minutes is an intolerable stretch of waiting, as if she were hanging by her thumbs. At last it was reasonable to imagine Adam stepping out of the house in the early darkness into the blue mist that might soon be rain, he would be on the way, and there was nothing to think about him, after all. There was only the wish to see him and the fear, the present threat, of not seeing him again; for every step they took towards each other seemed perilous, drawing them apart instead of together, as a swimmer in spite of his most determined strokes is yet drawn slowly backward by the tide. "I don't want to love," she would think in spite of herself, "not Adam, there is no time and we are not ready for it and yet this is all we have—"

And there he was on the sidewalk, with his foot on the first step, and Miranda almost ran down to meet him. Adam, holding her hands, asked, "Do you feel well now? Are you hungry? Are you tired? Will you feel like dancing after the show?"

"Yes to everything," said Miranda, "yes, yes. . . ." Her head was like a feather, and she steadied herself on his arm. The mist was still mist that might be rain later, and though the air was sharp and clean in her mouth, it did not, she decided, make breathing any easier. "I hope the show is good, or at least funny," she told him, "but I promise nothing."

It was a long, dreary play, but Adam and Miranda sat very quietly together waiting patiently for it to be over. Adam carefully and seriously pulled off her glove and held her hand as if he were accustomed to holding her hand in theaters. Once they turned and their eyes met, but only once, and the two pairs of eyes were equally steady and noncommittal. A deep tremor set up in Miranda, and she set about resisting herself methodically as if she were closing windows and doors and fastening down curtains against a rising storm. Adam sat watching the monotonous play with a strange shining excitement, his face quite fixed and still.

When the curtain rose for the third act, the third act did not take place at once. There was instead disclosed a backdrop almost covered with an American flag improperly and disrespectfully exposed, nailed at each upper corner, gathered in the middle and nailed again, sagging dustily. Before it posed a local dollar-a-year man, now doing his bit as a Liberty Bond salesman. He was an ordinary man past middle life, with a neat little melon buttoned into his trousers and waistcoat, an opinionated tight mouth, a face and figure in which nothing could be read save the inept sensual record of fifty years. But for once in his life he was an important fellow in an impressive situation, and he reveled, rolling his words in an actorish tone.

"Looks like a penguin," said Adam. They moved, smiled at each other, Miranda reclaimed her hand, Adam folded his together and they prepared to wear their way again through the same old moldy speech with the same old dusty backdrop. Miranda tried not to listen, but she heard. These vile Huns—glorious Belleau Wood—our keyword is Sacrifice—Martyred Belgium—give till it hurts—our noble boys Over There—Big Berthas—the death of civilization—the Boche—

"My head aches," whispered Miranda. "Oh, why won't he hush?"

"He won't," whispered Adam. "I'll get you some aspirin."

"In Flanders Field the poppies grow, Between the crosses row on row"—"He's getting into the home stretch," whispered Adam—atrocities, innocent babes hoisted on Boche bayonets—your child and my child—if our children are spared these things, then let us say with all reverence that these dead have not died in vain—the war, the *war*, the WAR to end WAR, war for Democracy, for humanity, a safe world forever and ever—and to prove our faith in Democracy to each other, and to the world, let everybody get together and buy Liberty Bonds and do without sugar and wool socks—was that it? Miranda asked herself, Say that over, I didn't catch the last line. Did you mention Adam? If you didn't I'm not interested. What about Adam, you little pig? And what are we going to sing this time, "Tipperary" or "There's a Long, Long Trail"? Oh, please do let the show go on and get over with. I must write a piece about it before I can go dancing with Adam and we have no time. Coal, oil, iron, gold, international finance, why don't you tell us about them, you little liar?

The audience rose and sang, "There's a Long, Long Trail A-winding," their opened mouths black and faces pallid in the reflected footlights; some of the faces grimaced and wept and had shining streaks like snail's tracks on them. Adam and Miranda joined in at the tops of their voices, grinning shamefacedly at each other once or twice.

In the street, they lit their cigarettes and walked slowly as always. "Just another nasty old man who would like to see the young ones killed," said Miranda in a low voice; "the tomcats try to eat the little tom-kittens, you know. They don't fool you really, do they, Adam?"

The young people were talking like that about the business by then. They felt they were seeing pretty clearly through that game. She went on, "I hate these potbellied baldheads, too fat, too old, too cowardly, to go to war themselves, they know they're safe; it's you they are sending instead—"

Adam turned eyes of genuine surprise upon her. "Oh, *that* one," he said. "Now what could the poor sap do if they did take him? It's not his fault," he explained, "he can't do anything but talk." His pride in his youth, his forbearance and tolerance and contempt for that unlucky being breathed out of his very pores as he strolled, straight and relaxed in his strength. "What *could* you expect of him, Miranda?"

She spoke his name often, and he spoke hers rarely. The little shock of pleasure the sound of her name in his mouth gave her stopped her answer. For a moment she hesitated, and began at another point of attack. "Adam," she said, "the worst of war is the fear and suspicion and the awful expression in all the eyes you meet . . . as if they had pulled down the shutters over their minds and their hearts and were peering out at you, ready to leap if you make one gesture or say one word they do not understand instantly. It frightens me; I live in fear too, and no one should have to live in fear. It's the skulking about, and the lying. It's what war does to the mind and the heart, Adam, and you can't separate these two—what it does to them is worse than what it can do to the body."

Adam said soberly, after a moment, "Oh, yes, but suppose one comes back whole? The mind and the heart sometimes get another chance, but if anything happens to the poor old human frame, why, it's just out of luck, that's all."

"Oh, yes," mimicked Miranda. "It's just out of luck, that's all."

"If I didn't go," said Adam, in a matter-of-fact voice, "I couldn't look myself in the face."

So that's all settled. With her fingers flattened on his arm, Miranda was silent, thinking about Adam. No, there was no resentment or revolt in him. Pure, she thought, all the way through, flawless, complete, as the sacrificial lamb must be. The sacrificial lamb strode along casually, accommodating his long pace to hers, keeping her on the inside of the walk in the good American style, helping her across street corners as if she were a cripple—"I hope we don't come to a mud puddle, he'll carry me over it"—giving off whiffs of tobacco smoke, a manly smell of scentless soap, freshly cleaned leather and freshly washed skin, breathing through his nose and carrying his chest easily. He threw back his head and smiled into the sky which still misted, promising rain. "Oh, boy," he said, "what a night. Can't you hurry that review of yours so we can get started?"

He waited for her before a cup of coffee in the restaurant next to the pressroom, nicknamed The Greasy Spoon. When she came down at last, freshly washed and combed and powdered, she saw Adam first, sitting near the dingy big window, face turned to the street, but looking down. It was an extraordinary face, smooth and fine and golden in the shabby light, but now set in a blind melancholy, a look of pained suspense and disillusion. For just one split second she got a glimpse of Adam when he would have been older, the face of the man he would not live to be. He saw her then, rose, and the bright glow was there.

Adam pulled their chairs together at their table; they drank hot tea and listened to the orchestra jazzing "Pack Up Your Troubles."

"In an old kit bag, and smoil, smoil, smoil," shouted half a dozen boys under the draft age, gathered around a table near the orchestra. They yelled incoherently, laughed in great hysterical bursts of something that appeared to be merriment, and passed around under the tablecloth flat bottles containing a clear liquid—for in this western city founded and built by roaring drunken miners, no one was allowed to take his alcohol openly—splashed it into their tumblers of ginger ale, and went on singing, "It's a Long Way to Tipperary." When the tune changed to "Madelon," Adam said, "Let's dance." It was a tawdry little place, crowded and hot and full of smoke, but there was nothing better. The music was gay; and life is completely crazy anyway, thought Miranda, so what does it matter? This is what we have, Adam and I, this is all we're going to get, this is the way it is with us. She wanted to say, "Adam, come out of your dream and listen to me. I have pains in my chest and my head and my heart and they're real. I am in pain all over, and you are in such danger as I can't bear to think about, and why can we not save each other?" When her hand tightened on his shoulder his arm tightened about her waist instantly, and stayed there, holding firmly. They said nothing but smiled continually at each other, odd changing smiles as though they had found a new language. Miranda, her face near Adam's shoulder, noticed a dark young pair sitting at a corner table, each with an arm around the waist of the other, their heads together, their eyes staring at the same thing, whatever it was, that hovered in space before them. Her right hand lay on the table, his hand over it, and her face was a blur with weeping. Now and then he raised her hand and kissed it, and set it down and held it, and her eyes would fill again. They were not shameless, they had merely forgotten where they were, or they had no other place to go, perhaps. They said not a word, and the small pantomime repeated itself, like a melancholy short film running monotonously over and over again. Miranda envied them. She envied that girl. At least she can weep if that helps, and he does not even have to ask, What is the matter? Tell me. They had cups of coffee before them, and after a long while—Miranda and Adam had danced and sat down again twice—when the coffee was quite cold, they drank it suddenly, then embraced as before, without a word and scarcely a glance at each other. Something was done and settled between them, at least; it was enviable, enviable, that they could sit quietly together and have the same expression on their faces while they looked into the hell they shared, no matter what kind of hell, it was theirs, they were together.

At the table nearest Adam and Miranda a young woman was leaning on her elbow, telling her young man a story. "And I don't like him because he's too fresh. He kept on asking me to take a drink and I kept telling him, I don't drink and he said, Now look here, I want a drink the worst way and I think it's mean of you not to drink with me, I can't sit up here and drink by myself, he said. I told him, You're not by yourself in the first place; I like that, I said, and if you want a drink go ahead and have it, I told him, why drag *me* in? So he called the waiter and ordered ginger ale and two glasses and I drank straight ginger ale like I always do but he poured a shot of hooch in his. He was awfully proud of that hooch, said he made it himself out of potatoes. Nice homemade likker, warm from the pipe, he told me, three drops of this and your ginger ale will taste like Mumm's Extry. But I said, No, and I mean no, can't you get that through your bean? He took another drink and said, Ah, come on, honey, don't be so stubborn, this'll make your shimmy shake. So I just got tired of the argument, and I said, I don't need to drink, to shake my shimmy, I can strut my stuff on tea, I said. Well, why don't you then, he wanted to know, and I just told him—"

She knew she had been asleep for a long time when all at once without even a warning footstep or creak of the door hinge, Adam was in the room turning on the light, and she knew it was he, though at first she was blinded and turned her head away. He came over at once and sat on the side of the bed and began to talk as if he were going on with something they had been talking about before. He crumpled a square of paper and tossed it in the fireplace.

"You didn't get my note," he said. "I left it under the door. I was called back suddenly to camp for a lot of inoculations. They kept me longer than I expected, I was late. I called the office and they told me you were not coming in today. I called Miss Hobbe here and she said you were in bed and couldn't come to the telephone. Did she give you my message?"

"No," said Miranda drowsily, "but I think I have been asleep all day. Oh, I do remember. There was a doctor here. Bill sent him. I was at the telephone once, for Bill told me he would send an ambulance and have me taken to the hospital. The doctor tapped my chest and left a prescription and said he would be back, but he hasn't come."

"Where is it, the prescription?" asked Adam.

"I don't know. He left it, though, I saw him."

Adam moved about searching the tables and the mantelpiece. "Here it is," he said. "I'll be back in a few minutes. I must look for an all-night drugstore. It's after one o'clock, Good-by."

Good-by, good-by. Miranda watched the door where he had disappeared for quite a while, then closed her eyes, and thought, When I am not here I cannot remember anything about this room where I lived for nearly a year, except that the curtains are too thin and there was never any way of shutting out the morning light. Miss Hobbe had promised heavier curtains, but they had never appeared. When Miranda in her dressing gown had been at the telephone that morning, Miss Hobbe had passed through, carrying a tray. She was a little red-haired nervously friendly creature, and her manner said all too plainly that the place was not paying and she was on the ragged edge.

"My dear *child,*" she said sharply, with a glance at Miranda's attire, "what is the matter?"

Miranda, with the receiver to her ear, said, "Influenza, I think."

"*Horrors,*" said Miss Hobbe, in a whisper, and the tray wavered in her hands. "Go back to bed at once . . . go at *once!*"

"I must talk to Bill first," Miranda had told her, and Miss Hobbe had hurried on and had not returned. Bill had shouted directions at her, promising everything, doctor, nurse, ambulance, hospital, her check every week as usual, everything, but she was to get back to bed and stay there. She dropped into bed, thinking that Bill was the only person she had ever seen who actually tore his own hair when he was excited enough . . . I suppose I should ask to be sent home, she thought, it's a respectable old custom to inflict your death on the family if you can manage it. No, I'll stay here, this is my business, but not in this room, I hope. . . . I wish I were in the cold mountains in the snow, that's what I should like best; and all about her rose the measured ranges of the Rockies wearing their perpetual snow, their majestic blue laurels of cloud, chilling her to the bone with their sharp breath. Oh, no, I must have warmth—and her memory turned and roved after another place she had known first and loved best, that now she could see only in drifting fragments of palm and cedar, dark shadows and a sky that warmed without dazzling, as this strange sky had dazzled without warming her; there was the long slow wavering of gray moss in the drowsy oak shade, the spacious hovering of buzzards overhead, the smell of crushed water herbs along a bank, and without warning a broad tranquil river into which flowed all the rivers she had known. The walls shelved away in one deliberate silent movement on either side, and a tall sailing ship was moored near by, with a gangplank weathered to blackness touching the foot of her bed. Back of the ship was jungle, and even as it appeared before her, she knew it was all she had ever read or had been told or felt or thought about jungles; a writhing terribly alive and secret place of death, creeping with tangles of spotted serpents, rainbow-colored birds with malign eyes, leopards with humanly wise faces and extravagantly crested lions; screaming long-armed monkeys tumbling among broad fleshy leaves that glowed with sulphur-colored light and exuded the ichor of death, and rotting trunks of unfamiliar trees sprawled in crawling slime. Without surprise, watching from her pillow, she saw herself run swiftly down this gangplank to the slanting deck, and standing there, she leaned on the rail and waved gaily to herself in bed, and the slender ship spread its wings and sailed away into the jungle. The air trembled with the shattering scream and the hoarse bellow of voices all crying together, rolling and colliding above her like ragged stormclouds, and the words became two words only rising and falling and clamoring about her head. Danger, danger, danger, the voices said, and War, war, war. There was her door half open, Adam standing with his hand on the knob, and Miss Hobbe with her face all out of shape with terror was crying shrilly, "I tell you, they must come for her *now*, or I'll put her on the sidewalk. . . . I tell you, this is a plague, a plague, my God, and I've got a houseful of people to think about!"

Adam said, "I know that. They'll come for her tomorrow morning."

"Tomorrow morning, my God, they'd better come now!"

"They can't get an ambulance," said Adam, "and there aren't any beds. And we can't find a doctor or a nurse. They're all busy. That's all there is to it. You stay out of the room, and I'll look after her."

"Yes, you'll look after her, I can see that," said Miss Hobbe, in a particularly unpleasant tone.

"Yes, that's what I said," answered Adam, drily, "and you keep out."

He closed the door carefully. He was carrying an assortment of misshapen packages, and his face was astonishingly impassive.

"Did you hear that?" he asked, leaning over and speaking very quietly.

"Most of it," said Miranda, "it's a nice prospect, isn't it?"

"I've got your medicine," said Adam, "and you're to begin with it this minute. She can't put you out."

"So it's really as bad as that," said Miranda.

"It's as bad as anything can be," said Adam, "all the theaters and nearly all the shops and restaurants are closed, and the streets have been full of funerals all day and ambulances all night—"

"But not one for me," said Miranda, feeling hilarious and lightheaded. She sat up and beat her pillow into shape and reached for her robe. "I'm glad you're here, I've been having a nightmare. Give me a cigarette, will you, and light one for yourself and open all the windows and sit near one of them. You're running a risk," she told him, "don't you know that? Why do you do it?"

"Never mind," said Adam, "take your medicine," and offered her two large cherry-colored pills. She swallowed them promptly and instantly vomited them up. "*Do* excuse me," she said, beginning to laugh. "I'm so sorry." Adam without a word and with a very concerned expression washed her face with a wet towel, gave her some cracked ice from one of the packages, and firmly offered her two more pills. "That's what they always did at home," she explained to him, "and it worked." Crushed with humiliation, she put her hands over her face and laughed again, painfully.

"There are two more kinds yet," said Adam, pulling her hands from her face and lifting her chin. "You've hardly begun. And I've got other things, like orange juice and ice cream—they told me to feed you ice cream—and coffee in a thermos bottle, and a thermometer. You have to work through the whole lot so you'd better take it easy."

"This time last night we were dancing," said Miranda, and drank something from a spoon. Her eyes followed him about the room, as he did things for her with an absentminded face, like a man alone; now and again he would come back, and slipping his hand under her head, would hold a cup or a tumbler to her mouth, and she drank, and followed him with her eyes again, without a clear notion of what was happening.

"Adam," she said, "I've just thought of something. Maybe they forgot St. Luke's Hospital. Call the sisters there and ask them not to be so selfish with their silly old rooms. Tell them I only want a very small dark ugly one for three days, or less. Do try them, Adam."

He believed, apparently, that she was still more or less in her right mind, for she heard him at the telephone explaining in his deliberate voice. He was back again almost at once, saying, "This seems to be my day for getting mixed up with peevish old maids. The sister said that even if they had a room you couldn't have it without doctor's orders. But they didn't have one, anyway. She was pretty sour about it."

"Well," said Miranda in a thick voice, "I think that's abominably rude and mean, don't you?" She sat up with a wide gesture of both arms, and began to retch again, violently.

"Hold it, as you were," called Adam, fetching the basin. He held her head, washed her face and hands with ice water, put her head straight on the pillow, and went over and looked out of the window. "Well," he said at last, sitting beside her again, "they haven't got a room. They haven't got a bed. They haven't even got a baby crib, the way she talked. So I think that's straight enough, and we may as well dig in."

"Isn't the ambulance coming?"

"Tomorrow, maybe."

He took off his tunic and hung it on the back of a chair. Kneeling before the fireplace, he began carefully to set kindling sticks in the shape of an Indian teepee, with a little paper in the center for them to lean upon. He lighted this and placed other sticks upon them, and larger bits of wood. When they were going nicely he added still heavier wood, and coal a few lumps at a time, until there was a good blaze, and a fire that would not need rekindling. He rose and dusted his hands together, the fire illuminated him from the back and his hair shone.

"Adam," said Miranda, "I think you're very beautiful." He laughed out at this, and shook his head at her. "What a hell of a word," he said, "for me." "It was the first that occurred to me," she said, drawing up on her elbow to catch the warmth of the blaze. "That's a good job, that fire."

He sat on the bed again, dragging up a chair and putting his feet on the rungs. They smiled at each other for the first time since he had come in that night. "How do you feel now?" he asked.

"Better, much better," she told him. "Let's talk. Let's tell each other what we meant to do."

"You tell me first," said Adam, "I want to know about you."

"You'd get the notion I had a very sad life," she said, "and perhaps it was, but I'd be glad enough to have it now. If I could have it back, it would be easy to be happy about almost anything at all. That's not true, but that's the way I feel now." After a pause, she said, "There's nothing to tell, after all, if it ends now, for all this time I was getting ready for something that was going to happen later, when the time came. So now it's nothing much."

"But it must have been worth having until now, wasn't it?" he asked seriously as if it were something important to know.

"Not if this is all," she repeated obstinately.

"Weren't you ever—happy?" asked Adam, and he was plainly afraid of the word; he was shy of it as he was of the word *love*, he seemed never to have spoken it before, and was uncertain of its sound or meaning.

"I don't know," she said, "I just lived and never thought about it. I remember things I liked, though, and things I hoped for."

"I was going to be an electrical engineer," said Adam. He stopped short. "And I shall finish up when I get back," he added, after a moment.

"Don't you love being alive?" asked Miranda. "Don't you love weather and the colors at different times of the day, and all the sounds and noises like children screaming in the next lot, and automobile horns and little bands playing in the street and the smell of food cooking?"

"I love to swim, too," said Adam.

"So do I," said Miranda; "we never did swim together."

"Do you remember any prayers?" she asked him suddenly. "Did you ever learn anything at Sunday School?"

"Not much," confessed Adam without contrition. "Well, the Lord's Prayer."

"Yes, and there's Hail Mary," she said, "and the really useful one beginning, I confess to Almighty God and to blessed Mary ever virgin and to the holy Apostles Peter and Paul—"

"Catholic," he commented.

"Prayers just the same, you big Methodist. I'll bet you *are* a Methodist."

"No, Presbyterian."

"Well, what others do you remember?"

"Now I lay me down to sleep—" said Adam.

"Yes, that one, and Blessed Jesus meek and mild—you see that my religious education wasn't neglected either. I even know a prayer beginning O Apollo. Want to hear it?"

"No," said Adam, "you're making fun."

"I'm not," said Miranda, "I'm trying to keep from going to sleep. I'm afraid to go to sleep, I may not wake up. Don't let me go to sleep, Adam. Do you know Matthew, Mark, Luke and John? Bless the bed I lie upon?"

"If I should die before I wake, I pray the Lord my soul to take. Is that it?" asked Adam. "It doesn't sound right, somehow."

"Light me a cigarette, please, and move over and sit near the window. We keep forgetting about fresh air. You must have it." He lighted the cigarette and held it to her lips. She took it between her fingers and dropped it under the edge of her pillow. He found it and crushed it out in the saucer under the water tumbler. Her head swam in darkness for an instant, cleared, and she sat up in panic, throwing off the covers and breaking into a sweat. Adam leaped up with an alarmed face, and almost at once was holding a cup of hot coffee to her mouth.

"You must have some too," she told him, quiet again, and they sat huddled together on the edge of the bed, drinking coffee in silence.

Adam said, "You must lie down again. You're awake now."

"Let's sing," said Miranda. "I know an old spiritual, I can remember some of the words." She spoke in a natural voice. "I'm fine now." She began in a hoarse whisper, " 'Pale horse, pale rider, done taken my lover away. . . .' Do you know that song?"

"Yes," said Adam, "I heard Negroes in Texas sing it, in an oil field."

"I heard them sing it in a cotton field," she said; "it's a good song."

They sang that line together. "But I can't remember what comes next," said Adam.

" 'Pale horse, pale rider,' " said Miranda. "(We really need a good banjo) 'done taken my lover away—' " Her voice cleared and she said, "But we ought to get on with it. What's the next line?"

"There's a lot more to it than that," said Adam, "about forty verses, the rider done taken away a mammy, pappy, brother, sister, the whole family besides the lover—"

"But not the singer, not yet," said Miranda. "Death always leaves one singer to mourn. 'Death,' " she sang, " 'oh, leave one singer to mourn—' "

" 'Pale horse, pale rider,' " chanted Adam, coming in on the beat, " 'done taken my lover away!' (I think we're good, I think we ought to get up an act—)"

"Go in Hut Service," said Miranda, "entertain the poor defenseless heroes Over There."

"We'll play banjos," said Adam; "I always wanted to play the banjo."

Miranda sighed, and lay back on the pillow and thought, I must give up, I can't hold out any longer. There was only that pain, only that room, and only Adam. There were no longer any multiple planes of living, no tough filaments of memory and hope pulling taut backwards and forwards holding her upright between them. There was only this one moment and it was a dream of time, and Adam's face, very near hers, eyes still and intent, was a shadow, and there was to be nothing more. . . .

"Adam," she said out of the heavy soft darkness that drew her down, down, "I love you, and I was hoping you would say that to me, too."

He lay down beside her with his arm under her shoulder, and pressed his smooth face against hers, his mouth moved towards her mouth and stopped. "Can you hear what I am saying? . . . What do you think I have been trying to tell you all this time?"

She turned towards him, the cloud cleared and she saw his face for an instant. He pulled the covers about her and held her, and said, "Go to sleep, darling, darling, if you will go to sleep now for one hour I will wake you up and bring you hot coffee and tomorrow we will find somebody to help. I love you, go to sleep—"

Almost with no warning at all, she floated into the darkness, holding his hand, in sleep that was not sleep but clear evening light in a small green wood, an angry dangerous wood full of inhuman concealed voices singing sharply like the whine of arrows and she saw Adam transfixed by a flight of these singing arrows that struck him in the heart and passed shrilly cutting their path through the leaves. Adam fell straight back before her eyes, and rose again unwounded and alive; another flight of arrows loosed from the invisible bow struck him again and he fell, and yet he was there before her untouched in a perpetual death and resurrection. She threw herself before him, angrily and selfishly she interposed between him and the track of the arrow, crying, No, no, like a child cheated in a game, It's my turn now, why must you always be the one to die? and the arrows struck her cleanly through the heart and through his body and he lay dead, and she still lived, and the wood whistled and sang and shouted, every branch and leaf and blade of grass had its own terrible accusing voice. She ran then, and Adam caught her in the middle of the room, running, and said, "Darling, I must have been asleep too. What happened, you screamed terribly?"

After he had helped her to settle again, she sat with her knees drawn up under her chin, resting her head on her folded arms and began carefully searching for her words because it was important to explain clearly. "It was a very odd sort of dream, I don't know why it could have frightened me. There was something about an old-fashioned valentine. There were two hearts carved on a tree, pierced by the same arrow—you know, Adam—"

"Yes, I know, honey," he said in the gentlest sort of way, and sat kissing her on the cheek and forehead with a kind of accustomedness, as if he had been kissing her for years, "one of those lace paper things."

"Yes, and yet they were alive, and were us, you understand—this doesn't seem to be quite the way it was, but it was something like that. It was in a wood—"

"Yes," said Adam. He got up and put on his tunic and gathered up the thermos bottle. "I'm going back to that little stand and get us some ice cream and hot coffee," he told her, "and I'll be back in five minutes, and you keep quiet. Good-by for five mintues," he said, holding her chin in the palm of his hand and trying to catch her eye, "and you be very quiet."

"Good-by," she said. "I'm awake again." But she was not, and the two alert young internes from the County hospital who had arrived, after frantic urgings from the noisy city editor of the Blue Mountain *News*, to carry her away in a police ambulance, decided that they had better go down and get the stretcher. Their voices roused her, she sat up, got out of bed at once and stood glancing about brightly. "Why, you're all right," said the darker and stouter of the two young men, both extremely fit and competent-looking in their white clothes, each with a flower in his buttonhole. "I'll just carry you." He unfolded a white blanket and wrapped it around her. She gathered up the folds and asked, "But where is Adam?" taking hold of the doctor's arm. He laid a hand on her drenched forehead, shook his head, and gave her a shrewd look. "Adam?"

"Yes," Miranda told him, lowering her voice confidentially, "he was here and now he is gone."

"Oh, he'll be back," the interne told her easily, "he's just gone round the block to get cigarettes. Don't worry about Adam. He's the least of your troubles."

"Will he know where to find me?" she asked, still holding back.

"We'll leave him a note," said the interne. "Come now, it's time we got out of here."

He lifted and swung her up to his shoulder. "I feel very badly," she told him; "I don't know why."

"I'll bet you do," said he, stepping out carefully, the other doctor going before them, and feeling for the first step of the stairs. "Put your arms around my neck," he instructed her. "It won't do you any harm and it's a great help to me."

"What's your name?" Miranda asked as the other doctor opened the front door and they stepped out into the frosty sweet air.

"Hildesheim," he said, in the tone of one humoring a child.

"Well, Dr. Hildesheim, aren't we in a pretty mess?"

"We certainly are," said Dr. Hildesheim.

The second young interne, still quite fresh and dapper in his white coat, though his carnation was withering at the edges, was leaning over listening to her breathing through a stethoscope, whistling thinly, "There's a Long, Long Trail—" From time to time he tapped her ribs smartly with two fingers, whistling. Miranda observed him for a few moments until she fixed his bright busy hazel eye not four inches from hers. "I'm not unconscious," she explained, "I know what I want to say." Then to her horror she heard herself babbling nonsense, knowing it was nonsense though she could not hear what she was saying. The flicker of attention in the eye near her vanished, the second interne went on tapping and listening, hissing softly under his breath.

"I wish you'd stop whistling," she said clearly. The sound stopped. "It's a beastly tune," she added. Anything, anything at all to keep her small hold on the life of human beings, a clear line of communication, no matter what, between her and the receding world. "Please let me see Dr. Hildesheim," she said, "I have something important to say to him. I must say it now." The second interne vanished. He did not walk away, he fled into the air without a sound, and Dr. Hildesheim's face appeared in his stead.

"Dr. Hildesheim, I want to ask you about Adam."

"That young man? He's been here, and left you a note, and has gone again," said Dr. Hildesheim, "and he'll be back tomorrow and the day after." His tone was altogether too merry and flippant.

"I don't believe you," said Miranda bitterly, closing her lips and eyes and hoping she might not weep.

"Miss Tanner," called the doctor, "have you got that note?"

Miss Tanner appeared beside her, handed her an unsealed envelope, took it back, unfolded the note and gave it to her.

"I can't see it," said Miranda, after a pained search of the page full of hasty scratches in black ink.

"Here, I'll read it," said Miss Tanner. "It says, 'They came and took you while I was away and now they will not let me see you. Maybe tomorrow they will, with my love, Adam,' " read Miss Tanner in a firm dry voice, pronouncing the words distinctly. "Now, do you see?" she asked soothingly.

Miranda, hearing the words one by one, forgot them one by one. "Oh, read it again, what does it say?" she called out over the silence that pressed upon her, reaching towards the dancing words that just escaped as she almost touched them. "That will do," said Dr. Hildesheim, calmly authoritarian. "Where is that bed?"

"There is no bed yet," said Miss Tanner, as if she said, We are short of oranges. Dr. Hildesheim said, "Well, we'll manage something," and Miss Tanner drew the narrow trestle with bright crossed metal supports and small rubbery wheels into a deep jut of the corridor, out of the way of the swift white figures darting about, whirling and skimming like water flies all in silence. The white walls rose sheer as cliffs, a dozen frosted moons followed each other in perfect self-possession down a white lane and dropped mutely one by one into a snowy abyss.

What is this whiteness and silence but the absence of pain? Miranda lay lifting the nap of her white blanket softly between eased fingers, watching a dance of tall deliberate shadows moving behind a wide screen of sheets spread upon a frame. It was there, near her, on her side of the wall where she could see it clearly and enjoy it, and it was so beautiful she had no curiosity as to its meaning. Two dark figures nodded, bent, curtsied to each other, retreated and bowed again, lifted long arms and spread great hands against the white shadow of the screen; then with a single round movement, the sheets were folded back, disclosing two speechless men in white, standing, and another speechless man in white, lying on the bare springs of a white iron bed. The man on the springs was swathed smoothly from head to foot in white, with folded bands across the face, and a large stiff bow like merry rabbit ears dangled at the crown of his head.

The two living men lifted a mattress standing hunched against the wall, spread it tenderly and exactly over the dead man. Wordless and white they vanished down the corridor, pushing the wheeled bed before them. It had been an entrancing and leisurely spectacle, but now it was over. A pallid white fog rose in their wake insinuatingly and floated before Miranda's eyes, a fog in which was concealed all terror and all weariness, all the wrung faces and twisted backs and broken feet of abused, outraged living things, all the shapes of their confused pain and their estranged hearts; the fog might part at any moment and loose the horde of human torments. She put up her hands and said, Not yet, not yet, but it was too late. The fog parted and two executioners, white clad, moved towards her pushing between them with marvelously deft and practiced hands the misshapen figure of an old man in filthy rags whose scanty beard waggled under his opened mouth as he bowed his back and braced his feet to resist and delay the fate they had prepared for him. In a high weeping voice he was trying to explain to them that the crime of which he was accused did not merit the punishment he was about to receive; and except for this whining cry there was silence as they advanced. The soiled cracked bowls of the old man's hands were held before him beseechingly as a beggar's as he said, "Before God I am not guilty," but they held his arms and drew him onward, passed, and were gone.

The road to death is a long march beset with all evils, and the heart fails little by little at each new terror, the bones rebel at each step, the mind sets up its own bitter resistance and to what end? The barriers sink one by one, and no covering of the eyes shuts out the landscape of disaster, nor the sight of crimes committed there. Across the field came Dr. Hildesheim, his face a skull beneath his German helmet, carrying a naked infant writhing on the point of his bayonet, and a huge stone pot marked Poison in Gothic letters. He stopped before the well that Miranda remembered in a pasture on her father's farm, a well once dry but now bubbling with living water, and into its pure depths he threw the child and the poison, and the violated water sank back soundlessly into the earth. Mirinda, screaming, ran with her arms above her head; her voice echoed and came back to her like a wolf's howl, Hildesheim is a Boche, a spy, a Hun, kill him, kill him before he kills you. . . . She woke howling, she heard the foul words accusing Dr. Hildesheim tumbling from her mouth; opened her eyes and knew she was in a bed in a small white room, with Dr. Hildesheim sitting beside her, two firm fingers on her pulse. His hair was brushed sleekly and his buttonhole flower was fresh. Stars gleamed through the window, and Dr. Hildesheim seemed to be gazing at them with no particular expression, his stethoscope dangling around his neck. Miss Tanner stood at the foot of the bed writing something on a chart.

"Hello," said Dr. Hildesheim, "at least you take it out in shouting. You don't try to get out of bed and go running around." Miranda held her eyes open with a terrible effort, saw his rather heavy, patient face clearly even as her mind tottered and slithed again, broke from its foundation and spun like a cast wheel in a ditch. "I didn't mean it, I never believed it, Dr. Hildesheim, you mustn't remember it—" and was gone again, not being able to wait for an answer.

The wrong she had done followed her and haunted her dream: this wrong took vague shapes of horror she could not recognize or name, though her heart cringed at sight of them. Her mind, split in two, acknowledged and denied what she saw in the one instant, for across an abyss of complaining darkness her reasoning coherent self watched the strange frenzy of the other coldly, reluctant to admit the truth of its visions, its tenacious remorses and despairs.

"I know those are your hands," she told Miss Tanner, "I know it, but to me they are white tarantulas, don't touch me."

"Shut your eyes," said Miss Tanner.

"Oh, no," said Miranda, "for then I see worse things," but her eyes closed in spite of her will, and the midnight of her internal torment closed about her.

Oblivion, thought Miranda, her mind feeling among her memories of words she had been taught to describe the unseen, the unknowable, is a whirlpool of gray water turning upon itself for all eternity . . . eternity is perhaps more than the distance to the farthest star. She lay on a narrow ledge over a pit that she knew to be bottomless, though she could not comprehend it; the ledge was her childhood dream of danger, and she strained back against a reassuring wall of granite at her shoulders, staring into the pit, thinking, There it is, there it is at last, it is very simple; and soft carefully shaped words like oblivion and eternity are curtains hung before nothing at all. I shall not know when it happens, I shall not feel or remember, why can't I consent now, I am lost, there is no hope for me. Look, she told herself, there it is, that is death and there is nothing to fear. But she could not consent, still shrinking stiffly against the granite wall that was her childhood dream of safety, breathing slowly for fear of squandering breath, saying desperately, Look, don't be afraid, it is nothing, it is only eternity.

Granite walls, whirlpools, stars, are things. None of them is death, nor the image of it. Death is death, said Miranda, and for the dead it has no attributes. Silenced, she sank easily through deeps under deeps of darkness until she lay like a stone at the farthest bottom of life, knowing herself to be blind, deaf, speechless, no longer aware of the members of her own body, entirely withdrawn from all human concerns, yet alive with a peculiar lucidity and coherence; all notions of the mind, the reasonable inquiries of doubt, all ties of blood and the desires of the heart, dissolved and fell away from her, and there remained of her only a minute fiercely burning particle of being that knew itself alone, that relied upon nothing beyond itself for its strength; not susceptible to any appeal or inducement, being itself composed entirely of one single motive, the stubborn will to live. This fiery motionless particle set itself unaided to resist destruction, to survive and to be in its own madness of being, motiveless and planless beyond that one essential end. Trust me, the hard unwinking angry point of light said. Trust me. I stay.

At once it grew, flattened, thinned to a fine radiance, spread like a great fan and curved out into a rainbow through which Miranda, enchanted, altogether believing, looked upon a deep clear landscape of sea and sand, of soft meadow and sky, freshly washed and glistening with transparencies of blue. Why, of course, of course, said Miranda, without surprise but with serene rapture as if some promise made to her had been kept long after she had ceased to hope for it. She rose from her narrow ledge and ran lightly through the tall portals of the great bow that arched in its splendor over the burning blue of the sea and the cool green of the meadow on either hand.

The small waves rolled in and over unhurriedly, lapped upon the sand in silence and retreated; the grasses flurried before a breeze that made no sound. Moving towards her leisurely as clouds through the shimmering air came a great company of human beings, and Miranda saw in an amazement of joy that they were all the living she had known. Their faces were transfigured, each in its own beauty, beyond what she remembered of them, their eyes were clear and untroubled as good weather, and they cast no shadows. They were pure identities and she knew them every one without calling their names or remembering what relation she bore to them. They surrounded her smoothly on silent feet, then turned their entranced faces again towards the sea, and she moved among them easily as a wave among waves. The drifting circle widened, separated, and each figure was alone but not solitary; Miranda, alone too, questioning nothing, desiring nothing, in the quietude of her ecstasy, stayed where she was, eyes fixed on the overwhelming deep sky where it was always morning.

Lying at ease, arms under her head, in the prodigal warmth which flowed evenly from sea and sky and meadow, within touch but not touching the serenely smiling familiar beings about her, Miranda felt without warning a vague tremor of apprehension, some small flick of distrust in her joy; a thin frost touched the edges of this confident tranquility; something, somebody, was missing, she had lost something, she had left something valuable in another country, oh, what could it be? There are no trees, no trees here, she said in fright, I have left something unfinished. A thought struggled at the back of her mind, came clearly as a voice in her ear. Where are the dead? We have forgotten the dead, oh, the dead, where are they? At once as if a curtain had fallen, the bright landscape faded, she was alone in a strange stony place of bitter cold, picking her way along a steep path of slippery snow, calling out, Oh, I must go back! But in what direction? Pain returned, a terrible compelling pain running through her veins like a heavy fire, the stench of corruption filled her nostrils, the sweetish sickening smell of rotting flesh and pus; she opened her eyes and saw pale light through a coarse white cloth over her face, knew that the smell of death was in her own body, and struggled to lift her hand. The cloth was drawn away; she saw Miss Tanner filling a hypodermic needle in her methodical expert way, and heard Dr. Hildesheim saying, "I think that will do the trick. Try another." Miss Tanner plucked firmly at Miranda's arm near the shoulder, and the unbelievable current of agony ran burning through her veins again. She struggled to cry out, saying, Let me go, let me go; but heard only incoherent sounds of animal suffering. She saw doctor and nurse glance at each other with the glance of initiates at a mystery, nodding in silence, their eyes alive with knowledgeable pride. They looked briefly at their handiwork and hurried away.

Bells screamed all off key, wrangling together as they collided in midair, horns and whistles mingled shrilly with cries of human distress; sulphur-colored light exploded through the black windowpane and flashed away in darkness. Miranda waking from a dreamless sleep asked without expecting an answer, "What is happening?" for there was a bustle of voices and footsteps in the corridor, and a sharpness in the air; the far clamor went on, a furious exasperated shrieking like a mob in revolt.

The light came on, and Miss Tanner said in a furry voice, "Hear that? They're celebrating. It's the Armistice. The war is over, my dear." Her hands trembled. She rattled a spoon in a cup, stopped to listen, held the cup out to Miranda. From the ward for old bedridden women down the hall floated a ragged chorus of cracked voices singing, "My country, 'tis of thee . . ."

Sweet land . . . oh, terrible land of this bitter world where the sound of rejoicing was a clamor of pain, where ragged tuneless old women, sitting up waiting for their evening bowl of cocoa, were singing, "Sweet land of Liberty—"

"Oh, say, can you see?" their hopeless voices were asking next, the hammer strokes of metal tongues drowning them out. "The war is over," said Miss Tanner, her underlip held firmly, her eyes blurred. Miranda said, "Please open the window, please, I smell death in here."

Now if real daylight such as I remember having seen in this world would only come again, but it is always twilight or just before morning, a promise of day that is never kept. What has become of the sun? That was the longest and loneliest night and yet it will not end and let the day come. Shall I ever see light again?

Sitting in a long chair, near a window, it was in itself a melancholy wonder to see the colorless sunlight slanting on the snow, under a sky drained of its blue. "Can this be my face?" Miranda asked her mirror. "Are these my own hands?" she asked Miss Tanner, holding them up to show the yellow tint like melted wax glimmering between the closed fingers. The body is a curious monster, no place to live in, how could anyone feel at home there? Is it possible I can ever accustom myself to this place? she asked herself. The human faces around her seemed dulled and tired, with no radiance of skin and eyes as Miranda remembered radiance; the once white walls of her room were now a soiled gray. Breathing slowly, falling asleep and waking again, feeling the splash of water on her flesh, taking food, talking in bare phrases with Dr. Hildesheim and Miss Tanner, Miranda looked about her with the covertly hostile eyes of an alien who does not like the country in which he finds himself, does not understand the language nor wish to learn it, does not mean to live there and yet is helpless, unable to leave it at his will.

"It is morning," Miss Tanner would say, with a sigh, for she had grown old and weary once for all in the past month, "morning again, my dear," showing Miranda the same monotonous landscape of dulled evergreens and leaden snow. She would rustle about in her starched skirts, her face bravely powdered, her spirit unbreakable as good steel, saying, "Look, my dear, what a heavenly morning, like a crystal," for she had an affection for the salvaged creature before her, the silent ungrateful human being whom she, Cornelia Tanner, a nurse who knew her business, had snatched back from death with her own hands. "Nursing is nine-tenths, just the same," Miss Tanner would tell the other nurses; "keep that in mind." Even the sunshine was Miss Tanner's own prescription for the further recovery of Miranda, this patient the doctors had given up for lost, and who yet sat here, visible proof of Miss Tanner's theory. She said, "Look at the sunshine, now," as she might be saying, "I ordered this for you, my dear, do sit up and take it."

"It's beautiful," Miranda would answer, even turning her head to look, thanking Miss Tanner for her goodness, most of all her goodness about the weather, "beautiful, I always loved it." And I might love it again if I saw it, she thought, but truth was, she could not see it. There was no light, there might never be light again, compared as it must always be with the light she had seen beside the blue sea that lay so tranquilly along the shore of her paradise. That was a child's dream of the heavenly meadow, the vision of repose that comes to a tired body in sleep, she thought, but I have seen it when I did not know it was a dream. Closing her eyes she would rest for a moment remembering that bliss which had repaid all the pain of the journey to reach it; opening them again she saw with a new anguish the dull world to which she was condemned, where the light seemed filmed over with cobwebs, all the bright surfaces corroded, the sharp planes melted and formless, all objects and beings meaningless, ah, dead and withered things that believed themselves alive!

At night, after the long effort of lying in her chair, in her extremity of grief for what she had so briefly won, she folded her painful body together and wept silently, shamelessly, in pity for herself and her lost rapture. There was no escape. Dr. Hildesheim, Miss Tanner, the nurses in the diet kitchen, the chemist, the surgeon, the precise machine of the hospital, the whole humane conviction and custom of society, conspired to pull her inseparable rack of bones and wasted flesh to its feet, to put in order her disordered mind, and to set her once more safely in the road that would lead her again to death.

Chuck Rouncivale and Mary Townsend came to see her, bringing her a bundle of letters they had guarded for her. They brought a basket of delicate small hothouse flowers, lilies of the valley with sweet peas and feathery fern, and above these blooms their faces were merry and haggard.

Mary said, "You *have* had a tussle, haven't you?" and Chuck said, "Well, you made it back, didn't you?" Then after an uneasy pause, they told her that everybody was waiting to see her again at her desk. "They've put me back on sports already, Miranda," said Chuck. For ten minutes Miranda smiled and told them how gay and what a pleasant surprise it was to find herself alive. For it will not do to betray the conspiracy and tamper with the courage of the living; there is nothing better than to be alive, everyone has agreed on that; it is past argument, and who attempts to deny it is justly outlawed. "I'll be back in no time at all," she said; "this is almost over."

Her letters lay in a heap in her lap and beside her chair. Now and then she turned one over to read the inscription, recognizing this handwriting or that, examined the blotted stamps and postmarks, and let them drop again. For two or three days they lay upon the table beside her, and she continued to shrink from them. "They will all be telling me again how good it is to be alive, they will say again they love me, they are glad I am living too, and what can I answer to that?" and her hardened, indifferent heart shuddered in despair at itself, because before it had been tender and capable of love.

Dr. Hildesheim said, "What, all these letters not opened yet?" and Miss Tanner said, "Read your letters, my dear, I'll open them for you." Standing beside the bed, she slit them cleanly with a paper knife. Miranda, cornered, picked and chose until she found a thin one in an unfamiliar handwriting. "Oh, no, now," said Miss Tanner, "take them as they come. Here, I'll hand them to you." She sat down, prepared to be helpful to the end.

What a victory, what triumph, what happiness to be alive, sang the letters in a chorus. The names were signed with flourishes like the circles in air of bugle notes, and they were the names of those she had loved best; some of those she had known well and pleasantly; and a few who meant nothing to her, then or now. The thin letter in the unfamiliar handwriting was from a strange man at the camp where Adam had been, telling her that Adam had died of influenza in the camp hospital. Adam had asked him, in case anything happened, to be sure to let her know.

If anything happened. To be sure to let her know. If anything happened. "Your friend, Adam Barclay," wrote the strange man. It had happened—she looked at the date—more than a month ago.

"I've been here a long time, haven't I?" she asked Miss Tanner, who was folding letters and putting them back in their proper envelopes.

"Oh, quite a while," said Miss Tanner, "but you'll be ready to go soon now. But you must be careful of yourself and not overdo, and you should come back now and then and let us look at you, because sometimes the aftereffects are very—"

Miranda, sitting up before the mirror, wrote carefully: "One lipstick, medium, one ounce flask Bois d'Hiver perfume, one pair of gray suede gauntlets without straps, two pair gray sheer stockings without clocks—"

Towney, reading after her, said, "Everything without something so that it will be almost impossible to get?"

"Try it, though," said Miranda, "they're nicer without. One walking stick of silvery wood with a silver knob."

"That's going to be expensive," warned Towney. "Walking is hardly worth it."

"You're right," said Miranda, and wrote in the margin, "a nice one to match my other things. Ask Chuck to look for this, Mary. Good-looking and not too heavy." Lazarus, come forth. Not unless you bring me my top hat and stick. Stay where you are then, you snob. Not at all, I'm coming forth. "A jar of cold cream," wrote Miranda, "a box of apricot powder—and, Mary, I don't need eye shadow, do I?" She glanced at her face in the mirror and away again. "Still, no one need pity this corpse if we look properly to the art of the thing."

Mary Townsend said, "You won't recognize yourself in a week."

"Do you suppose, Mary," asked Miranda, "I could have my old room back again?"

"That should be easy," said Mary. "We stored away all your things there with Miss Hobbe." Miranda wondered again at the time and trouble the living took to be helpful to the dead. But not quite dead now, she reassured herself, one foot in either world now; soon I shall cross back and be at home again. The light will seem real and I shall be glad when I hear that someone I know has escaped from death. I shall visit the escaped ones and help them dress and tell them how lucky they are, and how lucky I am to still have them. Mary will be back soon with my gloves and my walking stick, I must go now, I must begin saying good-by to Miss Tanner and Dr. Hildesheim. Adam, she said, now you need not die again, but still I wish you were here; I wish you had come back, what do you think I came back for, Adam, to be deceived like this?

At once he was there beside her, invisible but urgently present, a ghost but more alive than she was, the last intolerable cheat of her heart; for knowing it was false she still clung to the lie, the unpardonable lie of her bitter desire. She said, "I love you," and stood up trembling, trying by the mere act of her will to bring him to sight before her. If I could call you up from the grave I would, she said, if I could see your ghost I would say, I believe. . . . "I believe," she said aloud. "Oh, let me see you once more." The room was silent, empty, the shade was gone from it, struck away by the sudden violence of her rising and speaking aloud. She came to herself as if out of sleep. Oh, no, that is not the way, I must never do that, she warned herself. Miss Tanner said, "Your taxicab is waiting, my dear," and there was Mary. Ready to go.

No more war, no more plague, only the dazed silence that follows the ceasing of the heavy guns; noiseless houses with the shades drawn, empty streets, the dead cold light of tomorrow. Now there would be time for everything.

JOSEPH CONRAD

Heart of Darkness, *by common consent, is one of the very great short novels in the English language. Yet the author, a Pole by birth, did not begin to learn his adopted tongue until he was twenty years old. In* Heart of Darkness *Conrad so truthfully unveiled the conscience and consciousness of man that the recognition can be painful. The thesis of this short novel is that complete self-knowledge—including knowledge of human capacity for evil—is a concomitant of the capacity for good.*

Kurtz's final gasp, "The horror! The horror!" can be taken as a haunting indictment of civilized man. The rapaciousness, indeed, the barbaric savagery covered in the civilized breast by a thin veneer is exposed. A more stunning portrait of moral deterioration, in a story of comparable length, does not exist in English literature. Kurtz is testimony to all the high-sounding hypocritical pretensions with which men may delude themselves. The Congo is a heart of darkness, suggests Conrad, but what of the human spirit? It is with these unspeakable potentialities of man's soul that the author is concerned.

To place an intermediary between himself and the reader, Conrad utilizes the raconteur Marlow, of whom Conrad also makes use in several other pieces. What the total experience in the Congo meant to Marlow is at the center of the story. Marlow's quest for knowledge of Kurtz is a journey after self-knowledge. To fathom Kurtz is, for Marlow, to see into the primitive nature of his own spirit. Through prolonged self-examination of his deepest inner recesses, Marlow, too, is brought to the edge of darkness. His moral education is attained, however, and he is able to assimilate this new knowledge and to live, a sadder man. The reader will want to answer for himself the question of Marlow's lie to Kurtz's fiancée.

Conrad's writing resists any easy classification. He once said, "I have been called a writer of the sea, of the tropics, a descriptive writer, a romantic writer—and also an idealist, but as a matter of fact all my concern has been with the 'ideal' value of things, events, and people. That and nothing else." His work constitutes a valid statement of his faithful integrity and of the high value he placed on man. Moral honesty and moral corruption are the contending forces that animate his work. The grandeur of his accomplishment is suggested by the several levels on which Heart of Darkness *may be read: as an exciting narration of Marlow's trip into the Congo and of his adventures there; as a psychological-philosophic revelation of how Marlow matured through his recognition of the basic evil in man; and as a historical view of the process by which imperialism corrupts man's ethical values.*

Heart of Darkness

I

THE NELLIE, a cruising yawl, swung to her anchor without a flutter of the sail, and was at rest. The flood had made, the wind was nearly calm, and being bound down the river, the only thing for it was to come to and wait for the turn of the tide.

The sea-reach of the Thames stretched before us like the beginning of an interminable waterway. In the offing the sea and the sky were welded together without a joint, and in the luminous space the tanned sails of the barges drifted up with the tide seemed to stand still in red clusters of canvas sharply peaked, with gleams of varnished sprits. A haze rested on the low shores that ran out to sea in vanishing flatness. The air was dark above Gravesend, and farther back still seemed condensed into a mournful gloom, brooding motionless over the biggest, and the greatest, town on earth.

The Director of Companies was our captain and our host. We four affectionately watched his back as he stood in the bows looking to seaward. On the whole river there was nothing that looked half so nautical. He resembled a pilot, which to a seaman is trustworthiness personified. It was difficult to realize his work was not out there in the luminous estuary, but behind him, within the brooding gloom.

Between us there was, as I have already said somewhere, the bond of the sea. Besides holding our hearts together through long periods of separation, it had the effect of making us tolerant of each other's yarns —and even convictions. The Lawyer—the best of old fellows—had, because of his many years and many virtues, the only cushion on deck, and was lying on the only rug. The Accountant had brought out already a box of dominoes, and was toying architecturally with the bones. Marlow sat cross-legged right aft, leaning against the mizzen-mast. He had sunken cheeks, a yellow complexion, a straight back, an ascetic aspect, and, with his arms dropped, the palms of hands outwards, resembled an idol. The director, satisfied the anchor had good hold, made his way aft and sat down amongst us. We exchanged a few words lazily. Afterwards there was silence on board the yacht. For some reason or other we did not begin that game of dominoes. We felt meditative, and fit for nothing but placid staring. The day was ending in a serenity of still and exquisite brilliance. The water shone pacifically; the sky, without a speck, was a benign immensity of unstained light; the very mist on the Essex marsh was like a gauzy and radiant fabric, hung from the wooded rises inland, and draping the low shores in diaphanous folds. Only the gloom to the west, brooding over the upper reaches, became more sombre every minute, as if angered by the approach of the sun.

And at last, in its curved and imperceptible fall, the sun sank low, and from glowing white changed to a dull red without rays and without heat, as if about to go out suddenly, stricken to death by the touch of that gloom brooding over a crowd of men.

Forthwith a change came over the waters, and the serenity became less brilliant but more profound. The old river in its broad reach rested unruffled at the decline of day, after ages of good service done to the race that peopled its banks, spread out in the tranquil dignity of a waterway leading to the uttermost ends of the earth. We looked at the venerable stream not in the vivid flush of a short day that comes and departs for ever, but in the august light of abiding memories. And indeed nothing is easier for a man who has, as the phrase goes, "followed the sea" with reverence and affection, than to evoke the great spirit of the past upon the lower reaches of the Thames. The tidal current runs to and fro in its unceasing service, crowded with memories of men and ships it had borne to the rest of home or to the battles of the sea. It had known and served all the men of whom the nation is proud, from Sir Francis Drake to Sir John Franklin, knights all, titled and untitled—the great knights-errant of the sea. It had borne all the ships whose names are like jewels flashing in the night of time, from the *Golden Hind* returning with her round flanks full of treasure, to be visited by the Queen's Highness and thus pass out of the gigantic tale, to the *Erebus* and *Terror*, bound on other conquests—and that never returned. It had known the ships and the men. They had sailed from Deptford, from Greenwich, from Erith—the adventurers and the settlers; kings' ships and the ships of men on 'Change; captains, admirals, the dark "interlopers" of the Eastern trade, and the commissioned "generals" of East India fleets. Hunters for gold or pursuers of fame, they all had gone out on that stream, bearing the sword, and often the torch, messengers of the might within the land, bearers of a spark from the sacred fire. What greatness had not floated on the ebb of that river into the mystery of an unknown earth! . . . The dreams of men, the seed of commonwealths, the germs of empires.

The sun set; the dusk fell on the stream, and lights began to appear along the shore. The Chapman lighthouse, a three-legged thing erect on a mud-flat, shone strongly. Lights of ships moved in the fairway—a great stir of lights going up and going down. And farther west on the upper reaches the place of the monstrous town was still marked ominously on the sky, a brooding gloom in sunshine, a lurid glare under the stars.

"And this also," said Marlow suddenly, "has been one of the dark places of the earth."

He was the only man of us who still "followed the sea." The worst that could be said of him was that he did not represent his class. He was a seaman, but he was a wanderer, too, while most seamen lead, if one may so express it, a sedentary life. Their minds are of the stay-at-home order, and their home is always with them—the ship; and so is their country—the sea. One ship is very much like another, and the sea is always the same. In the immutability of their surroundings the foreign shores, the foreign faces, the changing immensity of life, glide past, veiled not by a sense of mystery but by a slightly disdainful ignorance; for there is nothing mysterious to a seaman unless it be the sea itself, which is the mistress of his existence and as inscrutable as

Destiny. For the rest, after his hours of work, a casual stroll or a casual spree on shore suffices to unfold for him the secret of a whole continent, and generally he finds the secret not worth knowing. The yarns of seamen have a direct simplicity, the whole meaning of which lies within the shell of a cracked nut. But Marlow was not typical (if his propensity to spin yarns be excepted), and to him the meaning of an episode was not inside like a kernel but outside, enveloping the tale which brought it out only as a glow brings out a haze, in the likeness of one of these misty halos that sometimes are made visible by the spectral illumination of moonshine.

His remark did not seem at all surprising. It was just like Marlow. It was acepted in silence. No one took the trouble to grunt even; and presently he said, very slow—

"I was thinking of very old times, when the Romans first came here, nineteen hundred years ago—the other day.... Light came out of this river since—you say Knights? Yes; but it is like a running blaze on a plain, like a flash of lightning in the clouds. We live in the flicker —may it last as long as the old earth keeps rolling! But darkness was here yesterday. Imagine the feelings of a commander of a fine—what d'ye call 'em?—trireme in the Mediterranean, ordered suddenly to the north; run overland across the Gauls in a hurry; put in charge of one of these craft the legionaries—a wonderful lot of handy men they must have been, too—used to build, apparently by the hundred, in a month or two, if we may believe what we read. Imagine him here—the very end of the world, a sea the colour of lead, a sky the colour of smoke, a kind of ship about as rigid as a concertina—and going up this river with stores, or orders, or what you like. Sand-banks, marshes, forests, savages,—precious little to eat fit for a civilized man, nothing but Thames water to drink. No Falernian wine here, no going ashore. Here and there a military camp lost in a wilderness, like a needle in a bundle of hay—cold, fog, tempests, disease, exile, and death—death skulking in the air, in the water, in the bush. They must have been dying like flies here. Oh, yes—he did it. Did it very well, too, no doubt, and without thinking much about it either, excepts afterwards to brag of what he had gone through in his time, perhaps. They were men enough to face the darkness. And perhaps he was cheered by keeping his eye on a chance of promotion to the fleet at Ravenna by and by, if he had good friends in Rome and survived the awful climate. Or think of a decent young citizen in a toga—perhaps too much dice, you know—

coming out here in the train of some prefect, or tax-gatherer, or trader even, to mend his fortunes. Land in a swamp, march through the woods, and in some inland post feel the savagery, the utter savagery, had closed round him—all that mysterous life of the wilderness that stirs in the forest, in the jungles, in the hearts of wild men. There's no initiation either into such mysteries. He has to live in the midst of the incomprehensible, which is also detestable. And it has a fascination, too, that goes to work upon him. The fascination of the abomination—you know, imagine the growing regrets, the longing to escape, the powerless disgust, the surrender, the hate."

He paused.

"Mind," he began again, lifting one arm from the elbow, the palm of the hand outwards, so that, with his legs folded before him, he had the pose of a Buddha preaching in European clothes and without a lotus-flower—"Mind, none of us would feel exactly like this. What saves us is efficiency—the devotion to efficiency. But these chaps were not much account, really. They were no colonists; their administration was merely a squeeze, and nothing more, I suspect. They were conquerors, and for that you want only brute force—nothing to boast of, when you have it, since your strength is just an accident arising from the weakness of others. They grabbed what they could get for the sake of what was to be got. It was just robbery with violence, aggravated murder on a great scale, and men going at it blind—as is very proper for those who tackle a darkness. The conquest of the earth, which mostly means the taking it away from those who have a different complexion or slightly flatter noses than ourselves, is not a pretty thing when you look into it too much. What redeems it is the idea only. An idea at the back of it; not a sentimental pretence but an idea; and an unselfish belief in the idea—something you can set up, and bow down before, and offer a sacrifice to. . . ."

He broke off. Flames glided in the river, small green flames, red flames, white flames, pursuing, overtaking, joining, crossing each other—then separating slowly or hastily. The traffic of the great city went on in the deepening night upon the sleepless river. We looked on, waiting patiently—there was nothing else to do till the end of the flood; but it was only after a long silence, when he said, in a hesitating voice, "I suppose you fellows remember I did once turn fresh-water sailor for a bit," that we knew we were fated, before the ebb began to run, to hear about one of Marlow's inconclusive experiences.

"I don't want to bother you much with what happened to me personally," he began, showing in this remark the weakness of many tellers of tales who seem so often unaware of what their audience would best like to hear; "yet to understand the effect of it on me you ought to know how I got out there, what I saw, how I went up that river to the place where I first met the poor chap. It was the farthest point of navigation and the culminating point of my experience. It seemed somehow to throw a kind of light on everything about me—and into my thoughts. It was sombre enough, too—and pitiful—not extraordinary in any way —not very clear either. No, not very clear. And yet it seemed to throw a kind of light.

"I had then, as you remember, just returned to London after a lot of Indian Ocean, Pacific, China Seas—a regular dose of the East—six years or so, and I was loafing about, hindering you fellows in your work and invading your homes, just as though I had got a heavenly mission to civilize you. It was very fine for a time, but after a bit I did get tired of resting. Then I began to look for a ship—I should think the hardest work on earth. But the ships wouldn't even look at me. And I got tired of that game, too.

"Now when I was a little chap I had a passion for maps. I would look for hours at South America, or Africa, or Australia, and lose myself in all the glories of exploration. At that time there were many blank spaces on the earth, and when I saw one that looked particularly inviting on a map (but they all look that) I would put my finger on it and say, 'When I grow up I will go there.' The North Pole was one of these places, I remember. Well, I haven't been there yet, and shall not try now. The glamour's off. Other places were scattered about the Equator, and in every sort of latitude all over the two hemispheres. I have been in some of them, and . . . well, we won't talk about that. But there was one yet—the biggest, the most blank, so to speak—that I had a hankering after.

"True, by this time it was not a blank space any more. It had got filled since my boyhood with rivers and lakes and names. It had ceased to be a blank space of delightful mystery—a white patch for a boy to dream gloriously over. It had become a place of darkness. But there was in it one river especially, a mighty big river, that you could see on the map, resembling an immense snake uncoiled, with its head in the sea, its body at rest curving afar over a vast country, and its tail lost in the depths of the land. And as I looked at the map of it in a shop-window, it fascinated me as a snake would a bird—a silly little bird. Then I remembered there was a big concern, a Company for trade on that river. Dash it all! I thought to myself, they can't trade without using some kind of craft on that lot of fresh water—steamboats! Why shouldn't I try to get charge of one? I went on along Fleet Street, but could not shake off the idea. The snake had charmed me.

"You understand it was a Continental concern, that Trading society; but I have a lot of relations living on the Continent, because it's cheap and not so nasty as it looks, they say.

"I am sorry to own I began to worry them. This was already a fresh departure for me. I was not used to get things that way, you know. I always went my own road and on my own legs where I had a mind to go. I wouldn't have believed it of myself; but, then—you see—I felt somehow I must get there by hook or by crook. So I worried them. The men said 'My dear fellow,' and did nothing. Then—would you believe it?—I tried the women. I, Charlie Marlow, set the women to work—to get a job. Heavens! Well, you see, the notion drove me. I had an aunt, a dear enthusiastic soul. She wrote: 'It will be delightful. I am ready to do anything, anything for you. It is a glorious idea. I know the wife of a very high personage in the Administration, and also a man who has lots of influence with,' etc., etc. She was determined to make no end of fuss to get me appointed skipper of a river steamboat, if such was my fancy.

"I got my appointment—of course; and I got it very quick. It appears the Company had received news that one of their captains had been killed in a scuffle with the natives. This was my chance, and it made me the more anxious to go. It was only months and months afterwards, when I made the attempt to recover what was left of the body, that I heard the original quarrel arose from a misunderstanding about some hens. Yes, two black hens. Fresleven—that was the fellow's name, a Dane—thought himself wronged somehow in the bargain, so he went ashore and started to hammer the chief of the village with a stick. Oh, it didn't surprise me in the least to hear this, and at the same time to be told that Fresleven was the gentlest, quietest creature that ever walked on two legs. No doubt he was; but he had been a couple of years already out there engaged in the noble cause, you know, and he probably felt the need at last of asserting his self-respect in some way. Therefore he whacked the old nigger mercilessly, while a big crowd of his people watched him, thunderstruck, till some man

—I was told the chief's son—in desperation at hearing the old chap yell, made a tentative jab with a spear at the white man—and of course it went quite easy between the shoulder-blades. Then the whole population cleared into the forest, expecting all kinds of calamities to happen, while, on the other hand, the steamer Fresleven commanded left also in a bad panic, in charge of the engineer, I believe. Afterwards nobody seemed to trouble much about Fresleven's remains, till I got out and stepped into his shoes. I couldn't let it rest, though; but when an opportunity offered at last to meet my predecessor, the grass growing through his ribs was tall enough to hide his bones. They were all there. The supernatural being had not been touched after he fell. And the village was deserted, the huts gaped black, rotting, all askew within the fallen enclosures. A calamity had come to it, sure enough. The people had vanished. Mad terror had scattered them, men, women, and children, through the bush, and they had never returned. What became of the hens I don't know either. I should think the cause of progress got them, anyhow. However, through this glorious affair I got my appointment, before I had fairly begun to hope for it.

"I flew around like mad to get ready, and before forty-eight hours I was crossing the Channel to show myself to my employers, and sign the contract. In a very few hours I arrived in a city that always makes me think of a whited sepulchre. Prejudice no doubt. I had no difficulty in finding the Company's offices. It was the biggest thing in the town, and everybody I met was full of it. They were going to run an over-sea empire, and make no end of coin by trade.

"A narrow and deserted street in deep shadow, high houses, innumerable windows with venetian blinds, a dead silence, grass sprouting between the stones, imposing carriage archways right and left, immense double doors standing ponderously ajar. I slipped through one of these cracks, went up a swept and ungarnished staircase, as arid as a desert, and opened the first door I came to. Two women, one fat and the other slim, sat on straw-bottomed chairs, knitting black wool. The slim one got up and walked straight at me—still knitting with downcast eyes—and only just as I began to think of getting out of her way, as you would for a somnambulist, stood still, and looked up. Her dress was as plain as an umbrella-cover, and she turned round without a word and preceded me into a waiting-room. I gave my name, and looked about. Deal table in the middle, plain chairs all round the walls, on one end a large shining map, marked with all the colours of a rainbow. There was a vast amount of red—good to see at any time, because one knows that some real work is done in there, a deuce of a lot of blue, a little green, smears of orange, and, on the East Coast, a purple patch, to show where the jolly pioneers of progress drink the jolly lager-beer. However, I wasn't going into any of these. I was going into the yellow. Dead in the centre. And the river was there—fascinating—deadly—like a snake. Ough! A door opened, a white-haired secretarial head, but wearing a compassionate expression, appeared, and a skinny forefinger beckoned me into the sanctuary. Its light was dim, and a heavy writing-desk squatted in the middle. From behind that structure came out an impression of pale plumpness in a frock-coat. The great man himself. He was five feet six, I should judge, and had his grip on the handle-end of ever so many millions. He shook hands, I fancy, murmured vaguely, was satisfied with my French. *Bon voyage.*

"In about forty-five seconds I found myself again in the waiting-room with the compassionate secretary, who, full of desolation and sympathy, made me sign some document. I believe I undertook amongst other things not to disclose any trade secrets. Well, I am not going to.

"I began to feel slightly uneasy. You know I am not used to such ceremonies, and there was something ominous in the atmosphere. It was just as though I had been let into some conspiracy—I don't know—something not quite right; and I was glad to get out. In the outer room the two women knitted black wool feverishly. People were arriving, and the younger one was walking back and forth introducing them. The old one sat on her chair. Her flat cloth slippers were propped up on a foot-warmer, and a cat reposed on her lap. She wore a starched white affair on her head, had a wart on one cheek, and silver-rimmed spectacles hung on the tip of her nose. She glanced at me above the glasses. The swift and indifferent placidity of that look troubled me. Two youths with foolish and cheery countenances were being piloted over, and she threw at them the same quick glance of unconcerned wisdom. She seemed to know all about them and about me, too. An eerie feeling came over me. She seemed uncanny and fateful. Often far away there I thought of these two, guarding the door of Darkness, knitting black wool as for a warm pall, one introducing, introducing continuously to the unknown, the other scrutinizing the cheery and foolish faces with unconcerned old eyes. *Ave!* Old knitter of black wool. *Morituri te salutant.* Not many of those she looked at ever saw her again—not half, by a long way.

"There was yet a visit to the doctor. 'A simple formality,' assured me the secretary, with an air of taking an immense part in all my sorrows. Accordingly a young chap wearing his hat over the left eyebrow, some clerk I suppose—there must have been clerks in the business, though the house was as still as a house in a city of the dead—came from somewhere up-stairs, and led me forth. He was shabby and careless, with inkstains on the sleeves of his jacket, and his cravat was large and billowy, under a chin shaped like the toe of an old boot. It was a little too early for the doctor, so I proposed a drink, and thereupon he developed a vein of joviality. As we sat over our vermouths he glorified the Company's business, and by and by I expressed casually my surprise at him not going out there. He became very cool and collected all at once. 'I am not such a fool as I look, quoth Plato to his disciples,' he said sententiously, emptied his glass with great resolution, and we rose.

"The old doctor felt my pulse, evidently thinking of something else the while. 'Good, good for there,' he mumbled, and then with a certain eagerness asked me whether I would let him measure my head. Rather surprised, I said Yes, when he produced a thing like calipers and got the dimensions back and front and every way, taking notes carefully. He was an unshaven little man in a threadbare coat like a gaberdine, with his feet in slippers, and I thought him a harmless fool. 'I always ask leave, in the interests of science, to measure the crania of those going out there,' he said. 'And when they come back, too?' I asked. 'Oh, I never see them,' he remarked; 'and, moreover, the changes take place inside, you know.' He smiled, as if at some quiet joke. 'So you are going out there. Famous. Interesting, too.' He gave me a searching glance, and made another note. 'Ever any madness in your family?' he asked, in a matter-of-fact tone. I felt very annoyed. 'Is that question in the interests of science, too?' 'It would be,' he said, without taking notice of my irritation, 'interesting for science to watch the mental changes of individuals, on the spot, but . . .' 'Are you an alienist?' I interrupted. 'Every doctor should be—a little,' answered that original, imperturbably. 'I have a little theory which you messieurs who go out there must help me to prove. This is my share in the advantages my country shall reap from the possession of such a magnificent dependency. The mere wealth I leave to others. Pardon my questions, but you are the first Englishman coming under my observation . . .' I hastened to assure him I was not in the least typical. 'If I were,' said I, 'I wouldn't be talking like this with you.' 'What you say is rather profound, and probably erroneous,' he said, with a laugh. 'Avoid irritation more than exposure to the sun. Adieu. How do you English say, eh? Good-bye. Ah! Good-bye. Adieu. In the tropics one must before everything keep calm.' . . . He lifted a warning forefinger. . . . '*Du calme, du calme. Adieu.*'

"One thing more remained to do—say good-bye to my excellent aunt. I found her triumphant. I had a cup of tea—the last decent cup of tea for many days—and in a room that most soothingly looked just as you would expect a lady's drawing-room to look, we had a long quiet chat by the fireside. In the course of these confidences it became quite plain to me I had been represented to the wife of the high dignitary, and goodness knows to how many more people besides, as an exceptional and gifted creature—a piece of good fortune for the Company —a man you don't get hold of every day. Good heavens! and I was going to take charge of a two-penny-half-penny river-steamboat with a penny whistle attached! It appeared, however, I was also one of the Workers, with a capital—you know. Something like an emissary of light, something like a lower sort of apostle. There had been a lot of such rot let loose in print and talk just about that time, and the excellent woman, living right in the rush of all that humbug, got carried off her feet. She talked about 'weaning those ignorant millions from their horrid ways,' till, upon my word, she made me quite uncomfortable. I ventured to hint that the Company was run for profit.

" 'You forget, dear Charlie, that the labourer is worthy of his hire,' she said, brightly. It's queer how out of touch with truth women are. They live in a world of their own, and there has never been anything like it, and never can be. It is too beautiful altogether, and if they were to set it up it would go to pieces before the first sunset. Some confounded fact we men have been living contentedly with ever since the day of creation would start up and knock the whole thing over.

"After this I got embraced, told to wear flannel, be sure to write often, and so on—and I left. In the street—I don't know why—a queer feeling came to me that I was an impostor. Odd thing that I, who used to clear out for any part of the world at twenty-four hours' notice, with less thought than most men give to the crossing of a street, had a moment—I won't say of hesitation, but of startled pause, before this commonplace affair. The best way I can explain it to you is by saying that, for a second or two, I felt as though, instead of going to the centre of a continent, I were about to set off for the centre of the earth.

"I left in a French steamer, and she called in every blamed port they have out there, for, as far as I could see, the sole purpose of landing soldiers and custom-house officers. I watched the coast. Watching a coast as it slips by the ship is like thinking about an enigma. There it is before you—smiling, frowning, inviting, grand, mean, insipid, or savage, and always mute with an air of whispering, 'Come and find out.' This one was almost featureless, as if still in the making, with an aspect of monotonous grimness. The edge of a colossal jungle, so dark-green as to be almost black, fringed with white surf, ran straight, like a ruled line, far, far away along a blue sea whose glitter was blurred by a creeping mist. The sun was fierce, the land seemed to glisten and drip with steam. Here and there greyish-whitish specks showed up clustered inside the white surf, with a flag flying above them perhaps. Settlements some centuries old, and still no bigger than pinheads on the untouched expanse of their background. We pounded along, stopped, landed soldiers; went on, landed custom-house clerks to levy toll in what looked like a God-forsaken wilderness, with a tin shed and a flag-pole lost in it; landed more soldiers—to take care of the custom-house clerks, presumably. Some, I heard, got drowned in the surf; but whether they did or not, nobody seemed particularly to care. They were just flung out there, and on we went. Every day the coast looked the same, as though we had not moved; but we passed various places —trading places—with names like Gran' Bassam, Little Popo; names that seemed to belong to some sordid farce acted in front of a sinister back-cloth. The idleness of a passenger, my isolation amongst all these men with whom I had no point of contact, the oily and languid sea, the uniform sombreness of the coast, seemed to keep me away from the truth of things, within the toil of a mournful and senseless delusion. The voice of the surf heard now and then was a positive pleasure, like the speech of a brother. It was something natural, that had its reason, that had a meaning. Now and then a boat from the shore gave one a momentary contact with reality. It was paddled by black fellows. You could see from afar the white of their eyeballs glistening. They shouted, sang; their bodies streamed with perspiration; they had faces like grotesque masks—these chaps; but they had bone, muscle, a wild vitality, an intense energy of movement, that was as natural and true as the surf along their coast. They wanted no excuse for being there. They were a great comfort to look at. For a time I would feel I belonged still to a world of straightforward facts; but the feeling would not last long. Something would turn up to scare it away. Once, I remember, we came upon a man-of-war anchored off the coast. There wasn't even a shed there, and she was shelling the bush. It appears the French had one of their wars going on thereabouts. Her ensign dropped limp like a rag; the muzzles of the long six-inch guns stuck out all over the low hull; the greasy, slimy swell swung her up lazily and let her down, swaying her thin masts. In the empty immensity of earth, sky, and water, there she was, incomprehensible, firing into a continent. Pop, would go one of the six-inch guns; a small flame would dart and vanish, a little white smoke would disappear, a tiny projectile would give a feeble screech—and nothing happened. Nothing could happen. There was a touch of insanity in the proceeding, a sense of lugubrious drollery in the sight; and it was not dissipated by somebody on board assuring me earnestly there was a camp of natives—he called them enemies!—hidden out of sight somewhere.

"We gave her her letters (I heard the men in that lonely ship were dying of fever at the rate of three a day) and went on. We called at some more places with farcical names, where the merry dance of death and trade goes on in a still and earthy atmosphere as of an overheated catacomb; all along the formless coast bordered by dangerous surf, as if Nature herself had tried to ward off intruders; in and out of rivers, streams of death in life, whose banks were rotting into mud, whose waters, thickened into slime, invaded the contorted mangroves, that seemed to writhe at us in the extremity of an impotent despair. Nowhere did we stop long enough to get a particularized impression, but the general sense of vague and oppressive wonder grew upon me. It was like a weary pilgrimage amongst hints for nightmares.

"It was upward of thirty days before I saw the mouth of the big river. We anchored off the seat of the government. But my work would not begin till some two hundred miles farther on. So as soon as I could I made a start for a place thirty miles higher up.

"I had my passage on a little sea-going steamer. Her captain was a Swede, and knowing me for a seaman, invited me on the bridge. He was a young man, lean, fair, and morose, with lanky hair and a shuffling gait. As we left the miserable little wharf, he tossed his head contemptuously at the shore. 'Been living there?' he asked. I said, 'Yes.' 'Fine lot these government chaps—are they not?' he went on, speaking English with great precision and considerable bitterness. 'It is funny what some people will do for a few francs a month. I wonder what becomes of that kind when it goes upcountry?' I said to him I expected to see that soon. 'So-o-o!' he exclaimed. He shuffled athwart, keeping one eye ahead vigilantly. 'Don't be too sure,' he continued. 'The other day I took up a man who hanged himself on the road. He was a Swede, too.' 'Hanged himself! Why, in God's name?' I cried. He kept on looking out watchfully. 'Who knows? The sun too much for him, or the country perhaps.'

"At last we opened a reach. A rocky cliff appeared, mounds of turned-up earth by the shore, houses on a hill, others with iron roofs, amongst a waste of excavations, or hanging to the declivity. A continuous noise of the rapids above hovered over this scene of inhabited devastation. A lot of people, mostly black and naked, moved about like ants. A jetty projected into the river. A blinding sunlight drowned all this at times in a sudden recrudescence of glare. 'There's your Company's station,' said the Swede, pointing to three wooden barrack-like structures on the rocky slope. 'I will send your things up. Four boxes did you say? So. Farewell.'

"I came upon a boiler wallowing in the grass, then found a path leading up the hill. It turned aside for the boulders, and also for an undersized railway-truck lying there on its back with its wheels in the air. One was off. The thing looked as dead as the carcass of some animal. I came upon more pieces of decaying machinery, a stack of rusty rails. To the left a clump of trees made a shady spot, where dark things seemed to stir feebly. I blinked, the path was steep. A horn tooted to the right, and I saw the black people run. A heavy and dull detonation shook the ground, a puff of smoke came out of the cliff, and that was all. No change appeared on the face of the rock. They were building a railway. The cliff was not in the way or anything; but this objectless blasting was all the work going on.

"A slight clinking behind me made me turn my head. Six black men advanced in a file, toiling up the path. They walked erect and slow, balancing small baskets full of earth on their heads, and the clink kept time with their footsteps. Black rags were wound round their loins, and the short ends behind waggled to and fro like tails. I could see every rib, the joints of their limbs were like knots in a rope; each had an iron collar on his neck, and all were connected together with a chain whose bights swung between them, rhythmically clinking. Another report from the cliff made me think suddenly of that ship of war I had seen firing into a continent. It was the same kind of ominous voice; but these men could by no stretch of imagination be called enemies. They were called criminals, and the outraged law, like the bursting shells, had come to them, an insoluble mystery from the sea. All their meagre breasts panted together, the violently dilated nostrils quivered, the eyes stared stonily uphill. They passed me within six inches, without a glance, with that complete, deathlike indifference of unhappy savages. Behind this raw matter one of the reclaimed, the product of the new forces at work, strolled despondently, carrying a rifle by its middle. He had a uniform jacket with one button off, and seeing a white man on the path, hoisted his weapon to his shoulder with alacrity. This was simple prudence, white men being so much alike at a distance that he could not tell who I might be. He was speedily reassured, and with a large, white, rascally grin, and a glance at his charge, seemed to take me into partnership in his exalted trust. After all, I also was a part of the great cause of these high and just proceedings.

"Instead of going up, I turned and descended to the left. My idea was to let that chain-gang get out of sight before I climbed the hill. You know I am not particularly tender; I've had to strike and to fend off. I've had to resist and to attack sometimes—that's only one way of resisting—without counting the exact cost, according to the demands of such sort of life as I had blundered into. I've seen the devil of violence, and the devil of greed, and the devil of hot desire; but, by all the stars! these were strong, lusty, red-eyed devils, that swayed and drove men—men, I tell you. But as I stood on this hillside, I foresaw that in the blinding sunshine of that land I would become acquainted with a flabby, pretending, weak-eyed devil of a rapacious and pitiless folly. How insidious he could be, too, I was only to find out several months later and a thousand miles farther. For a moment I stood appalled, as though by a warning. Finally I descended the hill, obliquely, towards the trees I had seen.

"I avoided a vast artificial hole somebody had been digging on the slope, the purpose of which I found it impossible to divine. It wasn't a quarry or a sandpit, anyhow. It was just a hole. It might have been connected with the philanthropic desire of giving the criminals something to do. I don't know. Then I nearly fell into a very narrow ravine, almost no more than a scar in the hillside. I discovered that a lot of imported drainage-pipes for the settlement had been tumbled in there. There wasn't one that was not broken. It was a wanton smash-up. At last I got under the trees. My purpose was to stroll into the shade for a moment; but no sooner within than it seemed to me I had stepped into the gloomy circle of some Inferno. The rapids were near, and an uninterrupted, uniform, headlong, rushing noise filled the mournful stillness of the grove, where not a breath stirred, not a leaf moved, with a mysterious sound—as though the tearing pace of the launched earth had suddenly become audible.

"Black shapes crouched, lay, sat between the trees leaning against the trunks, clinging to the earth, half coming out, half effaced within the dim light, in all the attitudes of pain, abandonment, and despair. Another mine on the cliff went off, followed by a slight shudder of the soil under my feet. The work was going on. The work! And this was the place where some of the helpers had withdrawn to die.

"They were dying slowly—it was very clear. They were not enemies, they were not criminals, they were nothing earthly now—nothing but black shadows of disease and starvation, lying confusedly in the greenish gloom. Brought from all the recesses of the coast in all the legality of time contracts, lost in uncongenial surroundings, fed on unfamiliar food, they sickened, became inefficient, and were then allowed to crawl away and rest. These moribund shapes were free as air—and nearly as thin. I began to distinguish the gleam of the eyes under the trees. Then, glancing down, I saw a face near my hand. The black bones reclined at full length with one shoulder against the tree, and slowly the eyelids rose and the sunken eyes looked up at me, enormous and vacant, a kind of blind, white flicker in the depths of the orbs, which died out slowly. The man seemed young—almost a boy—but you know with them it's hard to tell. I found nothing else to do but to offer him one of my good Swede's ship's biscuits I had in my pocket. The fingers closed slowly on it and held—there was no other movement and no other glance. He had tied a bit of white worsted round his neck—Why? Where did he get it? Was it a badge—an ornament—a charm—a propitiatory act? Was there any idea at all connected with it? It looked startling round his black neck, this bit of white thread from beyond the seas.

"Near the same tree two more bundles of acute angles sat with their legs drawn up. One, with his chin propped on his knees, stared at nothing, in an intolerable and appalling manner: his brother phantom rested its forehead, as if overcome with a great weariness; and all about others were scattered in every pose of contorted collapse, as in some picture of a massacre or a pestilence. While I stood horror-struck, one of these creatures rose to his hands and knees, and went off on all-fours towards the river to drink. He lapped out of his hand, then sat up in the sunlight, crossing his shins in front of him, and after a time let his woolly head fall on his breastbone.

"I didn't want any more loitering in the shade, and I made haste towards the station. When near the buildings I met a white man, in such an unexpected elegance of get-up that in the first moment I took him for a sort of vision. I saw a high starched collar, white cuffs, a light alpaca jacket, snowy trousers, a clean necktie, and varnished boots. No hat. Hair parted, brushed, oiled, under a green-lined parasol held in a big white hand. He was amazing, and had a penholder behind his ear.

"I shook hands with this miracle, and I learned he was the Company's chief accountant, and that all the book-keeping was done at this station. He had come out for a moment, he said, 'to get a breath of fresh air.' The expression sounded wonderfully odd, with its suggestion of sedentary desk-life. I wouldn't have mentioned the fellow to you at all, only it was from his lips that I first heard the name of the man who is so indissolubly connected with the memories of that time. Moreover, I respected the fellow. Yes; I respected his collars, his vast cuffs, his brushed hair. His appearance was certainly that of a hairdresser's dummy; but in the great demoralization of the land he kept up his appearance. That's backbone. His starched collars and got-up shirt-fronts were achievements of character. He had been out nearly three years; and, later, I could not help asking him how he managed to sport such linen. He had just the faintest blush, and said modestly, 'I've been teaching one of the native women about the station. It was difficult. She had a distaste for the work.' Thus this man had verily accomplished something. And he was devoted to his books, which were in apple-pie order.

"Everything else in the station was in a muddle—heads, things, buildings. Strings of dusty niggers with splay feet arrived and departed; a stream of manufactured goods, rubbishy cottons, beads, and brass-wire set into the depths of darkness, and in return came a precious trickle of ivory.

"I had to wait in the station for ten days—an eternity. I lived in a hut in the yard, but to be out of the chaos I would sometimes get into the accountant's office. It was built of horizontal planks, and so badly put together that, as he bent over his high desk, he was barred from neck to heels with narrow strips of sunlight. There was no need to open the big shutter to see. It was hot there, too; big flies buzzed fiendishly, and did not sting, but stabbed. I sat generally on the floor, while, of faultless appearance (and even slightly scented), perching on a high stool, he wrote, he wrote. Sometimes he stood up for exercise. When a truckle-bed with a sick man (some invalid agent from upcountry) was put in there, he exhibited a gentle annoyance. 'The groans of this sick person,' he said, 'distract my attention. And without that it is extremely difficult to guard against clerical errors in this climate.'

"One day he remarked, without lifting his head, 'In the interior you will no doubt meet Mr. Kurtz.' On my asking who Mr. Kurtz was, he said he was a first-class agent; and seeing my disappointment at this information, he added slowly, laying down his pen, 'He is a very remarkable person.' Further questions elicited from him that Mr. Kurtz was at present in charge of a trading-post, a very important one, in the true ivory-country, at 'the very bottom of there. Sends in as much ivory as all the others put together. . . .' He began to write again. The sick man was too ill to groan. The flies buzzed in a great peace.

"Suddenly there was a growing murmur of voices and a great tramping of feet. A caravan had come in. A violent babble of uncouth sounds burst out on the other side of the planks. All the carriers were speaking together, and in the midst of the uproar the lamentable voice of the chief agent was heard 'giving it up' tearfully for the twentieth time that day. . . . He rose slowly. 'What a frightful row,' he said. He crossed the room gently to look at the sick man, and returning, said to me, 'He does not hear.' 'What! Dead?' I asked, startled. 'No, not yet,' he answered, with great composure. Then, alluding with a toss of the head to the tumult in the station-yard, 'When one has got to make correct entries, one comes to hate those savages—hate them to the death.' He remained thoughtful for a moment. 'When you see Mr. Kurtz,' he went on, 'tell him from me that everything here'—he glanced at the desk—'is very satisfactory. I don't like to write to him—with those messengers of ours you never know who may get hold of your letter—at that Central Station.' He stared at me for a moment with his mild, bulging eyes. 'Oh, he will go far, very far,' he began again. 'He will be a somebody in the Administration before long. They, above—the Council in Europe, you know—mean him to be.'

"He turned to his work. The noise outside had ceased, and presently in going out I stopped at the door. In the steady buzz of flies the homeward-bound agent was lying flushed and insensible; the other, bent over his books, was making correct entries of perfectly correct transactions; and fifty feet below the doorstep I could see the still treetops of the grove of death.

"Next day I left that station at last, with a caravan of sixty men, for a two-hundred-mile tramp.

"No use telling you much about that. Paths, paths, everywhere; a stamped-in network of paths spreading over the empty land, through the long grass, through burnt grass, through thickets, down and up chilly ravines, up and down stony hills ablaze with heat; and a solitude, a solitude, nobody, not a hut. The population had cleared out a long time ago. Well, if a lot of mysterious niggers armed with all kinds of fearful weapons suddenly took to travelling on the road between Deal and Gravesend, catching the yokels right and left to carry heavy loads for them, I fancy every farm and cottage thereabouts would get empty very soon. Only here the dwellings were gone, too. Still I passed through several abandoned villages. There's something pathetically childish in the ruins of grass walls. Day after day, with the stamp and shuffle of sixty pair of bare feet behind me, each pair under a 60-lb. load. Camp, cook, sleep, strike camp, march. Now and then a carrier dead in harness, at rest in the long grass near the path, with an empty water-gourd and his long staff lying by his side. A great silence around and above. Perhaps on some quiet night the tremor of far-off drums, sinking, swelling, a tremor vast, faint; a sound weird, appealing, suggestive, and wild—and perhaps with as profound a meaning as the sound of bells in a Christian country. Once a white man in an unbuttoned uniform, camping on the path with an armed escort of lank Zanzibaris, very hospitable and festive—not to say drunk. Was looking after the upkeep of the road, he declared. Can't say I saw any road or any upkeep, unless the body of a middle-aged negro, with a bullet-hole in the forehead, upon which I absolutely stumbled three miles farther on, may be considered as a permanent improvement. I had a white companion, too, not a bad chap, but rather too fleshy and with the exasperating habit of fainting on the hot hillsides, miles away from the least bit of shade and water. Annoying, you know, to hold your own coat like a parasol over a man's head while he is coming to. I couldn't help asking him once what he meant by coming there at all. 'To make money, of course. What do you think?' he said, scornfully. Then he got fever, and had to be carried in a hammock slung under a pole. As he weighed sixteen stone I had no end of rows with the carriers. They jibbed, ran away, sneaked off with their loads in the night—quite a mutiny. So, one evening, I made a speech in English with gestures, not one of which was lost to the sixty pairs of eyes before me, and the next morning I started the hammock off in front all right. An hour afterwards I came upon the whole concern wrecked in a bush—man, hammock, groans, blankets, horrors. The heavy pole had skinned his poor nose. He was very anxious for me to kill somebody, but there wasn't the shadow of a carrier near. I remembered the old doctor—'It would be interesting for science to watch the mental changes of individuals, on the spot.' I felt I was becoming scientifically interesting. However, all that is to no purpose. On the fifteenth day I came in sight of the big river again, and hobbled into the Central Station. It was on a back water surrounded by scrub and forest, with a pretty border of smelly mud on one side, and on the three others enclosed by a crazy fence of rushes. A neglected gap was all the gate it had, and the first glance at the place was enough to let you see the flabby devil was running that show. White men with long staves in their hands appeared languidly from amongst the buildings, strolling up to take a look at me, and then retired out of sight somewhere. One of them, a stout, excitable chap with black moustaches, informed me with great volubility and many digressions, as soon as I told him who I was, that my steamer was at the bottom of the river. I was thunderstruck. What, how, why? Oh, it was 'all right.' The 'manager himself' was there. All quite correct. 'Everybody had behaved splendidly! splendidly!'—'you must,' he said in agitation, 'go and see the general manager at once. He is waiting!'

"I did not see the real significance of that wreck at once. I fancy I see it now, but I am not sure—not at all. Certainly the affair was too stupid—when I think of it—to be altogether natural. Still . . . But at the moment it presented itself simply as a confounded nuisance. The steamer was sunk. They had started two days before in a sudden hurry up the river with the manager on board, in charge of some volunteer skipper, and before they had been out three hours they tore the bottom out of her on stones, and she sank near the south bank. I asked myself what I was to do there, now my boat was lost. As a matter of fact, I had plenty to do in fishing my command out of the river. I had to set about it the very next day. That, and the repairs when I brought the pieces to the station, took some months.

"My first interview with the manager was curious. He did not ask me to sit down after my twenty-mile walk that morning. He was commonplace in complexion, in feature, in manners, and in voice. He was of middle size and of ordinary build. His eyes, of the usual blue, were perhaps remarkably cold, and he certainly could make his glance fall on one as trenchant and heavy as an axe. But even at these times the rest of his person seemed to disclaim the intention. Otherwise there was only an indefinable, faint expression of his lips, something stealthy —a smile—not a smile—I remember it, but I can't explain. It was unconscious, this smile was, though just after he had said something it got intensified for an instant. It came at the end of his speeches like a seal applied on the words to make the meaning of the commonest phrase appear absolutely inscrutable. He was a common trader, from his youth up employed in these parts—nothing more. He was obeyed, yet he inspired neither love nor fear, nor even respect. He inspired uneasiness. That was it! Uneasiness. Not a definite mistrust—just uneasiness—nothing more. You have no idea how effective such a . . . a. . . . faculty can be. He had no genius for organizing, for initiative, or for order even. That was evident in such things as the deplorable state of the station. He had no learning, and no intelligence. His position had come to him—why? Perhaps because he was never ill. . . . He had served three terms of three years out there. . . . Because triumphant health in the general rout of constitutions is a kind of power in itself. When he went home on leave he rioted on a large scale—pompously. Jack ashore—with a difference—in externals only. This one could gather from his casual talk. He originated nothing, he could keep the routine going—that's all. But he was great. He was great by this little thing that it was impossible to tell what could control such a man. He never gave that secret away. Perhaps there was nothing within him. Such a suspicion made one pause—for out there there were no external checks. Once when various tropical diseases had laid low almost every 'agent' in the station, he was heard to say, 'Men who come out here should have no entrails.' He sealed the utterance with that smile of his, as though it had been a door opening into a darkness he had in his keeping. You fancied you had seen things—but the seal was on. When annoyed at meal-times by the constant quarrels of the white men about precedence, he ordered an immense round table to be made, for which a special house had to be built. This was the station's mess-room. Where he sat was the first place—the rest were nowhere. One felt this to be his unalterable conviction. He was neither civil nor uncivil. He was quiet. He allowed his 'boy'—an overfed young negro from the coast—to treat the white men, under his very eyes, with provoking insolence.

"He began to speak as soon as he saw me. I had been very long on the road. He could not wait. Had to start without me. The up-river stations had to be relieved. There had been so many delays already that he did not know who was dead and who was alive, and how they got on—and so on, and so on. He paid no attention to my explanations, and, playing with a stick of sealing-wax, repeated several times that the situation was 'very grave, very grave.' There were rumours that a very important station was in jeopardy, and its chief, Mr. Kurtz, was ill. Hoped it was not true. Mr. Kurtz was . . . I felt weary and irritable. Hang Kurtz, I thought. I interrupted him by saying I had heard of Mr. Kurtz on the coast. 'Ah! So they talk of him down there,' he murmured to himself. Then he began again, assuring me Mr. Kurtz was the best agent he had, an exceptional man, of the greatest importance to the Company; therefore I could understand his anxiety. He was, he said, 'very, very uneasy.' Certainly he fidgeted on his chair a good deal, exclaimed, 'Ah, Mr. Kurtz!' broke the stick of sealing-wax and seemed dumfounded by the accident. Next thing he wanted to know 'how long it would take to' . . . I interrupted him again. Being hungry, you know, and kept on my feet too, I was getting savage. 'How can I tell?' I said. 'I haven't even seen the wreck yet—some months, no doubt.' All this talk seemed to me so futile. 'Some months,' he said. 'Well, let us say three months before we can make a start. Yes. That ought to do the affair.' I flung out of his hut (he lived all alone in a clay hut with a sort of verandah) muttering to myself my opinion of him. He was a chattering idiot. Afterwards I took it back when it was borne in upon me startlingly with what extreme nicety he had estimated the time requisite for the 'affair.'

"I went to work the next day, turning, so to speak, my back on that station. In that way only it seemed to me I could keep my hold on the redeeming facts of life. Still, one must look about sometimes; and then I saw this station, these men strolling aimlessly about in the sunshine of the yard. I asked myself sometimes what it all meant. They wandered here and there with their absurd long staves in their hands, like a lot of faithless pilgrims bewitched inside a rotten fence. The word 'ivory' rang in the air, was whispered, was sighed. You would think they were praying to it. A taint of imbecile rapacity blew through it all, like a whiff from some corpse. By Jove! I've never seen anything so unreal in my life. And outside, the silent wilderness surrounding this cleared speck on the earth struck me as something great and invincible, like evil or truth, waiting impatiently for the passing away of this fantastic invasion.

"Oh, these months! Well, never mind. Various things happened. One evening a grass shed full of calico, cotton prints, beads, and I don't know what else, burst into a blaze so suddenly that you would have thought the earth had opened up to let an avenging fire consume all that trash. I was smoking my pipe quietly by my dismantled steamer, and saw them all cutting capers in the light, with their arms lifted high, when the stout man with moustaches came tearing down to the river, a tin pail in his hand, assured me that everybody was 'behaving splendidly, splendidly,' dipped about a quart of water and tore back again. I noticed there was a hole in the bottom of his pail.

"I strolled up. There was no hurry. You see the thing had gone off like a box of matches. It had been hopeless from the very first. The flame had leaped high, driven everybody back, lighted up everything—and collapsed. The shed was already a heap of embers glowing fiercely. A nigger was being beaten near by. They said he had caused the fire in some way; be that as it may, he was screeching most horribly. I saw him, later, for several days, sitting in a bit of shade looking very sick and trying to recover himself: afterwards he arose and went out—and the wilderness without a sound took him into its bosom again. As I approached the glow from the dark I found myself at the back of the two men, talking. I heard the name of Kurtz pronounced, then the words, 'take advantage of this unfortunate accident.' One of the men was the manager. I wished him a good evening. 'Did you ever see anything like it—eh? it is incredible,' he said, and walked off. The other man remained. He was a first-class agent, young, gentlemanly, a bit reserved, with a forked little beard and a hooked nose. He was stand-offish with the other agents, and they on their side said he was the manager's spy upon them. As to me, I had hardly ever spoken to him before. We got into talk, and by and by we strolled away from the hissing ruins. Then he asked me to his room, which was in the main building of the station. He struck a match, and I perceived that this young aristocrat had not only a silver-mounted dressing-case but also a whole candle all to himself. Just at that time the manager was the only man supposed to have any right to candles. Native mats covered the clay walls; a collection of spears, assegais, shields, knives was hung up in trophies. The business intrusted to this fellow was the making of bricks—so I had been informed; but there wasn't a fragment of a brick anywhere in the station, and he had been there more than a year—waiting. It seems he could not make bricks without something, I don't know what—straw maybe. Anyway, it could not be found there and as it was not likely to be sent from Europe, it did not appear clear to me what he was waiting for. An act of special creation perhaps. However, they were all waiting—all the sixteen or twenty pilgrims of them—for something; and upon my word it did not seem an uncongenial occupation, from the way they took it, though the only thing that ever came to them was disease—as far as I could see. They beguiled the time by backbiting and intriguing against each other in a foolish kind of way. There was an air of plotting about that station, but nothing came of it, of course. It was as unreal as everything else—as the philanthropic pretence of the whole concern, as their talk, as their government, as their show of work. The only real feeling was a desire to get appointed to a trading-post where ivory was to be had, so that they could earn percentages. They intrigued and slandered and hated each other only on that account—but as to effectually lifting a little finger—oh, no. By heavens! there is something after all in the world allowing one man to steal a horse while another must not look at a halter. Steal a horse straight out. Very well. He has done it. Perhaps he can ride. But there is a way of looking at a halter that would provoke the most charitable of saints into a kick.

"I had no idea why he wanted to be sociable, but as we chatted in there it suddenly occurred to me the fellow was trying to get at something—in fact, pumping me. He alluded constantly to Europe, to the people I was supposed to know there—putting leading questions as to my acquaintances in the sepulchral city, and so on. His little eyes glittered like mica discs—with curiosity—though he tried to keep up a bit of superciliousness. At first I was astonished, but very soon I became awfully curious to see what he would find out from me. I couldn't possibly imagine what I had in me to make it worth his while. It was very pretty to see how he baffled himself, for in truth my body was full only of chills, and my head had nothing in it but that wretched steamboat business. It was evident he took me for a perfectly shameless prevaricator. At last he got angry, and, to conceal a movement of furious annoyance, he yawned. I rose. Then I noticed a small sketch in oils, on a panel, representing a woman, draped and blindfolded, carrying a lighted torch. The background was sombre—almost black. The movement of the woman was stately, and the effect of the torchlight on the face was sinister.

"It arrested me, and he stood by civilly, holding an empty half-pint champagne bottle (medical comforts) with the candle stuck in it. To my question he said Mr. Kurtz had painted this—in this very station more than a year ago—while waiting for means to go to his trading-post. 'Tell me, pray,' said I, 'who is this Mr. Kurtz?'

" 'The chief of the Inner Station,' he answered in a short tone, looking away. 'Much obliged,' I said, laughing. 'And you are the brickmaker of the Central Station. Every one knows that.' He was silent for a while. 'He is a prodigy,' he said at last. 'He is an emissary of pity and science and progress, and devil knows what else. We want,' he began to disclaim suddenly, 'for the guidance of the cause intrusted to us by Europe, so to speak, higher intelligence, wide sympathies, a singleness of purpose.' 'Who says that?' I asked. 'Lots of them,' he replied. 'Some even write that; and so *he* comes here, a special being, as you ought to know.' 'Why ought I to know?' I interrupted, really surprised. He paid no attention. 'Yes. Today he is chief of the best station, next year he will be assistant-manager, two years more and . . . but I daresay you know what he will be in two years' time. You are of the new gang—the gang of virtue. The same people who sent him specially also recommended you. Oh, don't say no. I've my own eyes to trust.' Light dawned upon me. My dear aunt's influential acquaintances were producing an unexpected effect upon that young man. I nearly burst into a laugh. 'Do you read the Company's confidential correspondence?' I asked. He hadn't a word to say. It was great fun. 'When Mr. Kurtz,' I continued, severely, 'is General Manager, you won't have the opportunity.'

"He blew the candle out suddenly, and we went outside. The moon had risen. Black figures strolled about listlessly, pouring water on the glow, whence proceeded a sound of hissing; steam ascended in the moonlight, the beaten nigger groaned somewhere. 'What a row the brute makes!' said the indefatigable man with the moustaches, appearing near us. 'Serve him right. Transgression—punishment—bang! Pitiless, pitiless. That's the only way. This will prevent all conflagrations for the future. I was just telling the manager . . .' He noticed my companion, and became crestfallen all at once. 'Not in bed yet,' he said, with a kind of servile heartiness; 'it's so natural. Ha! Danger—agitation.' He vanished. I went on to the river-side, and the other followed me. I heard a scathing murmur at my ear, 'Heap of muffs—go to.' The pilgrims could be seen in knots gesticulating, discussing. Several had still their staves in their hands. I verily believe they took these sticks to bed with them. Beyond the fence the forest stood up spectrally in the moonlight, and through the dim stir, through the faint sounds of that lamentable courtyard, the silence of the land went home to one's very heart—its mystery, its greatness, the amazing reality of its concealed life. The hurt nigger moaned feebly somewhere near by, and then fetched a deep sigh that made me mend my pace away from there. I felt a hand introducing itself under my arm. 'My dear sir,' said the fellow, 'I don't want to be misunderstood, and especially by you, who will see Mr. Kurtz long before I can have that pleasure. I wouldn't like him to get a false idea of my disposition. . . .'

"I let him run on, this papier-maché Mephistopheles, and it seemed to me that if I tried I could poke my forefinger through him, and would find nothing inside but a little loose dirt, maybe. He, don't you see, had been planning to be assistant-manager by and by under the present man, and I could see that the coming of that Kurtz had upset them both not a little. He talked precipitately, and I did not try to stop him. I had my shoulders against the wreck of my steamer, hauled up on the slope like a carcass of some big river animal. The smell of mud, of primeval mud, by Jove! was in my nostrils, the high stillness of primeval forest was before my eyes; there were shiny patches on the black creek. The moon had spread over everything a thin layer of silver—over the rank grass, over the mud, upon the wall of matted vegetation standing higher than the wall of a temple, over the great river I could see through a sombre gap glittering, glittering, as it flowed broadly by without a murmur. All this was great, expectant, mute, while the man jabbered about himself. I wondered whether the stillness on the face of the immensity looking at us two were meant as an appeal or as a menace. What were we who had strayed in here? Could we handle that dumb thing, or would it handle us? I felt how big, how confoundedly big, was that thing that couldn't talk, and perhaps was deaf as well. What was in there? I could see a little ivory coming out from there, and I had heard Mr. Kurtz was in there. I had heard enough about it, too—God knows! Yet somehow it didn't bring any image with it—no more than if I had been told an angel or a fiend was in there. I believed it in the same way one of you might believe there are inhabitants in the planet Mars. I knew once a Scotch sailmaker who was certain, dead sure, there were people in Mars. If you asked him for some idea how they looked and behaved, he would get shy and mutter something about 'walking on all-fours.' If you as much as smiled, he would—though a man of sixty—offer to fight you. I would not have gone so far as to fight for Kurtz, but I went for him near enough to a lie. You know I hate, detest, and can't bear a lie, not because I am straighter than the rest of us, but simply because it appalls me. There is a taint of death, a flavour of mortality in lies—which is exactly what I hate and detest in the world—what I want to forget. It makes me miserable and sick, like biting something rotten would do. Temperament, I suppose. Well, I went near enough to it by letting the young fool there believe anything he liked to imagine as to my influence in Europe. I became in an instant as much of a pretence as the rest of the bewitched pilgrims. This simply because I had a notion it somehow would be of help to that Kurtz whom at the time I did not see—you understand. He was just a word for me. I did not see the man in the name any more than you do. Do you see him? Do you see the story? Do you see anything? It seems to me I am trying to tell you a dream—making a vain attempt, because no relation of a dream can convey the dream-sensation, that commingling of absurdity, surprise, and bewilderment in a tremor of struggling revolt, that notion of being captured by the incredible which is of the very essence of dreams. . . ."

He was silent for a while.

". . . No, it is impossible; it is impossible to convey the life-sensation of any given epoch of one's existence—that which makes its truth, its meaning—its subtle and penetrating essence. It is impossible. We live, as we dream—alone. . . ."

He paused again as if reflecting, then added:

"Of course in this you fellows see more than I could then. You see me, whom you know. . . ."

It had become so pitch dark that we listeners could hardly see one another. For a long time already he, sitting apart, had been no more to us than a voice. There was not a word from anybody. The others might have been asleep, but I was awake. I listened, I listened on the watch for the sentence, for the word, that would give me the clue to the faint uneasiness inspired by this narrative that seemed to shape itself without human lips in the heavy night-air of the river.

". . . Yes—I let him run on," Marlow began again, "and think what he pleased about the powers that were behind me. I did! And there was nothing behind me! There was nothing but that wretched, old, mangled steamboat I was leaning against, while he talked fluently about 'the necessity for every man to get on.' 'And when one comes out here, you conceive, it is not to gaze at the moon.' Mr. Kurtz was a 'universal genius,' but even a genius would find it easier to work with 'adequate tools—intelligent men.' He did not make bricks—why, there was a physical impossibility in the way—as I was well aware; and if he did secretarial work for the manager, it was because 'no sensible man rejects wantonly the confidence of his superiors.' Did I see it? I saw it. What more did I want? What I really wanted was rivets, by heaven! Rivets. To get on with the work—to stop the hole. Rivets I wanted. There were cases of them down at the coast—cases—piled up—burst—split! You kicked a loose rivet at every second step in that station-yard on the hillside. Rivets had rolled into the grove of death. You could fill your pockets with rivets for the trouble of stooping down—and there wasn't one rivet to be found where it was wanted. We had plates that would do, but nothing to fasten them with. And every week the messenger, a lone negro, letter-bag on shoulder and staff in hand, left our station for the coast. And several times a week a coast caravan came in with trade goods—ghastly glazed calico that made you shudder only to look at it, glass beads value about a penny a quart, confounded spotted cotton handkerchiefs. And no rivets. Three carriers could have brought all that was wanted to set that steamboat afloat.

"He was becoming confidential now, but I fancy my unresponsive attitude must have exasperated him at last, for he judged it necessary to inform me he feared neither God nor devil, let alone any mere man. I said I could see that very well, but what I wanted was a certain quantity of rivets—and rivets were what really Mr. Kurtz wanted, if he had only known it. Now letters went to the coast every week. . . . 'My dear sir,' he cried, 'I write from dictation.' I demanded rivets. There was a way—for an intelligent man. He changed his manner; became very cold, and suddenly began to talk about a hippopotamus; wondered whether sleeping on board the steamer (I stuck to my salvage night and day) I wasn't disturbed. There was an old hippo that had the bad habit of getting out on the bank and roaming at night over the station grounds. The pilgrims used to turn out in a body and empty every rifle they could lay hands on at him. Some even had sat up o' nights for him. All this energy was wasted, though. 'That animal has a charmed life,' he said; 'but you can say this only of brutes in this country. No man—you apprehend me?—no man here bears a charmed life.' He stood there for a moment in the moonlight with his delicate hooked nose set a little askew, and his mica eyes glittering without a wink, then, with a curt Good-night, he strode off. I could see he was disturbed and considerably puzzled, which made me feel more hopeful than I had been for days. It was a great comfort to turn from that chap to my influential friend, the battered, twisted, ruined, tin-pot steamboat. I clambered on board. She rang under my feet like an empty Huntley & Palmer biscuit-tin kicked along a gutter; she was nothing so solid in make, and rather less pretty in shape, but I had expended enough hard work on her to make me love her. No influential friend would have served me better. She had given me a chance to come out a bit—to find out what I could do. No, I don't like work. I had rather laze about and think of all the fine things that can be done. I don't like work—no man does—but I like what is in the work—the chance to find yourself. Your own reality—for yourself, not for others—what no other man can ever know. They can only see the mere show, and never can tell what it really means.

"I was not surprised to see somebody sitting aft, on the deck, with his legs dangling over the mud. You see I rather chummed with the few mechanics there were in that station, whom the other pilgrims naturally despised—on account of their imperfect manners, I suppose. This was the foreman—a boiler-maker by trade—a good worker. He was a lank, bony, yellow-faced man, with big intense eyes. His aspect was worried, and his head was as bald as the palm of my hand; but his hair in falling seemed to have stuck to his chin, and had prospered in the new locality, for his beard hung down to his waist. He was a widower with six young children (he had left them in charge of a sister of his to come out there), and the passion of his life was pigeon-flying. He was an enthusiast and a connoisseur. He would rave about pigeons. After work hours he used sometimes to come over from his hut for a talk about his children and his pigeons; at work, when he had to crawl in the mud under the bottom of the steamboat, he would tie up that beard of his in a kind of white serviette he brought for the purpose. It had loops to go over his ears. In the evening he could be seen squatted on the bank rinsing that wrapper in the creek with great care, then spreading it solemnly on a bush to dry.

"I slapped him on the back and shouted, 'We shall have rivets!' He scrambled to his feet exclaiming, 'No! Rivets!' as though he couldn't believe his ears. Then in a low voice, 'You . . . eh?' I don't know why we behaved like lunatics. I put my finger to the side of my nose and nodded mysteriously. 'Good for you!' he cried, snapped his fingers above his head, lifting one foot. I tried a jig. We capered on the iron deck. A frightful clatter came out of that hulk, and the virgin forest on the other bank of the creek sent it back in a thundering roll upon the sleeping station. It must have made some of the pilgrims sit up in their hovels. A dark figure obscured the lighted doorway of the manager's hut, vanished, then, a second or so after, the doorway itself vanished, too. We stopped, and the silence driven away by the stamping of our feet flowed back again from the recesses of the land. The great wall of vegetation, an exuberant and entangled mass of trunks, branches, leaves, boughs, festoons, motionless in the moonlight, was like a rioting invasion of soundless life, a rolling wave of plants, piled up, crested, ready to topple over the creek, to sweep every little man of us out of his little existence. And it moved not. A deadened burst of mighty splashes and snorts reached us from afar, as though an ichthyosaurus had been taking a bath of glitter in the great river. 'After all,' said the boiler-maker in a reasonable tone, 'why shouldn't we get the rivets?' Why not, indeed! I did not know of any reason why we shouldn't. 'They'll come in three weeks,' I said, confidently.

"But they didn't. Instead of rivets there came an invasion, an infliction, a visitation. It came in sections during the next three weeks, each section headed by a donkey carrying a white man in new clothes and tan shoes, bowing from that elevation right and left to the impressed pilgrims. A quarrelsome band of footsore sulky niggers trod on the heels of the donkey; a lot of tents, camp-stools, tin boxes, white cases, brown bales would be shot down in the court-yard, and the air of mystery would deepen a little over the muddle of the station. Five such instalments came, with their absurd air of disorderly flight with the loot of innumerable outfit shops and provision stores, that, one would think, they were lugging, after a raid, into the wilderness for equitable division. It was an inextricable mess of things decent in themselves but that human folly made look like the spoils of thieving.

"This devoted band called itself the Eldorado Exploring Expedition, and I believe they were sworn to secrecy. Their talk, however, was the talk of sordid buccaneers: it was reckless without hardihood, greedy without audacity, and cruel without courage; there was not an atom of foresight or of serious intention in the whole batch of them, and they did not seem aware these things are wanted for the work of the world. To tear treasure out of the bowels of the land was their desire, with no more moral purpose at the back of it than there is in burglars breaking into a safe. Who paid the expenses of the noble enterprise I don't know; but the uncle of our manager was leader of that lot.

"In exterior he resembled a butcher in a poor neighbourhood, and his eyes had a look of sleepy cunning. He carried his fat paunch with ostentation on his short legs, and during the time his gang infested the station spoke to no one but his nephew. You could see these two roaming about all day long with their heads close together in an everlasting confab.

"I had given up worrying myself about the rivets. One's capacity for that kind of folly is more limited than you would suppose. I said Hang!—and let things slide. I had plenty of time for meditation, and now and then I would give some thought to Kurtz. I wasn't very interested in him. No. Still, I was curious to see whether this man, who had come out equipped with moral ideas of some sort, would climb to the top after all and how he would set about his work when there."

II

"ONE EVENING as I was lying flat on the deck of my steamboat, I heard voices approaching—and there were the nephew and the uncle strolling along the bank. I laid my head on my arm again, and had nearly lost myself in a doze, when somebody said in my ear, as it were: 'I am as harmless as a child, but I don't like to be dictated to. Am I the manager—or am I not? I was ordered to send him there. It's incredible.' . . . I became aware that the two were standing on the shore alongside the forepart of the steamboat, just below my head. I did not move; it did not occur to me to move: I was sleepy. 'It *is* unpleasant,' grunted the uncle. 'He has asked the Administration to be sent there,' said the other, 'with the idea of showing what he could do; and I was instructed accordingly. Look at the influence that man must have. Is it not frightful?' They both agreed it was frightful, then made several bizarre remarks: 'Make rain and fine weather—one man—the Council—by the nose'—bits of absurd sentences that got the better of my drowsiness, so that I had pretty near the whole of my wits about me when the uncle said, 'The climate may do away with this difficulty for you. Is he alone there?' 'Yes,' answered the manager; 'he sent his assistant down the river with a note to me in these terms: "Clear this poor devil out of the country, and don't bother sending more of that sort. I had rather be alone than have the kind of men you can dispose of with me." It was more than a year ago. Can you imagine such impudence!' 'Anything since then?' asked the other hoarsely. 'Ivory,' jerked the nephew; 'lots of it—prime sort—lots—most annoying, from him.' 'And with that?' questioned the heavy rumble. 'Invoice,' was the reply fired out, so to speak. Then silence. They had been talking about Kurtz.

"I was broad awake by this time, but, lying perfectly at ease, remained still, having no inducement to change my position. 'How did that ivory come all this way?' growled the elder man, who seemed very vexed. The other explained that it had come with a fleet of canoes in charge of an English half-caste clerk Kurtz had with him; that Kurtz had apparently intended to return himself, the station being by that time bare of goods and stores, but after coming three hundred miles, had suddenly decided to go back, which he started to do alone in a small dugout with four paddlers, leaving the half-caste to continue down the river with the ivory. The two fellows there seemed astounded at anybody attempting such a thing. They were at a loss for an adequate motive. As to me, I seemed to see Kurtz for the first time. It was a distinct glimpse: the dugout, four paddling savages, and the lone white man turning his back suddenly on the headquarters, on relief, on thoughts of home—perhaps; setting his face towards the depths of the wilderness, towards his empty and desolate station. I did not know the motive. Perhaps he was just simply a fine fellow who stuck to his work for its own sake. His name, you understand, had not been pronounced once. He was 'that man.' The half-caste, who, as far as I could see, had conducted a difficult trip with great prudence and pluck, was invariably alluded to as 'that scoundrel.' The 'scoundrel' had reported that the 'man' had been very ill—had recovered imperfectly. . . . The two below me moved away then a few paces, and strolled back and forth at some little distance. I heard: 'Military post—doctor—two hundred miles—quite alone now—unavoidable delays—nine months—no news—strange rumours.' They approached again, just as the manager was saying, 'No one, as far as I know, unless a species of wandering trader—a pestilential fellow, snapping ivory from the natives.' Who was it they were talking about now? I gathered in snatches that this was some man supposed to be in Kurtz's district, and of whom the manager did not approve. 'We will not be free from unfair competition till one of these fellows is hanged for an example,' he said. 'Certainly,' grunted the other; 'get him hanged! Why not? Anything—anything can be done in this country. That's what I say; nobody here, you understand, *here,* can endanger your position. And why? You stand the climate—you outlast them all. The danger is in Europe; but therc before I left I took care to——' They moved off and whispered, then their voices rose again. 'The extraordinary series of delays is not my fault. I did my best.' The fat man sighed. 'Very sad.' 'And the pestiferous absurdity of his talk,' continued the other; 'he bothered me enough when he was her. "Each station should be like a beacon on the road towards better things, a centre for trade of course, but also for humanizing, improving, instructing." Conceive you—that ass! And he wants to be manager; No, it's——' Here he got choked by excessive indignation, and I lifted my head the least bit. I was surprised to see how near they were—right under me. I could have spat upon their hats. They were looking on the ground, absorbed in thought. The manager was switching his leg with a slender twig: his sagacious relative lifted his head. 'You have been well since you came out this time?' he asked. The other gave a start. 'Who? I? Oh? Like a charm—like a charm. But the rest—oh, my goodness! All sick. They die so quick, too, that I haven't the time to send them out of the country—it's incredible!' 'H'm. Just so,' grunted the uncle. 'Ah! my boy, trust to this—I say, trust to this.' I saw him extend his short flipper of an arm for a gesture that took in the forest, the creek, the mud, the river—seemed to beckon with a dishonouring flourish before the sunlit face of the land a treacherous appeal to the lurking death, to the hidden evil, to the profound darkness of its heart. It was so startling that I leaped to my feet and looked back at the edge of the forest, as though I had expected an answer of some sort to that black display of confidence. You know the foolish notions that come to one sometimes. The high stillness confronted these two figures with its ominous patience, waiting for the passing away of a fantastic invasion.

"They swore aloud together—out of sheer fright, I believe—then pretending not to know anything of my existence, turned back to the station. The sun was low; and leaning forward side by side, they seemed to be tugging painfully uphill their two ridiculous shadows of unequal length, that trailed behind them slowly over the tall grass without bending a single blade.

"In a few days the Eldorado Expedition went into the patient wilderness, that closed upon it as the sea closes over a diver. Long afterwards the news came that all the donkeys were dead. I know nothing as to the fate of the less valuable animals. They, no doubt, like the rest of us, found what they deserved. I did not inquire. I was then rather excited at the prospect of meeting Kurtz very soon. When I say very soon I mean it comparatively. It was just two months from the day we left the creek when we came to the bank below Kurtz's station.

"Going up that river was like travelling back to the earliest beginnings of the world, when vegetation rioted on the earth and the big trees were kings. An empty stream, a great silence, an impenetrable forest. The air was warm, thick, heavy, sluggish. There was no joy in the brilliance of sunshine. The long stretches of the waterway ran on, deserted, into the gloom of over-shadowed distances. On silvery sandbanks hippos and alligators sunned themselves side by side. The broadening waters flowed through a mob of wooded islands; you lost your way on that river as you would in a desert, and butted all day long against shoals, trying to find the channel, till you thought yourself bewitched and cut off for ever from everything you had known once—somewhere—far away—in another existence perhaps. There were moments when one's past came back to one, as it will sometimes when you have not a moment to spare to yourself; but it came in the shape of an unrestful and noisy dream, remembered with wonder amongst the overwhelming realities of this strange world of plants, and water, and silence. And this stillness of life did not in the least resemble a peace. It was the stillness of an implacable force brooding over an inscrutable intention. It looked at you with a vengeful aspect. I got used to it afterwards; I did not see it any more; I had no time. I had to keep guessing at the channel; I had to discern, mostly by inspiration, the signs of hidden banks; I watched for sunken stones; I was learning to clap my teeth smartly before my heart flew out, when I shaved by a fluke some infernal sly old snag that would have ripped the life out of the tin-pot steamboat and drowned all the pilgrims; I had to keep a lookout for the signs of dead wood we could cut up in the night for next day's steaming. When you have to attend to things of that sort, to the mere incidents of the surface, the reality—the reality, I tell you—fades. The inner truth is hidden—luckily, luckily. But I felt it all the same; I felt often its mysterious stillness watching me at my monkey tricks, just as it watches you fellows performing on your respective tight-ropes for—what is it? half-a-crown a tumble——"

"Try to be civil, Marlow," growled a voice, and I knew there was at least one listener awake besides myself.

"I beg your pardon. I forgot the heartache which makes up the rest of the price. And indeed what does the price matter, if the trick be well done? You do your tricks very well. And I didn't do badly either, since I managed not to sink that steamboat on my first trip. It's a wonder to me yet. Imagine a blindfolded man set to drive a van over a bad road. I sweated and shivered over that business considerably, I can tell you. After all, for a seaman, to scrape the bottom of the thing that's supposed to float all the time under his care is the unpardonable sin. No one may know of it, but you never forget the thump—eh? A blow on the very heart. You remember it, you dream of it, you wake up at night and think of it—years after—and go hot and cold all over. I don't pretend to say that steamboat floated all the time. More than once she had to wade for a bit, with twenty cannibals splashing around and pushing. We had enlisted some of these chaps on the way for a crew. Fine fellows—cannibals—in their place. They were men one could work with, and I am grateful to them. And, after all, they did not eat each other before my face: they had brought along a provision of hippo-meat which went rotten, and made the mystery of the wilderness stink in my nostrils. Phoo! I can sniff it now. I had the manager on board and three or four pilgrims with their staves—all complete. Sometimes we came upon a station close by the bank, clinging to the skirts of the unknown, and the white men rushing out of a tumble-down hovel, with great gestures of joy and surprise and welcome, seemed very strange—had the appearance of being held there captive by a spell. The word ivory would ring in the air for a while—and on we went again into the silence, along empty reaches, round the still bends, between the high walls of our winding way, reverberating in hollow claps the ponderous beat of the stern-wheel. Trees, trees, millions of trees, massive, immense, running up high; and at their foot, hugging the bank against the stream, crept the little begrimed steamboat, like a sluggish beetle crawling on the floor of a lofty portico. It made you feel very small, very lost, and yet it was not altogether depressing, that feeling. After all, if you were small, the grimy beetle crawled on—which was just what you wanted it to do. Where the pilgrims imagined it crawled to I don't know. To some place where they expected to get something. I bet! For me it crawled towards Kurtz—exclusively; but when the steam-pipes started leaking we crawled very slow. The reaches opened before us and closed behind, as if the forest had stepped leisurely across the water to bar the way for our return. We penetrated deeper and deeper into the heart of darkness. It was very quiet there. At night sometimes the roll of drums behind the curtain of trees would run up the river and remain sustained faintly, as if hovering in the air high over our heads, till the first break of day. Whether it meant war, peace, or prayer we could not tell. The dawns were heralded by the descent of a chill stillness; the wood-cutters slept, their fires burned low; the snapping of a twig would make you start. We were wanderers on a prehistoric earth, on an earth that wore the aspect of an unknown planet. We could have fancied ourselves the first of men taking possession of an accursed inheritance, to be subdued at the cost of profound anguish and of excessive toil. But suddenly, as we struggled round a bend, there would be a glimpse of rush walls, of peaked grass-roofs, a burst of yells, a whirl of black limbs, a mass of hands clapping, of feet stamping, of bodies swaying, of eyes rolling, under the droop of heavy and motionless foliage. The steamer toiled along slowly on the edge of a black and incomprehensible frenzy. The prehistoric man was cursing us, praying to us, welcoming us—who could tell? We were cut off from the comprehension of our surroundings; we glided past like phantoms, wondering and secretly appalled, as sane men would be before an enthusiastic outbreak in a madhouse. We could not understand because we were too far and could not remember because we were travelling in the night of first ages, of those ages that are gone, leaving hardly a sign—and no memories.

"The earth seemed unearthly. We are accustomed to look upon the shackled form of a conquered monster, but there—there you could look at a thing monstrous and free. It was unearthly, and the men were—— No, they were not inhuman. Well, you know, that was the worst of it—this suspicion of their not being inhuman. It would come slowly to one. They howled and leaped, and spun, and made horrid faces; but what thrilled you was just the thought of their humanity—like yours—the thought of your remote kinship with this wild and passionate uproar. Ugly. Yes, it was ugly enough; but if you were man enough you would admit to yourself that there was in you just the faintest trace of a response to the terrible frankness of that noise, a dim suspicion of there being a meaning in it which you—you so remote from night of first ages—could comprehend. And why not? The mind of man is capable of anything—because everything is in it, all the past as well as all the future. What was there after all? Joy, fear, sorrow, devotion, valour, rage—who can tell?—but truth—truth stripped of its cloak of time. Let the fool gape and shudder—the man knows, and can look on without a wink. But he must at least be as much of a man as these on the shore. He must meet that truth with his own true stuff—with his own inborn strength. Principles won't do. Acquisitions, clothes, pretty rags—rags that would fly off at the first good shake. No; you want a deliberate belief. An appeal to me in this fiendish row—is there? Very well; I hear; I admit, but I have a voice, too, and for good or evil mine is the speech that cannot be silenced. Of course, a fool, what with sheer fright and fine sentiments, is always safe. Who's that grunting? You wonder I didn't go ashore for a howl and a dance? Well, no—I didn't. Fine sentiments, you say? Fine sentiments, be hanged! I had no time. I had to mess about with white-lead and strips of woolen blanket helping to put bandages on those leaky steam-pipes—I tell you. I had to watch the steering, and circumvent those snags, and get the tin-pot along by hook or by crook. There was surface-truth enough in these things to save a wiser man. And between whiles I had to look after the savage who was fireman. He was an improved specimen; he could fire up a vertical boiler. He was there below me, and, upon my word, to look at him was as edifying as seeing a dog in a parody of breeches and a feather hat, walking on his hind-legs. A few months of training had done for that really fine chap. He squinted at the steam-gauge and at the water-gauge with an evident effort of intrepidity—and he had filed teeth, too, the poor devil, and the wool of his pate shaved into queer patterns, and three ornamental scars on each of his cheeks. He ought to have been clapping his hands and stamping his feet on the bank, instead of which he was hard at work, a thrall to strange witchcraft, full of improving knowledge. He was useful because he had been instructed; and what he knew was this—that should the water in that transparent thing disappear, the evil spirit inside the boiler would get angry through the greatness of his thirst, and take a terrible vengeance. So he sweated and fired up and watched the glass fearfully (with an impromptu charm, made of rags, tied to his arm, and a piece of polished bone, as big as a watch, stuck flatways through his lower lip), while the wooded banks slipped past us slowly, the short noise was left behind, the interminable miles of silence—and we crept on, towards Kurtz. But the snags were thick, the water was treacherous and shallow, the boiler seemed indeed to have a sulky devil in it, and thus neither that fireman nor I had any time to peer into our creepy thoughts.

"Some fifty miles below the Inner Station we came upon a hut of reeds, an inclined and melancholy pole, with the unrecognizable tatters of what had been a flag of some sort flying from it, and a neatly stacked wood-pile. This was unexpected. We came to the bank, and on the stack of firewood found a flat piece of board with some faded pencil-writing on it. When deciphered it said: 'Wood for you. Hurry up. Approach cautiously.' There was a signature, but it was illegible—not Kurtz—a much longer word. 'Hurry up.' Where? Up the river? 'Approach cautiously.' We had not done so. But the warning could not have been meant for the place where it could be only found after approach. Something was wrong above. But what—and how much? That was the question. We commented adversely upon the imbecility of that telegraphic style. The bush around said nothing, and would not let us look very far, either. A torn curtain of red twill hung in the doorway of the hut, and flapped sadly in our faces. The dwelling was dismantled; but we could see a white man had lived there not very long ago. There remained a rude table—a plank on two posts; a heap of rubbish reposed in a dark corner, and by the door I picked up a book. It had lost its covers, and the pages had been thumbed into a state of extremely dirty softness; but the back had been lovingly stitched afresh with white cotton thread, which looked clean yet. It was an extraordinary find. Its title was, *An Inquiry into some Points of Seamanship*, by a man Towser, Towson—some such name—Master in his Majesty's Navy. The matter looked dreary reading enough, with illustrative diagrams and repulsive tables of figures, and the copy was sixty years old. I handled this amazing antiquity with the greatest possible tenderness, lest it should dissolve in my hands. Within, Towson or Towser was inquiring earnestly into the breaking strain of ships' chains and tackle, and other such matters. Not a very enthralling book; but at the first glance you could see there a singleness of intention, an honest concern for the right way of going to work, which made these humble pages, thought out so many years ago, luminous with another than a professional light. The simple old sailor, with his talk of chains and purchases, made me forget the jungle and the pilgrims in a delicious sensation of having come upon something unmistakably real. Such a book being there was wonderful enough; but still more astounding were the notes pencilled in the margin, and plainly referring to the text. I couldn't believe my eyes! They were in cipher! Yes, it looked like cipher. Fancy a man lugging with him a book of that description into this nowhere and studying it —and making notes—in cipher at that! It was an extravagant mystery.

"I had been dimly aware for some time of a worrying noise, and when I lifted my eyes I saw the wood-pile was gone, and the manager, aided by all the pilgrims, was shouting at me from the riverside. I slipped the book into my pocket. I assure you to leave off reading was like tearing my self away from the shelter of an old and solid friendship.

"I started the lame engine ahead. 'It must be this miserable trader—this intruder,' exclaimed the manager, looking back malevolently at the place we had left. 'He must be English,' I said. 'It will not save him from getting into trouble if he is not careful,' muttered the manager darkly. I observed with assumed innocence that no man was safe from trouble in this world.

"The current was more rapid now, the steamer seemed at her last gasp, the stern-wheel flopped languidly, and I caught myself listening on tiptoe for the next beat of the boat, for in sober truth I expected the wretched thing to give up every moment. It was like watching the last flickers of a life. But still we crawled. Sometimes I would pick out a tree a little way ahead to measure our progress towards Kurtz by, but I lost it invariably before we got abreast. To keep the eyes so long on one thing was too much for human patience. The manager displayed a beautiful resignation. I fretted and fumed and took to arguing with myself whether or no I would talk openly with Kurtz; but before I could come to any conclusion it occurred to me that my speech or my silence, indeed any action of mine, would be a mere futility. What did it matter what any one knew or ignored? What did it matter who was manager? One gets sometimes such a flash of insight. The essentials of this affair lay deep under the surface, beyond my reach, and beyond my power of meddling.

"Towards the evening of the second day we judged ourselves about eight miles from Kurtz's station. I wanted to push on; but the manager looked grave, and told me the navigation up there was so dangerous that it would be advisable, the sun being low already, to wait where we were till next morning. Moreover, he pointed out that if the warning to approach cautiously were to be followed, we must approach in daylight—not at dusk or in the dark. This was sensible enough. Eight miles meant nearly three hours' steaming for us, and I could also see suspicious ripples at the upper end of the reach. Nevertheless, I was annoyed beyond expression at the delay, and most unreasonably, too, since one night more could not matter much after so many months. As we had plenty of wood, and caution was the word, I brought up in the middle of the stream. The reach was narrow, straight, with high sides like a railway cutting. The dusk came gliding into it long before the sun had set. The current ran smooth and swift, but a dumb immobility sat on the banks. The living trees, lashed together by the creepers and every living bush of the undergrowth, might have been changed into stone, even to the slenderest twig, to the lightest leaf. It was not sleep—it seemed unnatural, like a state of trance. Not the faintest sound of any kind could be heard. You looked on amazed, and began to suspect yourself of being deaf—then the night came suddenly, and struck you blind as well. About three in the morning some large fish leaped, and the loud splash made me jump as though a gun had been fired. When the sun rose there was a white fog, very warm and clammy, and more blinding than the night. It did not shift or drive; it was just there, standing all round you like something solid. At eight or nine, perhaps, it lifted as a shutter lifts. We had a glimpse of the towering multitude of trees, of the immense matted jungle, with the blazing little ball of the sun hanging over it—all perfectly still—and then the white shutter came down again, smoothly, as if sliding in greased grooves. I ordered the chain, which we had begun to heave in, to be paid out again. Before it stopped running with a muffled rattle, a cry, a very loud cry as of infinite desolation, soared slowly in the opaque air. It ceased. A complaining clamour, modulated in savage discords, filled our ears. The sheer unexpectedness of it made my hair stir under my cap. I don't know how it struck the others: to me it seemed as though the mist itself had screamed, so suddenly, and apparently from all sides at once, did this tumultuous and mournful uproar arise. It culminated in a hurried outbreak of almost intolerably excessive shrieking, which stopped short, leaving us stiffened in a variety of silly attitudes, and obstinately listening to the nearly as appalling and excessive silence. 'Good God! What is the meaning——' stammered at my elbow one of the pilgrims—a little fat man, with sandy hair and red whiskers, who wore sidespring boots, and pink pyjamas tucked into his socks. Two others remained open-mouthed a whole minute, then dashed into the little cabin, to rush out incontinently and stand darting scared glances, with Winchesters at 'ready' in their hands. What we could see was just the steamer we were on, her outlines blurred as though she had been on the point of dissolving, and a misty strip of water, perhaps two feet broad, around her—and that was all. The rest of the world was nowhere, as far as our eyes and ears were concerned. Just nowhere. Gone, disappeared; swept off without leaving a whisper or a shadow behind.

"I went forward, and ordered the chain to be hauled in short, so as to be ready to trip the anchor and move the steamboat at once if necessary. 'Will they attack?' whispered an awed voice. 'We will be all butchered in this fog,' murmured another. The faces twitched with the strain, the hands trembled slightly, the eyes forgot to wink. It was very curious to see the contrast of expressions of the white men and of the black fellows of our crew, who were as much strangers to that part of the river as we, though their homes were only eight hundred miles away. The whites, of course greatly discomposed, had besides a curious look of being painfully shocked by such an outrageous row. The others had an alert, naturally interested expression; but their faces were essentially quiet, even those of the one or two who grinned as they hauled at the chain. Several exchanged short, grunting phrases, which seemed to settle the matter to their satisfaction. Their headman, a young, broad-chested black, severely draped in dark blue fringed cloths, with fierce nostrils and his hair all done up artfully in oily ringlets, stood near me. 'Aha!' I said, just for good fellowship's sake. 'Catch 'im,' he snapped, with a bloodshot widening of his eyes and a flash of sharp teeth—'catch 'im. Give 'im to us.' 'To you, eh?' I asked; 'what would you do with them?' 'Eat 'im!' he said curtly, and, leaning his elbow on the rail, looked out into the fog in a dignified and profoundly pensive attitude. I would no doubt have been properly horrified, had it not occurred to me that he and his chaps must be very hungry: that they must have been growing increasingly hungry for at least this month past. They had been engaged for six months (I don't think a single one of them had any clear idea of time, as we at the end of countless ages have. They still belonged to the beginnings of time—had no inherited experience to teach them as it were), and of course, as long as there was a piece of paper written over in accordance with some farcical law or other made down the river, it didn't enter anybody's head to trouble how they would live. Certainly they had brought with them some rotten hippo-meat, which couldn't have lasted very long, anyway, even if the pilgrims hadn't, in the midst of a shocking hullabaloo, thrown a considerable quantity of it overboard. It looked like a high-handed proceeding; but it was really a case of legitimate self-defence. You can't breathe dead hippo waking, sleeping, and eating, and at the same time keep your precarious grip on existence. Besides that, they had given them every week three pieces of brass wire, each about nine inches long;

and the theory was they were to buy their provisions with that currency in riverside villages. You can see how *that* worked. There were either no villages, or the people were hostile, or the director, who like the rest of us fed out of tins, with an occasional old he-goat thrown in, didn't want to stop the steamer for some more or less recondite reason. So, unless they swallowed the wire itself, or made loops of it to snare the fishes with, I don't see what good their extravagant salary could be to them. I must say it was paid with a regularity worthy of a large and honourable trading company. For the rest, the only thing to eat—though it didn't look eatable in the least—I saw in their possession was a few lumps of some stuff like half-cooked dough, of a dirty lavender colour, they kept wrapped in leaves, and now and then swallowed a piece of, but so small that it seemed done more for the looks of the thing than for any serious purpose of sustenance. Why in the name of all the gnawing devils of hunger they didn't go for us—they were thirty to five—and have a good tuck-in for once, amazes me now when I think of it. They were big powerful men, with not much capacity to weigh the consequences, with courage, with strength, even yet, though their skins were no longer glossy and their muscles no longer hard. And I saw that something restraining, one of those human secrets that baffle probability, had come into play there. I looked at them with a swift quickening of interest—not because it occurred to me I might be eaten by them before very long, though I own to you that just then I perceived—in a new light, as it were—how unwholesome the pilgrims looked, and I hoped, yes, I positively hoped, that my aspect was not so—what shall I say?—so—unappetizing: a touch of fantastic vanity which fitted well with the dream-sensation that pervaded all my days at that time. Perhaps I had a little fever, too. One can't live with one's finger everlastingly on one's pulse. I had often 'a little fever,' or a little touch of other things—the playful paw-strokes of the wilderness, the preliminary trifling before the more serious onslaught which came in due course. Yes; I looked at them as you would on any human being, with a curiosity of their impulses, motives, capacities, weaknesses, when brought to the test of an inexorable physical necessity. Restraint! What possible restraint? Was it superstition, disgust, patience, fear—or some kind of primitive honour? No fear can stand up to hunger, no patience can wear it out, disgust simply does not exist where hunger is; and as to superstition, beliefs, and what you may call principles, they are less than chaff in a breeze. Don't you know the devilry of lingering starvation, its exasperating torment, its black thoughts, its sombre and brooding ferocity? Well, I do. It takes a man all his inborn strength to fight hunger properly. It's really easier to face bereavement, dishonour, and the perdition of one's soul—than this kind of prolonged hunger. Sad, but true. And these chaps, too, had no earthly reason for any kind of scruple. Restraint! I would just as soon have expected restraint from a hyena prowling amongst the corpses of a battlefield. But there was the fact facing me—the fact dazzling, to be seen, like the foam on the depths of the sea, like a ripple on an unfathomable enigma, a mystery greater—when I thought of it—than the curious, inexplicable note of desperate grief in this savage clamour that had swept by us on the river-bank, behind the blind whiteness of the fog.

"Two pilgrims were quarrelling in hurried whispers as to which bank. 'Left.' 'No, no; how can you? Right, right, of course.' 'It is very serious,' said the manager's voice behind me; 'I would be desolated if anything should happen to Mr. Kurtz before we came up.' I looked at him, and had not the slightest doubt he was sincere. He was just the kind of man who would wish to preserve appearances. That was his restraint. But when he muttered something about going on at once, I did not even take the trouble to answer him. I knew, and he knew, that it was impossible. Were we to let go our hold of the bottom, we would be absolutely in the air—in space. We wouldn't be able to tell where we were going to—whether up or down stream, or across—till we fetched against one bank or the other—and then we wouldn't know at first which it was. Of course I made no move. I had no mind for a smash-up. You couldn't imagine a more deadly place for a shipwreck. Whether drowned at once or not, we were sure to perish speedily in one way or another. 'I authorize you to take all the risks,' he said, after a short silence. 'I refuse to take any,' I said shortly; which was just the answer he expected, though its tone might have surprised him. 'Well, I must defer to your judgment. You are captain,' he said with marked civility. I turned my shoulder to him in sign of my appreciation, and looked into the fog. How long would it last? It was the most hopeless lookout. The approach to this Kurtz grubbing for ivory in the wretched bush was beset by as many dangers as though he had been an enchanted princess sleeping in a fabulous castle. 'Will they attack, do you think?' asked the manager, in a confidential tone.

"I did not think they would attack, for several obvious reasons. The thick fog was one. If they left the bank in their canoes they would get lost in it, as we would be if we attempted to move. Still, I had also judged the jungle of both banks quite impenetrable—and yet eyes were in it, eyes that had seen us. The riverside bushes were certainly very thick; but the undergrowth behind was evidently penetrable. However, during the short life I had seen no canoes anywhere in the reach—certainly not abreast of the steamer. But what made the idea of attack inconceivable to me was the nature of the noise—of the cries we had heard. They had not the fierce character boding immediate hostile intention. Unexpected, wild, and violent as they had been, they had given me an irresistible impression of sorrow. The glimpse of the steamboat had for some reason filled those savages with unrestrained grief. The danger, if any, I expounded, was from our proximity to a great human passion let loose. Even extreme grief may ultimately vent inself in violence—but more generally takes the form of apathy. . . .

"You should have seen the pilgrims stare! They had no heart to grin, or even to revile me: but I believe they thought me gone mad—with fright, maybe. I delivered a regular lecture. My dear boys, it was no good bothering. Keep a lookout? Well, you may guess I watched the fog for the signs of lifting as a cat watches a mouse; but for anything else our eyes were of no more use to us than if we had been buried miles deep in a heap of cotton-wool. It felt like it, too—choking, warm, stifling. Besides, all I said, though it sounded extravagant, was absolutely true to fact. What we afterwards alluded to as an attack was really an attempt at repulse. The action was very far from being aggressive—it was not even defensive, in the usual sense: it was undertaken under the stress of desperation, and in its essence was purely protective.

"It developed itself, I should say, two hours after the fog lifted, and its commencement was at a spot, roughly speaking, about a mile and a half below Kurtz's station. We had just floundered and flopped round a bend, when I saw an islet, a mere grassy hummock of bright green, in the middle of the stream. It was the only thing of the kind; but as we opened the reach more, I perceived it was the head of a long sandbank, or rather of a chain of shallow patches stretching down the middle of the river. They were discoloured, just awash, and the whole lot was seen just under the water, exactly as a man's backbone is seen running down the middle of his back under the skin. Now, as far as I did see, I could go to the right or to the left of this. I didn't know either channel, of course. The banks looked pretty well alike, the depth appeared the same; but as I had been informed the station was on the west side, I naturally headed for the western passage.

"No sooner had we fairly entered it than I became aware it was much narrower than I had supposed. To the left of us there was the long uninterrupted shoal, and to the right a high, steep bank heavily overgrown with bushes. Above the bush the trees stood in serried ranks. The twigs overhung the current thickly, and from distance to distance a large limb of some tree projected rigidly over the stream. It was then well on in the afternoon, the face of the forest was gloomy, and a broad strip of shadow had already fallen on the water. In this shadow we steamed up—very slowly, as you may imagine. I sheered her well inshore—the water being deepest near the bank, as the sounding-pole informed me.

"One of my hungry and forbearing friends was sounding in the bows just below me. This steamboat was exactly like a decked scow. On the deck, there were two little teakwood houses, with doors and windows. The boiler was in the fore-end, and the machinery right astern. Over the whole there was a light roof, supported on stanchions. The funnel projected through that roof, and in front of the funnel a small cabin built of light planks served for a pilot-house. It contained a couch, two camp-stools, a loaded Martini-Henry leaning in one corner, a tiny table, and the steering-wheel. It had a wide door in front and a broad shutter at each side. All these were always thrown open, of course. I spent my days perched up there on the extreme fore-end of that roof, before the door. At night I slept, or tried to, on the couch. An athletic black belonging to some coast tribe and educated by my poor predecessor, was the helmsman. He sported a pair of brass earrings, wore a blue cloth wrapper from the waist to the ankles, and thought all the world of himself. He was the most unstable kind of fool I had ever seen. He steered with no end of a swagger while you were by; but if he lost sight of you, he became instantly the prey of an abject funk, and would let that cripple of a steamboat get the upper hand of him in a minute.

"I was looking down at the sounding-pole, and feeling much annoyed to see at each try a little more of it stick out of that river, when I saw my policeman give up the business suddenly, and stretch himself flat on the deck, without even taking the trouble to haul his pole in. He kept hold on it though, and it trailed in the water. At the same time the fireman, whom I could also see below me, sat down abruptly before his furnace and ducked his head. I was amazed. Then I had to look at the river mighty quick, because there was a snag in the fairway. Sticks, little sticks, were flying about—thick: they were whizzing before my nose, dropping below me, striking behind me against my pilot-house.

All this time the river, the shore, the woods, were very quiet—perfectly quiet. I could only hear the heavy splashing thump of the stern-wheel and the patter of these things. We cleared the snag clumsily. Arrows, by Jove! We were being shot at! I stepped in quickly to close the shutter on the land-side. That fool-helmsman, his hands on the spokes, was lifting his knees high, stamping his feet, champing his mouth, like a reined-in horse. Confound him! And we were staggering within ten feet of the bank. I had to lean right out to swing the heavy shutter, and I saw a face amongst the leaves on the level with my own, looking at me very fierce and steady; and then suddenly, as though a veil had been removed from my eyes, I made out, deep in the tangled gloom, naked breasts, arms, legs, glaring eyes—the bush was swarming with human limbs in movement, glistening, of bronze colour. The twigs shook, swayed, and rustled, the arrows flew out of them, and then the shutter came to. 'Steer her straight,' I said to the helmsman. He held his head rigid, face forward; but his eyes rolled, he kept on lifting and setting down his feet gently, his mouth foamed a little. 'Keep quiet!' I said in a fury. I might just as well have ordered a tree not to sway in the wind. I darted out. Below me there was a great scuffle of feet on the iron deck; confused exclamations; a voice screamed. 'Can you turn back?' I caught sight of a V-shaped ripple on the water ahead. What? Another snag! A fusillade burst out under my feet. The pilgrims had opened with their Winchesters, and were simply squirting lead into that bush. A deuce of a lot of smoke came up and drove slowly forward. I swore at it. Now I couldn't see the ripple or the snag either. I stood in the doorway, peering, and the arrows came in swarms. They might have been poisoned, but they looked as though they wouldn't kill a cat. The bush began to howl. Our wood-cutters raised a warlike whoop; the report of a rifle just at my back deafened me. I glanced over my shoulder, and the pilot-house was yet full of noise and smoke when I made a dash at the wheel. The fool-nigger had dropped everything, to throw the shutter open and let off that Martini-Henry. He stood before the wide opening, glaring, and I yelled at him to come back, while I straightened the sudden twist out of that steamboat. There was no room to turn even if I had wanted to, the snag was somewhere very near ahead in that confounded smoke, there was no time to lose, so I just crowded her into the bank—right into the bank, where I knew the water was deep.

"We tore slowly along the overhanging bushes in a whirl of broken twigs and flying leaves. The fusillade below stopped short, as I had foreseen it would when the squirts got empty. I threw my head back to a glinting whizz that traversed the pilot-house, in at one shutter-hole and out at the other. Looking past that mad helmsman, who was shaking the empty rifle and yelling at the shore, I saw vague forms of men running bent double, leaping, gliding, distinct, incomplete, evanescent. Something big appeared in the air before the shutter, the rifle went overboard, and the man stepped back swiftly, looked at me over his shoulder in an extraordinary, profound, familiar manner, and fell upon my feet. The side of his head hit the wheel twice, and the end of what appeared a long cane clattered round and knocked over a little camp-stool. It looked as though after wrenching that thing from somebody ashore he had lost his balance in the effort. The thin smoke had blown away, we were clear of the snag, and looking ahead I could see that in another hundred yards or so I would be free to sheer off, away from the bank; but my feet felt so very warm and wet that I had to look down. The man had rolled on his back and stared straight up at me; both his hands clutched that cane. It was the shaft of a spear that, either thrown or lunged through the opening, had caught him in the side just below the ribs; the blade had gone in out of sight, after making a frightful gash; my shoes were full; a pool of blood lay very still, gleaming dark-red under the wheel; his eyes shone with an amazing lustre. The fusillade burst out again. He looked at me anxiously, gripping the spear like something precious, with an air of being afraid I would try to take it away from him. I had to make an effort to free my eyes from his gaze and attend to the steering. With one hand I felt above my head for the line of the steam whistle, and jerked out screech after screech hurriedly. The tumult of angry and warlike yells was checked instantly, and then from the depths of the woods went out such a tremulous and prolonged wail of mournful fear and utter despair as may be imagined to follow the flight of the last hope from the earth. There was a great commotion in the bush; the shower of arrows stopped, a few dropping shots rang out sharply—then silence, in which the languid beat of the stern-wheel came plainly to my ears. I put the helm hard a-starboard at the moment when the pilgrim in pink pyjamas, very hot and agitated, appeared in the doorway. 'The manager sends me——' he began in an official tone, and stopped short. 'Good God!' he said, glaring at the wounded man.

"We two whites stood over him, and his lustrous and inquiring glance enveloped us both. I declare it looked as though he would presently put to us some question in an understandable language; but he died without uttering a sound, without moving a limb, without twitching a muscle. Only in the very last moment, as though in response to some sign we could not see, to some whisper we could not hear, he frowned heavily, and that frown gave to his black death-mask an inconceivably sombre, brooding, and menacing expression. The lustre of inquiring glance faded swiftly into vacant glassiness. 'Can you steer?' I asked the agent eagerly. He looked very dubious; but I made a grab at his arm, and he understood at once I meant him to steer whether or no. To tell you the truth, I was morbidly anxious to change my shoes and socks. 'He is dead,' murmured the fellow, immensely impressed. 'No doubt about it,' said I, tugging like mad at the shoe-laces. 'And by the way, I suppose Mr. Kurtz is dead as well by this time.'

"For the moment that was the dominant thought. There was a sense of extreme disappointment, as though I had found out I had been striving after something altogether without a substance. I couldn't have been more disgusted if I had travelled all this way for the sole purpose of talking with Mr. Kurtz. Talking with . . . I flung one shoe overboard, and became aware that that was exactly what I had been looking forward to—a talk with Kurtz. I made the strange discovery that I had never imagined him as doing, you know, but as discoursing. I didn't say to myself, 'Now I will never see him,' or 'Now I will never shake him by the hand,' but, 'Now I will never hear him.' The man presented himself as a voice. Not of course that I did not connect him with some sort of action. Hadn't I been told in all the tones of jealousy and admiration that he had collected, bartered, swindled, or stolen more ivory than all the other agents together? That was not the point. The point was in his being a gifted creature, and that of all his gifts the one that stood out preëminently, that carried with it a sense of real presence, was his ability to talk, his words—the gift of expression, the bewildering, the illuminating, the most exalted and the most contemptible, the pulsating stream of light, or the deceitful flow from the heart of an impenetrable darkness.

"The other shoe went flying unto the devil-god of that river. I thought, 'By Jove! it's all over. We are too late; he has vanished—the gift has vanished, by means of some spear, arrow, or club. I will never hear that chap speak after all'—and my sorrow had a startling extravagance of emotion, even such as I had noticed in the howling sorrow of these savages in the bush. I couldn't have felt more of lonely desolation somehow, had I been robbed of a belief or had missed my destiny in life. . . . Why do you sigh in this beastly way, somebody? Absurd? Well, absurd. Good Lord! mustn't a man ever—— Here, give me some tobacco." . . .

There was a pause of profound stillness, then a match flared, and Marlow's lean face appeared, worn, hollow, with downward folds and dropped eyelids, with an aspect of concentrated attention; and as he took vigorous draws at his pipe, it seemed to retreat and advance out of the night in the regular flicker of tiny flame. The match went out.

"Absurd!" he cried. "This is the worst of trying to tell. . . . Here you all are, each moored with two good addresses, like a hulk with two anchors, a butcher round one corner, a policeman round another, excellent appetites, and temperature normal—you hear—normal from year's end to year's end. And you say, Absurd! Absurd be—exploded! Absurd! My dear boys, what can you expect from a man who out of sheer nervousness had just flung overboard a pair of new shoes! Now I think of it, it is amazing I did not shed tears. I am, upon the whole, proud of my fortitude. I was cut to the quick at the idea of having lost the inestimable privilege of listening to the gifted Kurtz. Of course I was wrong. The privilege was waiting for me. Oh, yes, I heard more than enough. And I was right, too. A voice. He was very little more than a voice. And I heard—him—it—this voice—other voices—all of them were so little more than voices—and the memory of that time itself lingers around me, impalpable, like a dying vibration of one immense jabber, silly, atrocious, sordid, savage, or simply mean, without any kind of sense. Voices, voices—even the girl herself—now——"

He was silent for a long time.

"I laid the ghost of his gifts at last with a lie," he began, suddenly. "Girl! What? Did I mention a girl? Oh, she is out of it—completely. They—the women I mean—are out of it—should be out of it. We must help them to stay in that beautiful world of their own, lest ours gets worse. Oh, she had to be out of it. You should have heard the disinterred body of Mr. Kurtz saying, 'My Intended.' You would have perceived directly then how competely she was out of it. And the lofty frontal bone of Mr. Kurtz! They say the hair goes on growing sometimes, but this—ah—specimen, was impressively bald. The wilderness had patted him on the head, and, behold, it was like a ball—an ivory ball; it had caressed him, and—lo!—he had withered; it had taken him, loved him, embraced him, got into his veins, consumed his flesh, and sealed his soul to its own by the inconceivable ceremonies of some devilish initiation. He was its spoiled and pampered favourite. Ivory? I should think so. Heaps of it, stacks of it. The old mud shanty was bursting with it. You would think there was not a single tusk left either above or below the ground in the whole country. 'Mostly fossil,' the manager had remarked, disparagingly. It was no more fossil than I am; but they call it fossil when it is dug up. It appears these niggers do bury the tusks sometimes—but evidently they couldn't bury this parcel deep enough to save the gifted Mr. Kurtz from his fate. We filled the steamboat with it, and had to pile a lot on the deck. Thus he could see and enjoy as long as he could see, because the appreciation of this favour had remained with him to the last. You should have heard him say, 'My Ivory.' Oh, yes, I heard him. 'My Intended, my ivory, my station, my river, my——' everything belonged to him. It made me hold my breath in expectation of hearing the wilderness burst into a prodigious peal of laughter that would shake the fixed stars in their places. Everything belonged to him—but that was a trifle. The thing was to know what he belonged to, how many powers of darkness claimed him for their own. That was the reflection that made you creepy all over. It was impossible—it was not good for one either—trying to imagine. He had taken a high seat amongst the devils of the land—I mean literally. You can't understand. How could you—with solid pavement under your feet, surrounded by kind neighbours ready to cheer you or to fall on you, stepping delicately between the butcher and the policeman, in the holy terror of scandal and gallows and lunatic asylums—how can you imagine what particular region of the first ages a man's untrammelled feet may take him into by the way of solitude—utter solitude without a policeman—by the way of silence—utter silence, where no warning voice of a kind neighbour can be heard whispering of public opinion? These little things make all the great difference. When they are gone you must fall back upon your own innate strength, upon your own capacity for faithfulness. Of course you may be too much of a fool to go wrong—too dull even to know you are being assaulted by the powers of darkness. I take it, no fool ever made a bargain for his soul with the devil; the fool is too much of a fool, or the devil too much of a devil—I don't know which. Or you may be such a thunderingly exalted creature as to be altogether deaf and blind to anything but heavenly sights and sounds. Then the earth for you is only a standing place—and whether to be like this is your loss or your gain I won't pretend to say. But most of us are neither one nor the other. The earth for us is a place to live in, where we must put up with the sights, with sounds, with smells, too, by Jove!—breathe dead hippo, so to speak, and not be contaminated. And there, don't you see? Your strength comes in, the faith in your ability for the digging of unostentatious holes to bury the stuff in—your power of devotion, not to yourself, but to an obscure, back-breaking business. And that's difficult enough. Mind, I am not trying to excuse or even explain—I am trying to account to myself for—for—Mr. Kurtz—for the shade of Mr. Kurtz. This initiated wraith from the back of Nowhere honoured me with its amazing confidence before it vanished altogether. This was because it could speak English to me. The original Kurtz had been educated partly in England, and—as he was good enough to say himself—his sympathies were in the right place. His mother was half-English, his father was half-French. All Europe contributed to the making of Kurtz; and by and by I learned that, most appropriately, the International Society for the Suppression of Savage Customs had intrusted him with the making of a report, for its future guidance. And he had written it, too. I've seen it. I've read it. It was eloquent, vibrating with eloquence, but too high-strung, I think. Seventeen pages of close writing he had found time for! But this must have been before his—let us say—nerves, went wrong, and caused him to preside at certain midnight dances ending with unspeakable rites, which—as far as I reluctantly gathered from what I heard at various times—were offered up to him—do you understand?—to Mr. Kurtz himself. But it was a beautiful piece of writing. The opening paragraph, however, in the light of later information, strikes me now as ominous. He began with the argument that we whites, from the point of development we had arrived at, 'must necessarily appear to them [savages] in the nature of supernatural beings—we approach them with the might as of a deity,' and so on, and so on. 'By the simple exercise of our will we can exert a power for good practically unbounded,' etc., etc. From that point he soared and took me with him. The peroration was magnificent, though difficult to remember, you know. It gave me the notion of an exotic Immensity ruled by an august Benevolence. It made me tingle with enthusiasm. This was the unbounded power of eloquence—of words—of burning noble words. There were no practical hints to interrupt the magic current of phrases, unless a kind of note at the foot of the last page, scrawled evidently much later, in an unsteady hand, may be regarded as the exposition of a method. It was very simple, and at the end of that moving appeal to every altruistic sentiment it blazed at you, luminous and terrifying, like a flash of lightning in a serene sky: 'Exterminate all the brutes!' The curious part was that he had apparently forgotten all about that valuable postcriptum, because, later on, when he in a sense came to himself, he repeatedly entreated me to take good care of 'my pamphlet' (he called it), as it was sure to have in the future a good influence upon his career. I had full information about all these things, and, besides, as it turned out, I was to have the care of his memory. I've done enough for it to give me the indisputable right to lay it, if I choose, for an everlasting rest in the dust-bin of progress, amongst all the sweepings and, figuratively speaking, all the dead cats of civilization. But then, you see, I can't choose. He won't be forgotten. Whatever he was, he was not common. He had the power to charm or frighten rudimentary souls into an aggravated witch-dance in his honour; he could also fill the small souls of the pilgrims with bitter misgivings: he had one devoted friend at least, and he had conquered one soul in the world that was neither rudimentary nor tainted with self-seeking. No; I can't forget him, though I am not prepared to affirm the fellow was exactly worth the life we lost in getting to him. I missed my late helmsman awfully—I missed him even while his body was still lying in the pilot-house. Perhaps you will think it passing strange this regret for a savage who was of no more account than a grain of sand in a black Sahara. Well, don't you see, he had done something, he had steered; for months I had him at my back—a help—an instrument. It was a kind of partnership. He steered for me—I had to look after him, I worried about his deficiencies, and thus a subtle bond had been created, of which I only became aware when it was suddenly broken. And the intimate profundity of that look he gave me when he received his hurt remains to this day in my memory—like a claim of distant kinship affirmed in a supreme moment.

"Poor fool! If he had only left that shutter alone. He had no restraint, no restraint—just like Kurtz—a tree swayed by the wind. As soon as I had put on a dry pair of slippers, I dragged him out, after first jerking the spear out of his side, which operation I confess I performed with my eyes shut tight. His heels leaped together over the little doorstep; his shoulders were pressed to my breast; I hugged him from behind desperately. Oh! he was heavy, heavy; heavier than any man on earth,

I should imagine. Then without more ado I tipped him overboard. The current snatched him as though he had been a wisp of grass, and I saw the body roll over twice before I lost sight of it for ever. All the pilgrims and the manager were then congregated on the awning-deck about the pilot-house, chattering at each other like a flock of excited magpies, and there was a scandalized murmur at my heartless promptitude. What they wanted to keep that body hanging about for I can't guess. Embalm it, maybe. But I had also heard another, and a very ominous, murmur on the deck below. My friends the wood-cutters were likewise scandalized, and with a better show of reason—though I admit that the reason itself was quite inadmissible. Oh, quite! I had made up my mind that if my late helmsman was to be eaten, the fishes alone should have him. He had been a very second-rate helmsman while alive, but now he was dead he might have become a first-class temptation, and possibly cause some startling trouble. Besides, I was anxious to take the wheel, the man in pink pyjamas showing himself a hopeless duffer at the business.

"This I did directly the simple funeral was over. We were going half-speed, keeping right in the middle of the stream, and I listened to the talk about me. They had given up Kurtz, they had given up the station; Kurtz was dead, and the station had been burnt—and so on—and so on. The red-haired pilgrim was beside himself with the thought that at least this poor Kurtz had been properly avenged. 'Say! We must have made a glorious slaughter of them in the bush. Eh? What do you think? Say?' He positively danced, the bloodthirsty little gingery beggar. And he had nearly fainted when he saw the wounded man! I could not help saying, 'You made a glorious lot of smoke, anyhow.' I had seen, from the way the tops of the bushes rustled and flew, that almost all the shots had gone too high. You can't hit anything unless you take aim and fire from the shoulder; but these chaps fired from the hip with their eyes shut. The retreat, I maintained—and I was right—was caused by the screeching of the steam whistle. Upon this they forgot Kurtz, and began to howl at me with indignant protests.

"The manager stood by the wheel murmuring confidentially about the necessity of getting well away down the river before dark at all events, when I saw in the distance a clearing on the riverside and the outlines of some sort of building. 'What's this?' I asked. He clapped his hands in wonder. 'The station!' he cried. I edged in at once, still going half-speed.

"Through my glasses I saw the slope of a hill interspersed with rare trees and perfectly free from undergrowth. A long decaying building on the summit was half buried in the high grass; the large holes in the peaked roof gaped black from afar; the jungle and the woods made a background. There was no enclosure or fence of any kind; but there had been one apparently, for near the house half-a-dozen slim posts remained in a row, roughly trimmed, and with their upper ends ornamented with round carved balls. The rails, or whatever there had been between, had disappeared. Of course the forest surrounded all that. The river-bank was clear, and on the waterside I saw a white man under a hat like a cartwheel beckoning persistently with his whole arm. Examining the edge of the forest above and below, I was almost certain I could see movements—human forms gliding here and there. I steamed past prudently, then stopped the engines and let her drift down. The man on the shore began to shout, urging us to land. 'We have been attacked,' screamed the manager. 'I know—I know. It's all right,' yelled back the other, as cheerful as you please. 'Come along. It's all right. I am glad.'

"His aspect reminded me of something I had seen—something funny I had seen somewhere. As I manœuvered to get alongside, I was asking myself, 'What does this fellow look like?' Suddenly I got it. He looked like a harlequin. His clothes had been made of some stuff that was brown holland probably, but it was covered with patches all over, with bright patches, blue, red, and yellow—patches on the back, patches on the front, patches on elbows, on knees; coloured binding around his jacket, scarlet edging at the bottom of his trousers; and the sunshine made him look extremely gay and wonderfully neat withal, because you could see how beautifully all this patching had been done. A beardless, boyish face, very fair, no features to speak of, nose peeling, little blue eyes, smiles and frowns chasing each other over that open countenance like sunshine and shadow on a wind-swept plain. 'Look out, captain!' he cried; 'there's a snag lodged in here last night.' What! Another snag? I confess I swore shamefully. I had nearly holed my cripple, to finish off that charming trip. The harlequin on the bank turned his little pug-nose up to me. 'You English?' he asked, all smiles. 'Are you?' I shouted from the wheel. The smiles vanished, and he shook his head as if sorry for my disappointment. Then he brightened up. 'Never mind!' he cried encouragingly. 'Are we in time?' I asked. 'He is up there,' he replied, with a toss of the head up the hill, and becoming gloomy all of a sudden. His face was like the autumn sky, overcast one moment and bright the next.

"When the manager, escorted by the pilgrims, all of them armed to the teeth, had gone to the house this chap came on board. 'I say, I don't like this. These natives are in the bush,' I said. He assured me earnestly it was all right. 'They are simple people,' he added; 'well, I am glad you came. It took me all my time to keep them off.' 'But you said it was all right,' I cried. 'Oh, they meant no harm,' he said; and as I stared he corrected himself, 'Not exactly.' Then vivaciously, 'My faith, your pilot-house wants a clean-up!' In the next breath he advised me to keep enough steam on the boiler to blow the whistle in case of any trouble. 'One good screech will do more for you than all your rifles. They are simple people,' he repeated. He rattled away at such a rate he quite overwhelmed me. He seemed to be trying to make up for lots of silence, and actually hinted, laughing, that such was the case. 'Don't you talk with Mr. Kurtz?' I said. 'You don't talk with that man—you listen to him,' he exclaimed with severe exaltation. 'But now——' He waved his arm, and in the twinkling of an eye was in the uttermost depths of despondency. In a moment he came up again with a jump, possessed himself of both my hands, shook them continuously, while he gabbled: 'Brother sailor . . . honour . . . pleasure . . . delight . . . introduce myself . . . Russian . . . son of an arch-priest . . . Government of Tambov . . . What? Tobacco! English tobacco; the excellent English tobacco! Now, that's brotherly. Smoke? Where's a sailor that does not smoke?'

"The pipe soothed him, and gradually I made out he had run away from school, had gone to sea in a Russian ship; ran away again; served some time in English ships; was now reconciled with the arch-priest. He made a point of that. 'But when one is young one must see things, gather experience, ideas; enlarge the mind.' 'Here!' I interrupted. 'You can never tell! Here I met Mr. Kurtz,' he said, youthfully solemn and reproachful. I held my tongue after that. It appears he had persuaded a Dutch trading-house on the coast to fit him out with stores and goods, and had started for the interior with a light heart and no more idea of what would happen to him than a baby. He had been wandering about that river for nearly two years alone, cut off from everybody and everything. 'I am not so young as I look. I am twenty-five,' he said. 'At first old Van Shuyten would tell me to go to the devil,' he narrated with keen enjoyment; 'but I stuck to him, and talked and talked, till at last he got afraid I would talk the hind-leg off his favourite dog, so he gave me some cheap things and a few guns, and told me he hoped he would never see my face again. Good old Dutchman, Van Shuyten. I've sent him one small lot of ivory a year ago, so that he can't call me a little thief when I get back. I hope he got it. And for the rest I don't care. I had some wood stacked for you. That was my old house. Did you see?'

"I gave him Towson's book. He made as though he would kiss me, but restrained himself. 'The only book I had left, and I thought I had lost it,' he said, looking at it ecstatically. 'So many accidents happen to a man going about alone, you know. Canoes get upset sometimes—and sometimes you've got to clear out so quick when the people get angry.' He thumbed the pages. 'You made notes in Russian?' I asked. He nodded. "I thought they were written in cipher,' I said. He laughed, then became serious. 'I had lots of trouble to keep these people off,' he said. 'Did they want to kill you?' I asked. 'Oh, no!' he cried, and checked himself. 'Why did they attack us?' I pursued. He hesitated, then said shamefacedly, 'They don't want him to go.' 'Don't they?' I said curiously. He nodded a nod full of mystery and wisdom. 'I tell you,' he cried, 'this man has enlarged my mind.' He opened his arms wide, staring at me with his little blue eyes that were perfectly round."

III

"I LOOKED at him, lost in astonishment. There he was before me, in motley, as though he had absconded from a troupe of mimes, enthusiastic, fabulous. His very existence was improbable, inexplicable, and altogether bewildering. He was an insoluble problem. It was inconceivable how he had existed, how he had succeeded in getting so far, how he had managed to remain—why he did not instantly disappear. 'I went a little farther,' he said, 'then still a little farther—till I had gone so far that I don't know how I'll ever get back. Never mind. Plenty time. I can manage. You take Kurtz away quick—quick—I tell you.' The glamour of youth enveloped his parti-coloured rags, his destitution, his loneliness, the essential desolation of his futile wanderings. For months—for years—his life hadn't been worth a day's purchase; and there he was gallantly, thoughtlessly alive, to all appearance indestructible solely by the virtue of his few years and of his unreflecting audacity. I was seduced into something like admiration—like envy. Glamour urged him on, glamour kept him unscathed. He surely wanted nothing from the wilderness but space to breathe in and to push on through. His need was to exist, and to move onwards at the greatest possible risk, and with a maximum of privation. If the absolutely pure, uncalculating, unpractical spirit of adventure had ever ruled a human being, it ruled this bepatched youth. I almost envied him the possession of this modest and clear flame. It seemed to have consumed all thought of self so completely, that even while he was talking to you, you forgot that it was he—the man before your eyes—who had gone through these things. I did not envy him his devotion to Kurtz, though. He had not meditated over it. It came to him, and he accepted it with a sort of eager fatalism. I must say that to me it appeared about the most dangerous thing in every way he had come upon so far.

"They had come together unavoidably, like two ships becalmed near each other, and lay rubbing sides at last. I suppose Kurtz wanted an audience, because on a certain occasion, when encamped in the forest, they had talked all night, or more probably Kurtz had talked. 'We talked of everything,' he said, quite transported at the recollection. 'I forgot there was such a thing as sleep. The night did not seem to last an hour. Everything! Everything . . . Of love, too.' 'Ah, he talked to you of love!' I said, much amused. 'It isn't what you think,' he cried, almost passionately. 'It was in general. He made me see things—things.'

"He threw his arms up. We were on deck at the time, and the headman of my wood-cutters, lounging near by, turned upon him his heavy and glittering eyes. I looked around, and I don't know why, but I assure you that never, never before, did this land, this river, this jungle, the very arch of this blazing sky, appear to me so hopeless and so dark, so impenetrable to human thought, so pitiless to human weakness. 'And, ever since, you have been with him, of course?' I said.

"On the contrary. It appears their intercourse had been very much broken by various causes. He had, as he informed me proudly, managed to nurse Kurtz through two illnesses (he alluded to it as you would to some risky feat), but as a rule Kurtz wandered alone, far in the depths of the forest. 'Very often coming to this station, I had to wait days and days before he would turn up,' he said. 'Ah, it was worth waiting for! —sometimes' 'What was he doing? exploring or what?' I asked. 'Oh, yes, of course'; he had discovered lots of villages, a lake, too—he did not know exactly in what direction; it was dangerous to inquire too much—but mostly his expeditions had been for ivory. 'But he had no goods to trade with by that time,' I objected. 'There's a good lot of cartridges left even yet,' he answered, looking away. 'To speak plainly, he raided the country,' I said. He nodded. 'Not alone, surely!' He muttered something about the villages round that lake. 'Kurtz got the tribe to follow him, did he?' I suggested. He fidgeted a little. 'They adored him,' he said. The tone of these words was so extraordinary that I looked at him searchingly. It was curious to see his mingled eagerness and reluctance to speak of Kurtz. The man filled his life, occupied his thoughts, swayed his emotions. 'What can you expect?' he burst out; 'he came to them with thunder and lightning, you know—and they had never seen anything like it—and very terrible. He could be very terrible. You can't judge Mr. Kurtz as you would an ordinary man. No, no, no! Now—just to give you an idea—I don't mind telling you, he wanted to shoot me, too, one day—but I don't judge him.' 'Shoot you!' I cried 'What for?' 'Well, I had a small lot of ivory the chief of that village near my house gave me. You see I used to shoot game for them. Well, he wanted it, and wouldn't hear reason. He declared he would shoot me unless I gave him the ivory and then cleared out of the country, because he could do so, and had a fancy for it, and there was nothing on earth to prevent him killing whom he jolly well pleased. And it was true, too. I gave him the ivory. What did I care! But I didn't clear out. No, no. I couldn't leave him. I had to be careful, of course, till we got friendly again for a time. He had his second illness then. Afterwards I had to keep out of the way; but I didn't mind. He was living for the most part in those villages on the lake. When he came down to the river, sometimes he would take to me, and sometimes it was better for me to be careful. This man suffered too much. He hated all this, and somehow he couldn't get away. When I had a chance I begged him to try and leave while there was time; I offered to go back with him. And he would say yes, and then he would remain; go off on another ivory hunt; disappear for weeks; forget himself amongst these people—forget himself—you know.' 'Why! he's mad,' I said. He protested indignantly. Mr. Kurtz couldn't be mad. If I had heard him talk, only two days ago, I wouldn't dare hint at such a thing. . . . I had taken up my binoculars while we talked, and was looking at the shore, sweeping the limit of the forest at each side and at the back of the house. The consciousness of there being people in that bush, so silent, so quiet—as silent and quiet as the ruined house on the hill—made me uneasy. There was no sign on the face of nature of this amazing tale that was not so much told as suggested to me in desolate exclamations, completed by shrugs, in interrupted phrases, in hints ending in deep sighs. The woods were unmoved, like a mask—heavy, like the closed door of a prison—they looked with their air of hidden knowledge, of patient expectation, of unapproachable silence. The Russian was explaining to me that it was only lately that Mr. Kurtz had come down to the river, bringing along with him all the fighting men of that lake tribe. He had been absent for several months—getting himself adored, I suppose—and had come down unexpectedly, with the intention to all appearance of making a raid either across the river or down stream. Evidently the appetite for more ivory had got the better of the—what shall I say?—less material aspirations. However he had got much worse suddenly. 'I heard he was lying helpless, and so I came up—took my chance,' said the Russian. 'Oh, he is bad, very bad.' I directed my glass to the house. There were no signs of life, but there was the ruined roof, the long mud wall peeping above the grass, with three little square window-holes, no two of the same size; all this brought within reach of my hand, as it were. And then I made a brusque movement, and one of the remaining posts of that vanished fence leaped up in the field of my glass. You remember I told you I had been struck at the distance by certain attempts at ornamentation, rather remarkable in the ruinous aspect of the place. Now I had suddenly a nearer view, and its first result was to make me throw my head back as if before a blow. Then I went carefully from post to post with my glass, and I saw my mistake. These round knobs were not ornamental but symbolic; they were expressive and puzzling, striking and disturbing—food for thought and also for vultures if there had been any looking down from the sky; but at all events for such ants as were industrious enough to ascend the pole. They would have been even more impressive, those heads on the stakes, if their faces had not been turned to the house. Only one, the first I had made out, was facing my way. I was not so shocked as you may think. The start back I had given was really nothing but a movement of surprise. I had expected to see a knob of wood there, you know. I returned deliberately to the first I had seen—and there it was, black, dried, sunken, with closed eyelids—a head that seemed to sleep at the top of that pole, and, with the shrunken dry lips showing a narrow white line of the teeth, was smiling, too, smiling continuously at some endless and jocose dream of that eternal slumber.

"I am not disclosing any trade secrets. In fact, the manager said afterwards that Mr. Kurtz's methods had ruined the district. I have no opinion on that point, but I want you clearly to understand that there was nothing exactly profitable in these heads being there. They only showed that Mr. Kurtz lacked restraint in the gratification of his various lusts, that there was something wanting in him—some small matter which, when the pressing need arose, could not be found under his magnificent eloquence. Whether he knew of this deficiency himself I can't say. I think the knowledge came to him at last— only at the very last. But the wilderness had found him out early, and had taken on him a terrible vengeance for the fantastic invasion. I think it had whispered to him things about himself which he did not know, things of which he had no conception till he took counsel with this great solitude—and the whisper had proved irresistibly fascinating. It echoed loudly within him because he was hollow at the core. . . . I put down the glass, and the head that had appeared near enough to be spoken to seemed at once to have leaped away from me into inaccessible distance.

"The admirer of Mr. Kurtz was a bit crestfallen. In a hurried, indistinct voice he began to assure me he had not dared to take these—say, symbols—down. He was not afraid of the natives; they would not stir till Mr. Kurtz gave the word. His ascendancy was extraordinary. The camps of these people surrounded the place, and the chiefs came every day to see him. They would crawl. . . . 'I don't want to know anything of the ceremonies used when approaching Mr. Kurtz,' I shouted. Curious, this feeling that came over me that such details would be more intolerable than those heads drying on the stakes under Mr. Kurtz's windows. After all, that was only a savage sight, while I seemed at one bound to have been transported into some lightless region of subtle horrors, where pure, uncomplicated savagery was a positive relief, being something that had a right to exist—obviously—in the sunshine. The young man looked at me with surprise. I suppose it did not occur to him that Mr. Kurtz was no idol of mine. He forgot I hadn't heard any of these splendid monologues on, what was it? on love, justice, conduct of life—or what not. If it had come to crawling before Mr. Kurtz, he crawled as much as the veriest savage of them all. I had no idea of the conditions, he said: these heads were the heads of rebels.

I shocked him excessively by laughing. Rebels! What would be the next definition I was to hear? There had been enemies, criminals, workers—and these were rebels. Those rebellious heads looked very subdued to me on their sticks. 'You don't know how such a life tries a man like Kurtz,' cried Kurtz's last disciple. 'Well, and you?' I said. 'I! I! I am a simple man. I have no great thoughts. I want nothing from anybody. How can you compare me to . . . ?' His feelings were too much for speech, and suddenly he broke down. 'I don't understand,' he groaned. 'I've been doing my best to keep him alive, and that's enough. I had no hand in all this. I have no abilities. There hasn't been a drop of medicine or a mouthful of invalid food for months here. He was shamefully abandoned. A man like this, with such ideas. Shamefully! Shamefully! I—I—haven't slept for the last ten nights. . . .'

"His voice lost itself in the calm of the evening. The long shadows of the forest had slipped downhill while we talked, had gone far beyond the ruined hovel, beyond the symbolic row of stakes. All this was in the gloom, while we down there were yet in the sunshine, and the stretch of the river abreast of the clearing glittered in a still and dazzling splendour, with a murky and overshadowed bend above and below. Not a living soul was seen on the shore. The bushes did not rustle.

"Suddenly round the corner of the house a group of men appeared, as though they had come up from the ground. They waded waist-deep in the grass, in a compact body, bearing an improvised stretcher in their midst. Instantly, in the emptiness of the landscape, a cry arose whose shrillness pierced the still air like a sharp arrow flying straight to the very heart of the land; and, as if by enchantment, streams of human beings—of naked human beings—with spears in their hands, with bows, with shields, with wild glances and savage movements, were poured into the clearing by the dark-faced and pensive forest. The bushes shook, the grass swayed for a time, and then everything stood still in attentive immobility.

" 'Now, if he does not say the right thing to them we are all done for,' said the Russian at my elbow. The knot of men with the stretcher had stopped, too, halfway to the steamer, as if petrified. I saw the man on the stretcher sit up, lank and with an uplifted arm, above the shoulders of the bearers. 'Let us hope that the man who can talk so well of love in general will find some particular reason to spare us this time,' I said. I resented bitterly the absurd danger of our situation, as if to be at the mercy of that atrocious phantom had been a dishonouring necessity. I could not hear a sound, but through my glasses I saw the thin arm extended commandingly, the lower jaw moving, the eyes of that apparition shining darkly far in its bony head that nodded with grotesque jerks. Kurtz—Kurtz—that means short in German—don't it? Well, the name was as true as everything else in his life—and death. He looked at least seven feet long. His covering had fallen off, and his body emerged from it pitiful and appalling as from a winding-sheet. I could see the cage of his ribs all astir, the bones of his arm waving. It was as though an animated image of death carved out of old ivory had been shaking its hand with menaces at a motionless crowd of men made of dark and glittering bronze. I saw him open his mouth wide—it gave him a weirdly voracious aspect, as though he had wanted to swallow all the air, all the earth, all the men before him. A deep voice reached me faintly. He must have been shouting. He fell back suddenly. The stretcher shook as the bearers staggered forward again, and almost at the same time I noticed that the crowd of savages was vanishing without any perceptible movement of retreat, as if the forest that had ejected these beings so suddenly had drawn them in again as the breath is drawn in a long aspiration.

"Some of the pilgrims behind the stretcher carried his arms—two shot-guns, a heavy rifle, and a light revolver-carbine—the thunderbolts of that pitiful Jupiter. The manager bent over him murmuring as he walked beside his head. They laid him down in one of the little cabins —just a room for a bed place and a camp-stool or two, you know. We had brought his belated correspondence, and a lot of torn envelopes and open letters littered his bed. His hand roamed feebly amongst these papers. I was struck by the fire of his eyes and the composed languor of his expression. It was not so much the exhaustion of disease. He did not seem in pain. This shadow looked satiated and calm, as though for the moment it had had its fill of all the emotions.

"He rustled one of the letters, and looking straight in my face said, 'I am glad.' Somebody had been writing to him about me. These special recommendations were turning up again. The volume of tone he emitted without effort, almost without the trouble of moving his lips, amazed me. A voice! a voice! It was grave, profound, vibrating, while the man did not seem capable of a whisper. However, he had enough strength in him—factitious no doubt—to very nearly make an end of us, as you shall hear directly.

"The manager appeared silently in the doorway; I stepped out at once and he drew the curtain after me. The Russian, eyed curiously by the pilgrims, was staring at the shore. I followed the direction of his glance.

"Dark human shapes could be made out in the distance, flitting indistinctly against the gloomy border of the forest, and near the river two bronze figures, leaning on tall spears, stood in the sunlight under fantastic head-dresses of spotted skins, warlike and still in statuesque repose. And from right to left along the lighted shore moved a wild and gorgeous apparition of a woman.

"She walked with measured steps, draped in striped and fringed cloths, treading the earth proudly, with a slight jingle and flash of barbarous ornaments. She carried her head high; her hair was done in the shape of a helmet; she had brass leggings to the knee, brass wire gauntlets to the elbow, a crimson spot on her tawny cheek, innumerable necklaces of glass beads on her neck; bizarre things, charms, gifts of witch-men, that hung about her, glittered and trembled at every step. She must have had the value of several elephant tusks upon her. She was savage and superb, wild-eyed and magnificent; there was something ominous and stately in her deliberate progress. And in the hush that had fallen suddenly upon the whole sorrowful land, the immense wilderness, the colossal body of the fecund and mysterious life seemed to look at her, pensive, as though it had been looking at the image of its own tenebrous and passionate soul.

"She came abreast of the steamer, stood still, and faced us. Her long shadow fell to the water's edge. Her face had a tragic and fierce aspect of wild sorrow and of dumb pain mingled with the fear and of some struggling, half-shaped resolve. She stood looking at us without a stir, and like the wilderness itself, with the air of brooding over an inscrutable purpose. A whole minute passed, and then she made a step forward. There was a low jingle, a glint of yellow metal, a sway of fringed draperies, and she stopped as if her heart had failed her. The young fellow by my side growled. The pilgrims murmured at my back. She looked at us all as if her life had depended upon the unswerving steadiness of her glance. Suddenly she opened her bared arms and threw them up rigid above her head, as though in an uncontrollable desire to touch the sky, and at the same time the swift shadows darted out on the earth, swept around on the river, gathering the steamer into a shadowy embrace. A formidable silence hung over the scene.

"She turned away slowly, walked on, following the bank, and passed into the bushes to the left. Once only her eyes gleamed back at us in the dusk of the thickets before she disappeared.

" 'If she had offered to come aboard I really think I would have tried to shoot her,' said the man of patches, nervously. 'I have been risking my life every day for the last fortnight to keep her out of the house. She got in one day and kicked up a row about those miserable rags I picked up in the storeroom to mend my clothes with. I wasn't decent. At least it must have been that, for she talked like a fury to Kurtz for an hour, pointing at me now and then. I don't understand the dialect of this tribe. Luckily for me, I fancy Kurtz felt too ill that day to care, or there would have been mischief. I don't understand. . . . No—it's too much for me. Ah, well, it's all over now.'

"At this moment I heard Kurtz's deep voice behind the curtain: 'Save me!—save the ivory, you mean. Don't tell me. Save *me!* Why, I've had to save you. You are interrupting my plans now. Sick! Sick! Not so sick as you would like to believe. Never mind. I'll carry my ideas out yet—I will return. I'll show you what can be done. You with your little peddling notions—you are interfering with me. I will return. I. . . .'

"The manager came out. He did me the honour to take me under the arm and lead me aside. 'He is very low, very low,' he said. He considered it necessary to sigh, but neglected to be consistently sorrowful. 'We have done all we could for him—haven't we? But there is no disguising the fact, Mr. Kurtz has done more harm than good to the Company. He did not see the time was not ripe for vigorous action. Cautiously, cautiously—that's my principle. We must be cautious yet. The district is closed to us for a time. Deplorable! Upon the whole, the trade will suffer. I don't deny there is a remarkable quantity of ivory—mostly fossil. We must save it, at all events—but look how precarious the position is—and why? Because the method is unsound.' 'Do you,' said I, looking at the shore, 'call it "unsound method?" ' 'Without doubt,' he exclaimed hotly. 'Don't you?' . . . 'No method at all,' I murmured after a while. 'Exactly,' he exulted. 'I anticipated this. Shows a complete want of judgment. It is my duty to point it out in the proper quarter.' 'Oh,' said I, 'that fellow—what's his name?—the brickmaker, will make a readable report for you.' He appeared confounded for a moment. It seemed to me I had never breathed an atmosphere so vile, and I turned mentally to Kurtz for relief—positively for relief. 'Nevertheless I think Mr. Kurtz is a remarkable man,' I said with emphasis. He started, dropped on me a cold heavy glance, said very quietly, 'he *was*,' and turned his back on me. My hour of favour was over; I found myself lumped along with Kurtz as a partisan of methods for which the time was not ripe: I was unsound! Ah! but it was something to have at least a choice of nightmares.

"I had turned to the wilderness really, not to Mr. Kurtz, who, I was ready to admit, was as good as buried. And for a moment it seemed to me as if I also were buried in a vast grave full of unspeakable secrets. I felt an intolerable weight oppressing my breast, the smell of the damp earth, the unseen presence of victorious corruption, the darkness of an impenetrable night. . . . The Russian tapped me on the shoulder. I heard him mumbling and stammering something about 'brother seaman—couldn't conceal—knowledge of matters that would affect Mr. Kurtz's reputation.' I waited. For him evidently Mr. Kurtz was not in his grave; I suspect that for him Mr. Kurtz was one of the immortals. 'Well!' said I at last, 'speak out. As it happens, I am Mr. Kurtz's friend —in a way.'

"He stated with a good deal of formality that had we not been 'of the same profession,' he would have kept the matter to himself without regard to consequences. 'He suspected there was an active ill-will towards him on the part of these white men that——' 'You are right,' I said, remembering a certain conversation I had overheard. 'The manager thinks you ought to be hanged.' He showed a concern at this intelligence which amused me at first. 'I had better get out of the way quietly,' he said earnestly. 'I can do no more for Kurtz now, and they would soon find some excuse. What's to stop them? There's a military post three hundred miles from here.' 'Well, upon my word,' said I, 'perhaps you had better go if you have any friends amongst the savages near by.' 'Plenty,' he said. 'They are simple people—and I want nothing, you know.' He stood biting his lip, then: 'I don't want any harm to happen to these whites here, but of course I was thinking of Mr. Kurtz's reputation—but you are a brother seaman and——' 'All right,' said I, after a time. 'Mr. Kurtz's reputation is safe with me.' I did not know how truly I spoke.

"He informed me, lowering his voice, that it was Kurtz who had ordered the attack to be made on the steamer. 'He hated sometimes the idea of being taken away—and then again. . . . But I don't understand these matters. I am a simple man. He thought it would scare you away—that you would give it up, thinking him dead. I could not stop him. Oh, I had an awful time of it this last month.' 'Very well,' I said. 'He is all right now,' 'Ye-e-es,' he muttered, not very convinced apparently. 'Thanks,' said I; 'I shall keep my eyes open.' 'But quiet—eh?' he urged anxiously. 'It would be awful for his reputation if anybody here ——' I promised a complete discretion with great gravity. 'I have a canoe and three black fellows waiting not very far. I am off. Could you give me a few Martini-Henry cartridges?' I could, and did, with proper secrecy. He helped himself, with a wink at me, to a handful of my tobacco. 'Between sailors—you know—good English tobacco.' At the door of the pilot-house he turned round—'I say, haven't you a pair of shoes you could spare?' He raised one leg. 'Look.' The soles were tied with knotted strings sandalwise under his bare feet. I rooted out an old pair, at which he looked with admiration before tucking it under his left arm. One of his pockets (bright red) was bulging with cartridges, from the other (dark blue) peeped 'Towson's Inquiry,' etc., etc. He seemed to think himself excellently well equipped for a renewed encounter with the wilderness. 'Ah! I'll never, never meet such a man again. You ought to have heard him recite poetry—his own, too, it was, he told me. Poetry!' He rolled his eyes at the recollection of these delights. 'Oh, he enlarged my mind!' 'Good-bye,' said I. He shook hands and vanished in the night. Sometimes I ask myself whether I had ever really seen him—whether it was possible to meet such a phenomenon! . . .

"When I woke up shortly after midnight his warning came to my mind with its hint of danger that seemed, in the starred darkness, real enough to make me get up for the purpose of having a look round. On the hill a big fire burned, illuminating fitfully a crooked corner of the station-house. One of the agents with a picket of a few of our blacks, armed for the purpose, was keeping guard over the ivory; but deep within the forest, red gleams that wavered, that seemed to sink and rise from the ground amongst confused columnar shapes of intense blackness, showed the exact position of the camp where Mr. Kurtz's adorers were keeping their uneasy vigil. The monotonous beating of a big drum filled the air with muffled shocks and a lingering vibration. A steady droning sound of many men chanting each to himself some weird incantation came out from the black, flat wall of the woods as the humming of bees comes out of a hive, and had a strange narcotic effect upon my half-awake senses. I believe I dozed off leaning over the rail, till an abrupt burst of yells, an overwhelming outbreak of a pent-up and mysterious frenzy, woke me up in a bewildered wonder. It was cut short all at once, and the low droning went on with an effect of audible and soothing silence. I glanced casually into the little cabin. A light was burning within, but Mr. Kurtz was not there.

"I think I would have raised an outcry if I had believed my eyes. But I didn't believe them at first—the thing seemed so impossible. The fact is I was completely unnerved by a sheer blank fright, pure abstract terror, unconnected with any distinct shape of physical danger. What made this emotion so overpowering was—how shall I define it?—the moral shock I received, as if something altogether monstrous, intolerable to thought and odious to the soul, had been thrust upon me unexpectedly. This lasted of course the merest fraction of a second, and then the usual sense of commonplace, deadly danger, the possibility of a sudden onslaught and massacre, or something of the kind, which I saw impending, was positively welcome and composing. It pacified me, in fact, so much that I did not raise an alarm.

"There was an agent buttoned up inside an ulster and sleeping on a chair on deck within three feet of me. The yells had not awakened him; he snored very slightly; I left him to his slumbers and leaped ashore. I did not betray Mr. Kurtz—it was ordered I should never betray him—it was written I should be loyal to the nightmare of my choice. I was anxious to deal with this shadow by myself alone—and to this day I don't know why I was so jealous of sharing with any one the peculiar blackness of that experience.

"As soon as I got on the bank I saw a trail—a broad trail through the grass. I remembered the exaltation with which I said to myself, 'He can't walk—he is crawling on all-fours—I've got him.' The grass was wet with dew. I strode rapidly with clenched fists. I fancy I had some vague notion of falling upon him and giving him a drubbing. I don't know. I had some imbecile thoughts. The knitting old woman with the cat obtruded herself upon my memory as a most improper person to be sitting at the other end of such an affair. I saw a row of pilgrims squirting lead in the air out of Winchesters held to the hip. I thought I would never get back to the steamer, and imagined myself living alone and unarmed in the woods to an advanced age. Such silly things—you know. And I remember I confounded the beat of the drum with the beating of my heart, and was pleased at its calm regularity.

"I kept to the track though—then stopped to listen. The night was very clear; a dark blue space, sparkling with dew and starlight, in which black things stood very still. I thought I could see a kind of motion ahead of me. I was strangely cocksure of everything that night. I actually left the track and ran in a wide semicircle (I verily believe chuckling to myself) so as to get in front of that stir, of that motion I had seen—if indeed I had seen anything. I was circumventing Kurtz as though it had been a boyish game.

"I came upon him, and, if he had not heard me coming, I would have fallen over him, too, but he got up in time. He rose, unsteady, long, pale, indistinct, like a vapour exhaled by the earth, and swayed slightly, misty and silent before me; while at my back the fires loomed between the trees, and the murmur of many voices issued from the forest. I had cut him off cleverly; but when actually confronting him I seemed to come to my senses, I saw the danger in its right proportion. It was by no means over yet. Suppose he began to shout? Though he could hardly stand, there was still plenty of vigour in his voice. 'Go away—hide yourself,' he said, in that profound tone. It was very awful. I glanced back. We were within thirty yards from the nearest fire. A black figure stood up, strode on long black legs, waving long black arms, across the glow. It had horns—antelope horns, I think—on its head. Some sorcerer, some witch-man, no doubt: it looked fiendlike enough. 'Do you know what you are doing?' I whispered. 'Perfectly,' he answered, raising his voice for that single word: it sounded to me far off and yet loud, like a hail through a speaking-trumpet. 'If he makes a row we are lost,' I thought to myself. This clearly was not a case for fisticuffs, even apart from the very natural aversion I had to beat that Shadow—this wandering and tormented thing. 'You will be lost,' I said—'utterly lost.' One gets sometimes such a flash of inspiration, you know. I did say the right thing, though indeed he could not have been more irretrievably lost than he was at this very moment, when the foundations of our intimacy were being laid—to endure—to endure—even to the end—even beyond.

" 'I had immense plans,' he muttered irresolutely. 'Yes,' said I; 'but if you try to shout I'll smash your head with——' There was not a stick or a stone near. 'I will throttle you for good,' I corrected myself. 'I was on the threshold of great things,' he pleaded, in a voice of longing, with a wistfulness of tone that made my blood run cold. 'And now for this stupid scoundrel——' 'Your success in Europe is assured in any case,' I affirmed steadily. I did not want to have the throttling of him, you understand—and indeed it would have been very little use for any practical purpose. I tried to break the spell—the heavy, mute spell of the wilderness—that seemed to draw him to its pitiless breast by the awakening of forgotten and brutal instincts, by the memory of gratified and monstrous passions. This alone, I was convinced, had driven him out to the edge of the forest, to the bush, towards the gleam of fires, the throb of drums, the drone of weird incantations; this alone had beguiled his unlawful soul beyond the bounds of permitted aspirations. And, don't you see, the terror of the position was not in being knocked on the head—though I had a very lively sense of that danger, too—but in this, that I had to deal with a being to whom I could not appeal in the name of anything high or low. I had, even like the niggers, to invoke him—himself—his own exalted and incredible degradation. There was nothing either above or below him, and I knew it. He had kicked himself loose of the earth. Confound the man! he had kicked the very earth to pieces. He was alone, and I before him did not know whether I stood on the ground or floated in the air. I've been telling you what we said—repeating the phrases we pronounced—but what's the good? They were common everyday words—the familiar, vague sounds exchanged on every waking day of life. But what of that? They had behind them, to my mind, the terrific suggestiveness of words heard in dreams, of phrases spoken in nightmares. Soul! If anybody ever struggled with a soul, I am the man. And I wasn't arguing with a lunatic either. Believe me or not, his intelligence was perfectly clear—concentrated, it is true, upon himself with horrible intensity, yet clear; and therein was my only chance—barring, of course, the killing him there and then, which wasn't so good, on account of unavoidable noise. But his soul was mad. Being alone in the wilderness, it had looked within itself, and, by heavens! I tell you, it had gone mad. I had—for my sins, I suppose—to go through the ordeal of looking into it myself. No eloquence could have been so withering to one's belief in mankind as his final burst of sincerity. He struggled with himself, too. I saw it—I heard it. I saw the inconceivable mystery of a soul that knew no restraint, no faith, and no fear, yet struggling blindly with itself. I kept my head pretty well; but when I had him at last stretched on the couch, I wiped my forehead, while my legs shook under me as though I had carried half a ton on my back down that hill. And yet I had only supported him, his bony arm clasped round my neck—and he was not much heavier than a child.

"When next day we left at noon, the crowd, of whose presence behind the curtain of trees I had been acutely conscious all the time, flowed out of the woods again, filled the clearing, covered the slope with a mass of naked, breathing, quivering, bronze bodies. I steamed up a bit, then swung down stream, and two thousand eyes followed the evolutions of the splashing, thumping, fierce river demon beating the water with its terrible tail and breathing black smoke into the air. In front of the first rank, along the river, three men, plastered with bright red earth from head to foot, strutted to and fro restlessly. When we came abreast again, they faced the river, stamped their feet, nodded their horned heads, swayed their scarlet bodies; they shook towards the fierce river-demon a bunch of black feathers, a mangy skin with a pendent tail—something that looked like a dried gourd; they shouted periodically together strings of amazing words that resembled no sounds of human language; and the deep murmurs of the crowd, interrupted suddenly, were like the responses of some satanic litany.

"We had carried Kurtz into the pilot-house: there was more air there. Lying on the couch, he stared through the open shutter. There was an eddy in the mass of human bodies, and the woman with helmeted head and tawny cheeks rushed out to the very brink of the stream. She put out her hands, shouted something, and all that wild mob took up the shout in a roaring chorus of articulated, rapid, breathless utterance.

" 'Do you understand this?' I asked.

"He kept on looking out past me with fiery, longing eyes, with a mingled expression of wistfulness and hate. He made no answer, but I saw a smile, a smile of indefinable meaning, appear on his colourless lips that a moment after twitched convulsively. 'Do I not?' he said slowly, gasping, as if the words had been torn out of him by a supernatural power.

"I pulled the string of the whistle, and I did this because I saw the pilgrims on deck getting out their rifles with an air of anticipating a jolly lark. At the sudden screech there was a movement of abject terror through that wedged mass of bodies. 'Don't! don't you frighten them away,' cried some one on deck disconsolately. I pulled the string time after time. They broke and ran, they leaped, they crouched, they swerved, they dodged the flying terror of the sound. The three red chaps had fallen flat, face down on the shore, as though they had been shot dead. Only the barbarous and superb woman did not so much as flinch, and stretched tragically her bare arms after us over the sombre and glittering river.

"And then that imbecile crowd down on the deck started their little fun, and I could see nothing more for smoke.

"The brown current ran swiftly out of the heart of darkness, bearing us down towards the sea with twice the speed of our upward progress; and Kurtz's life was running swiftly, too, ebbing, ebbing out of his heart into the sea of inexorable time. The manager was very placid, he had no vital anxieties now, he took us both in with a comprehensive and satisfied glance: the 'affair' had come off as well as could be wished. I saw the time approaching when I would be left alone of the party of 'unsound method.' The pilgrims looked upon me with disfavour. I was, so to speak, numbered with the dead. It is strange how I accepted this unforseen partnership, this choice of nightmares forced upon me in the tenebrous land invaded by these mean and greedy phantoms.

"Kurtz discoursed. A voice! a voice! It rang deep to the very last. It survived his strength to hide in the magnificent folds of eloquence the barren darkness of his heart. Oh, he struggled! he struggled! The wastes of his weary brain were haunted by shadowy images now—images of wealth and fame revolving obsequiously round his unextinguishable gift of noble and lofty expression. My intended, my station, my career, my ideas—these were the subjects for the occasional utterances of elevated sentiments. The shade of the original Kurtz frequented the bedside of the hollow sham, whose fate it was to be buried presently in the mould of primeval earth. Both the diabolic love and the unearthly hate of the mysteries it had penetrated fought for the possession of that soul satiated with primitive emotions, avid of lying fame, of sham distinction, of all the appearances of success and power.

"Sometimes he was contemptibly childish. He desired to have kings meet him at railway-stations on his return from some ghastly Nowhere, where he intended to accomplish great things. 'You show them you have in you something that is really profitable, and then there will be no limits to the recognition of your ability,' he would say. 'Of course you must take care of the motives—right motives—always.' The long reaches that were like one and the same reach, monotonous bends that were exactly alike, slipped past the steamer with their multitude of secular trees looking patiently after this grimy fragment of another world, the forerunner of change, of conquest, of trade, of massacres, of blessings. I looked ahead—piloting. 'Close the shutter,' said Kurtz suddenly one day; 'I can't bear to look at this.' I did so. There was a silence. 'Oh, but I will wring your heart yet!' he cried at the invisible wilderness.

"We broke down—as I had expected—and had to lie up for repairs at the head of an island. This delay was the first thing that shook Kurtz's confidence. One morning he gave me a packet of papers and a photograph—the lot tied together with a shoe-string. 'Keep this for me,' he said. 'This noxious fool' (meaning the manager) 'is capable of prying into my boxes when I am not looking.' In the afternoon I saw him. He was lying on his back with closed eyes, and I withdrew quietly, but I heard him mutter, 'Live rightly, die, die . . .' I listened. There was nothing more. Was he rehearsing some speech in his sleep, or was it a fragment of a phrase from some newspaper article? He had been writing for the papers and meant to do so again, 'for the furthering of my ideas. It's a duty.'

"His was an impenetrable darkness. I looked at him as you peer down at a man who is lying at the bottom of a precipice where the sun never shines. But I had not much time to give him, because I was helping the engine-driver to take to pieces the leaky cylinders, to straighten a bent connecting-rod, and in other such matters. I lived in an infernal mess of rust, filings, nuts, bolts, spanners, hammers, ratchet-drills—things I abominate, because I don't get on with them. I tended the little forge we fortunately had aboard; I toiled wearily in a wretched scrap-heap—unless I had the shakes too bad to stand.

"One evening coming in with a candle I was startled to hear him say a little tremulously, 'I am lying here in the dark waiting for death.' The light was within a foot of his eyes. I forced myself to murmur, 'Oh, nonsense!' and stood over him as if transfixed.

"Anything approaching the change that came over his features I have never seen before, and hope never to see again. Oh, I wasn't touched. I was fascinated. It was as though a veil had been rent. I saw on that ivory face the expression of sombre pride, of ruthless power, of craven terror—of an intense and hopeless despair. Did he live his life again in every detail of desire, temptation, and surrender during that supreme moment of complete knowledge? He cried in a whisper at some image, at some vision—he cried out twice, a cry that was no more than a breath:

" 'The horror! The horror!'

"I blew the candle out and left the cabin. The pilgrims were dining in the mess-room, and I took my place opposite the manager, who lifted his eyes to give me a questioning glance, which I successfully ignored. He leaned back, serene, with that peculiar smile of his sealing the unexpressed depths of his meanness. A continuous shower of small flies streamed upon the lamp, upon the cloth, upon our hands and faces. Suddenly the manager's boy put his insolent black head in the doorway, and said in a tone of scathing contempt:

" 'Mistah Kurtz—he dead.'

"All the pilgrims rushed out to see. I remained, and went on with my dinner. I believe I was considered brutally callous. However, I did not eat much. There was a lamp in there—light, don't you know—and outside it was so beastly, beastly dark. I went no more near the remarkable man who had pronounced a judgment upon the adventures of his soul on this earth. The voice was gone. What else had been there? But I am of course aware that next day the pilgrims buried something in a muddy hole.

"And then they very nearly buried me.

"However, as you see, I did not go to join Kurtz there and then. I did not. I remained to dream the nightmare out to the end, and to show my loyalty to Kurtz once more. Destiny. My destiny! Droll thing life is—that mysterious arrangement of merciless logic for a futile purpose. The most you can hope from it is some knowledge of yourself—that comes too late—a crop of unextinguishable regrets. I have wrestled with death. It is the most unexciting contest you can imagine. It takes place in an impalpable greyness, with nothing underfoot, with nothing around, without spectators, without clamour, without glory, without the great desire of victory, without the great fear of defeat, in a sickly atmosphere of tepid scepticism, without much belief in your own right, and still less in that of your adversary. If such is the form of ultimate wisdom, then life is a greater riddle than some of us think it to be. I was within a hair's breadth of the last opportunity for pronouncement, and I found with humiliation that probably I would have nothing to say. This is the reason why I affirm that Kurtz was a remarkable man. He had something to say. He said it. Since I had peeped over the edge myself, I understand better the meaning of his stare, that could not see the flame of the candle, but was wide enough to embrace the whole universe, piercing enough to penetrate all the hearts that beat in the darkness. He had summed up—he had judged. 'The horror!' He was a remarkable man. After all, this was the expression of some sort of belief; it had candour, it had conviction, it had a vibrating note of revolt in its whisper, it had the appalling face of a glimpsed truth—the strange commingling of desire and hate. And it is not my own extremity I remember best—a vision of greyness without form filled with physical pain, and a careless contempt for the evanescence of all things—even of this pain itself. No! It is his extremity that I seem to have lived through. True, he had made that last stride, he had stepped over the edge, while I had been permitted to draw back my hesitating foot. And perhaps in this is the whole difference; perhaps all the wisdom, and all truth, and all sincerity, are just compressed into that inappreciable moment of time in which we step over the threshold of the invisible. Perhaps! I like to think my summing-up would not have been a word of careless contempt. Better his cry—much better. It was an affirmation, a moral victory paid for by innumerable defeats, by abominable terrors, by abominable satisfactions. But it was a victory! That is why I have remained loyal to Kurtz to the last, and even beyond, when a long time after I heard once more, not his own voice, but the echo of his magnificent eloquence thrown to me from a soul as translucently pure as a cliff of crystal.

"No, they did not bury me, though there is a period of time which I remember mistily, with a shuddering wonder, like a passage through some inconceivable world that had no hope in it and no desire. I found myself back in the sepulchral city resenting the sight of people hurrying through the streets to filch a little money from each other, to devour their infamous cookery, to gulp their unwholesome beer, to dream their insignificant and silly dreams. They trespassed upon my thoughts. They were intruders whose knowledge of life was to me an irritating pretence, because I felt so sure they could not possibly know the things I knew. Their bearing, which was simply the bearing of commonplace individuals going about their business in the assurance of perfect safety, was offensive to me like the outrageous flauntings of folly in the face of a danger it is unable to comprehend. I had no particular desire to enlighten them, but I had some difficulty in restraining myself from laughing in their faces so full of stupid importance. I daresay I was not very well at that time. I tottered about the streets—there were various affairs to settle—grinning bitterly at perfectly respectable persons. I admit my behaviour was inexcusable, but then my temperature was seldom normal in these days. My dear aunt's endeavours to 'nurse up my strength' seemed altogether beside the mark. It was not my strength that wanted nursing, it was my imagination that wanted soothing. I kept the bundle of papers given me by Kurtz, not knowing exactly what to do with it. His mother had died lately, watched over, as I was told, by his Intended. A clean-shaved man, with an official manner and wearing gold-rimmed spectacles, called on me one day and made inquiries, at first circuitous, afterwards suavely pressing, about what he was pleased to denominate certain 'documents.' I was not surprised, because I had had two rows with the manager on the subject out there. I had refused to give up the smallest scrap out of that package, and I took the same attitude with the spectacled man. He became darkly menacing at last, and with much heat argued that the Company had the right to every bit of information about its 'territories.' And said he, 'Mr. Kurtz's knowledge of unexplored regions must have been necessarily extensive and peculiar—owing to his great abilities and to the deplorable circumstances in which he had been placed: therefore——' I assured him Mr. Kurtz's knowledge, however extensive, did not bear upon the problems of commerce or administration. He invoked then the name of science. 'It would be an incalculable loss if,' etc., etc. I offered him the report on the 'Suppression of Savage Customs,' with the postscriptum torn off. He took it up eagerly, but ended by sniffing at it with an air of contempt. 'This is not what we had a right to expect,' he remarked. 'Expect nothing else,' I said. 'There are only private letters.' He withdrew upon some threat of legal proceedings, and I saw him no more; but another fellow, calling himself Kurtz's cousin, appeared two days later, and was anxious to hear all the details about his dear relative's last moments. Incidentally he gave me to understand that Kurtz had been essentially a great musician. 'There was the making of an immense success,' said the man, who was an organist, I believe, with lank grey hair flowing over a greasy coat-collar. I had no reason to doubt his statement; and to this day I am unable to say what was Kurtz's profession, whether he ever had any—which was the greatest of his talents. I had taken him for a painter who wrote for the papers, or else for a journalist who could paint—but even the cousin (who took snuff during the interview) could not tell me what he had been—exactly. He was a universal genius—on that point I agreed with the old chap, who thereupon blew his nose noisily into a large cotton handkerchief and withdrew in senile agitation, bearing off some family letters and memoranda without importance. Ultimately a journalist anxious to know something of the fate of his 'dear colleague' turned up. This visitor informed me Kurtz's proper sphere ought to have been politics 'on the popular side.' He had furry straight eyebrows, bristly hair cropped short, an eyeglass on a broad ribbon, and, becoming expansive, confessed his opinion that Kurtz really couldn't write a bit—'but heavens! how that man could talk. He electrified large meetings. He had faith—don't you see?—he had the faith. He could get himself to believe anything—anything. He would have been a splendid leader of an extreme party.' 'What party?' I asked. 'Any party,' answered the other. 'He was an—an—extremist.' Did I not think so? I assented. Did I know, he asked, with a sudden flash of curiosity, 'what it was that had induced him to go out there?' 'Yes,' said I, and forthwith handed him the famous Report for publication, if he thought fit. He glanced through it hurriedly, mumbling all the time, judged 'it would do,' and took himself off with this plunder.

"Thus I was left at last with a slim packet of letters and the girl's portrait. She struck me as beautiful—I mean she had a beautiful expression. I know that the sunlight can be made to lie, too, yet one felt that no manipulation of light and pose could have conveyed the delicate shade of truthfulness upon those features. She seemed ready to listen without mental reservation, without suspicion, without a thought for herself. I concluded I would go and give her back her portrait and those letters myself. Curiosity? Yes; and also some other feeling perhaps. All that had been Kurtz's had passed out of my hands: his soul, his body, his station, his plans, his ivory, his career. There remained only his memory and his Intended—and I wanted to give that up, too, to the past, in a way—to surrender personally all that remained of him with me to that oblivion which is the last word of our common fate. I don't defend myself. I had no clear perception of what it was I really wanted. Perhaps it was an impulse of unconscious loyalty, or the fulfilment of one of those ironic necessities that lurk in the facts of human existence. I don't know. I can't tell. But I went.

"I thought his memory was like the other memories of the dead that accumulate in every man's life—a vague impress on the brain of shadows that had fallen on it in their swift and final passage; but before the high and ponderous door, between the tall houses of a street as still and decorous as a well-kept alley in a cemetery, I had a vision of him on the stretcher, opening his mouth voraciously, as if to devour all the earth with all its mankind. He lived then before me; he lived as much as he had ever lived—a shadow insatiable of splendid appearances, of frightful realities; a shadow darker than the shadow of the night, and draped nobly in the folds of a gorgeous eloquence. The vision seemed to enter the house with me—the stretcher, the phantom-bearers, the wild crowd of obedient worshippers, the gloom of the forests, the glitter of the reach between the murky bends, the beat of the drum, regular and muffled like the beating of a heart—the heart of a conquering darkness. It was a moment of triumph for the wilderness, an invading and vengeful rush which, it seemed to me, I would have to keep back alone for the salvation of another soul. And the memory of what I had heard him say afar there, with the horned shapes stirring at my back, in the glow of fires, within the patient woods, those broken phrases came back to me, were heard again in their ominous and terrifying simplicity. I remembered his abject pleading, his abject threats, the colossal scale of his vile desires, the meanness, the torment, the tempestuous anguish of his soul. And later on I seemed to see his collected languid manner, when he said one day, 'This lot of ivory now is really mine. The Company did not pay for it. I collected it myself at a very great personal risk. I am afraid they will try to claim it as theirs though. H'm. It is a difficult case. What do you think I ought to do—resist? Eh? I want no more than justice.' . . . He wanted no more than justice—no more than justice. I rang the bell before a mahogany door on the first floor, and while I waited he seemed to stare at me out of the glassy panel—stare with that wide and immense stare embracing, condemning, loathing all the universe. I seemed to hear the whispered cry, 'The horror! The horror!'

"The dusk was falling. I had to wait in a lofty drawing-room with three long windows from floor to ceiling that were like three luminous and bedraped columns. The bent gilt legs and backs of the furniture shone in indistinct curves. The tall marble fireplace had a cold and monumental whiteness. A grand piano stood massively in a corner; with dark gleams on the flat surfaces like a sombre and polished sarcophagus. A high door opened—closed. I rose.

"She came forward, all in black, with a pale head, floating towards me in the dusk. She was in mourning. It was more than a year since his death, more than a year since the news came; she seemed as though she would remember and mourn forever. She took both my hands in hers and murmured, 'I had heard you were coming.' I noticed she was not very young—I mean not girlish. She had a mature capacity for fidelity, for belief, for suffering. The room seemed to have grown darker, as if all the sad light of the cloudy evening had taken refuge on her forehead. This fair hair, this pale visage, this pure brow, seemed surrounded by an ashy halo from which the dark eyes looked out at me. Their glance was guileless, profound, confident, and trustful. She carried her sorrowful head as though she were proud of that sorrow, as though she would say, 'I—I alone know how to mourn for him as he deserves.' But while we were still shaking hands, such a look of awful desolation came upon her face that I perceived she was one of those creatures that are not the playthings of Time. For her he had died only yesterday. And, by Jove! the impression was so powerful that for me, too, he seemed to have died only yesterday—nay, this very minute. I saw her and him in the same instant of time—his death and her sorrow—I saw her sorrow in the very moment of his death. Do you understand? I saw them together—I heard them together. She had said, with a deep catch of the breath, 'I have survived' while my strained ears seemed to hear distinctly, mingled with her tone of despairing regret, the summing up whisper of his eternal condemnation. I asked myself what I was doing there, with a sensation of panic in my heart as though I had blundered into a place of cruel and absurd mysteries not fit for a human being to behold. She motioned me to a chair. We sat down. I laid the packet gently on the little table, and she put her hand over it. . . . 'You knew him well,' she murmured, after a moment of mourning silence.

" 'Intimacy grows quickly out there,' I said. 'I knew him as well as it is possible for one man to know another.'

" 'And you admired him,' she said. 'It was impossible to know him and not to admire him. Was it?'

" 'He was a remarkable man,' I said, unsteadily. Then before the appealing fixity of her gaze, that seemed to watch for more words on my lips, I went on, 'It was impossible not to——'

" 'Love him,' she finished eagerly, silencing me into an appalled dumbness. 'How true! how true! But when you think that no one knew him so well as I! I had all his noble confidence. I knew him best.'

" 'You knew him best,' I repeated. And perhaps she did. But with every word spoken the room was growing darker, and only her forehead, smooth and white, remained illumined by the unextinguishable light of belief and love.

" 'You were his friend,' she went on. 'His friend,' she repeated, a little louder. 'You must have been, if he had given you this, and sent you to me. I feel I can speak to you—and oh! I must speak. I want you —you who have heard his last words—to know I have been worthy of him. . . . It is not pride. . . . Yes! I am proud to know I understood him better than any one on earth—he told me so himself. And since his mother died I have had no one—no one—to—to——'

"I listened. The darkness deepened. I was not even sure whether he had given me the right bundle. I rather suspect he wanted me to take care of another batch of his papers which, after his death, I saw the manager examining under the lamp. And the girl talked, easing her pain in the certitude of my sympathy; she talked as thirsty men drink. I had heard that her engagement with Kurtz had been disapproved by her people. He wasn't rich enough or something. And indeed I don't know whether he had not been a pauper all his life. He had given me some reason to infer that it was his impatience of comparative poverty that drove him out there.

" '. . . Who was not his friend who had heard him speak once?' she was saying. 'He drew men towards him by what was best in them.' She looked at me with intensity. 'It is the gift of the great,' she went on, and the sound of her low voice seemed to have the accompaniment of all the other sounds, full of mystery, desolation, and sorrow, I had ever heard—the ripple of the river, the soughing of the trees swayed by the wind, the murmurs of the crowds, the faint ring of incomprehensible words cried from afar, the whisper of a voice speaking from beyond the threshold of an eternal darkness. 'But you have heard him! You know!' she cried.

" 'Yes, I know,' I said with something like despair in my heart, but bowing my head before the faith that was in her, before that great and saving illusion that shone with an unearthly glow in the darkness, in the triumphant darkness from which I could not have defended her—from which I could not even defend myself.

" 'What a loss to me—to us!'—she corrected herself with beautiful generosity; then added in a murmur, 'To the world.' By the last gleams of twilight I could see the glitter of her eyes, full of tears—of tears that would not fall.

" 'I have been very happy—very fortunate—very proud,' she went on. 'Too fortunate. Too happy for a little while. And now I am unhappy for—for life.'

"She stood up; her fair hair seemed to catch all the remaining light in a glimmer of gold. I rose, too.

" 'And of all this,' she went on mournfully, 'of all his promise, and of all his greatness, of his generous mind, of his noble heart, nothing remains—nothing but a memory. You and I——'

" 'We shall always remember him,' I said hastily.

" 'No!' she cried. 'It is impossible that all this should be lost—that such a life should be sacrificed to leave nothing—but sorrow. You know what vast plans he had. I knew of them, too—I could not perhaps understand—but others knew of them. Something must remain. His words, at least, have not died.'

" 'His words will remain,' I said.

" 'And his example,' she whispered to herself. 'Men looked up to him—his goodness shone in every act. His example——'

" 'True,' I said; 'his example, too. Yes, his example. I forgot that.'

" 'But I do not. I cannot—I cannot believe—not yet. I cannot believe that I shall never see him again, that nobody will see him again, never, never, never.'

"She put out her arms as if after a retreating figure, stretching them back and with clasped pale hands across the fading and narrow sheen of the window. Never see him! I saw him clearly enough then. I shall see this eloquent phantom as long as I live, and shall see her, too, a tragic and familiar Shade, resembling in this gesture another one, tragic also, and bedecked with powerless charms, stretching bare brown arms over the glitter of the infernal stream, the stream of darkness. She said suddenly very low, 'He died as he lived.'

" 'His end,' said I, with dull anger stirring in me, 'was in every way worthy of his life.'

" 'And I was not with him,' she murmured. My anger subsided before a feeling of infinite pity.

" 'Everything that could be done——' I mumbled.

" 'Ah, but I believed in him more than any one on earth—more than his own mother, more than—himself. He needed me! Me! I would have treasured every sigh, every word, every sign, every glance.'

"I felt like a chill grip on my chest. 'Don't,' I said, in a muffled voice.

" 'Forgive me. I—I have mourned so long in silence—in silence. . . . You were with him—to the last? I think of his loneliness. Nobody near to understand him as I would have understood. Perhaps no one to hear. . . .'

" 'To the very end,' I said shakily. 'I heard his very last words. . . .' I stopped in a fright.

" 'Repeat them,' she murmured in a heart-broken tone. 'I want—I want—something—something—to—to live with.'

"I was on the point of crying at her, 'Don't you hear them?' The dusk was repeating them in a persistent whisper all around us, in a whisper that seemed to swell menacingly like the first whisper of a rising wind. 'The horror! The horror!'

" 'His last word—to live with,' she insisted. 'Don't you understand I loved him—I loved him—I loved him!'

"I pulled myself together and spoke slowly.

" 'The last word he pronounced was—your name.'

"I heard a light sigh and then my heart stood still, stopped dead short by an exulting and terrible cry, by the cry of inconceivable triumph and of unspeakable pain. 'I knew it—I was sure!' . . . She knew. She was sure. I heard her weeping; she had hidden her face in her hands. It seemed to me that the house would collapse before I could escape, that the heavens would fall upon my head. But nothing happened. The heavens do not fall for such a trifle. Would they have fallen, I wonder, if I had rendered Kurtz that justice which was his due? Hadn't he said he wanted only justice? But I couldn't. I could not tell her. It would have been too dark—too dark altogether. . . ."

Marlow ceased, and sat apart, indistinct and silent, in the pose of a meditating Buddha. Nobody moved for a time. "We have lost the first of the ebb," said the Director suddenly. I raised my head. The offing was barred by a black bank of clouds, and the tranquil waterway leading to the uttermost ends of the earth flowed sombre under an overcast sky—seemed to lead into the heart of an immense darkness.

HENRY JAMES

"It is a piece of ingenuity pure and simple, of cold artistic calculation, an amusette to catch those not easily caught (the 'fun' of the capture of the merely witless being ever but small), the jaded, the disillusioned, the fastidious." Henry James said this of The Turn of the Screw.

In the more than six decades since it appeared, despite James's disclaimer, the novella has invited profuse critical commentary of broadly two kinds, Freudian and "Archetypal." The Freudians, led by such eminent critics as Edmund Wilson, say that "the governess who is made to tell the story is a neurotic case of sex repression, and that the ghosts are not real ghosts, but hallucinations of the governess." Those who, like Professor Heilman, see in the tale a symbolic Christian myth, an archetype of religious experience, believe that the theme is "the struggle of evil to possess the human soul."

James's marvelous ambiguity incites our wonder. Though we do not believe in ghosts, the first literal reading presents an almost plausible case from the governess's point of view. True, we are utterly stunned by the depravity of the beautiful children, but we reluctantly accept it. But then the disquieting ambiguities become insistent. James's control of his material and of his reader is quite complete. The author does not fully tip his hand, and we are left stranded—bedeviled and bewildered.

If the piece is a study in psychopathology, it is the characterization of a Victorian young woman who is in love with her handsome employer but is afraid to admit her natural sexual impulses to herself. The apparitions become manifestations of her guilty repressions and a means of justifying her position. She has complete reign over the hapless children, communicates her hysteria to them, and finally frightens the little boy to death. In this light, The Turn of the Screw *is a horror story of unutterable power.*

If the novella, on the other hand, is an allegory of Good and Evil, then the apparitions are personifications of the dark power of Quint and Jessel over the innocent children they have corrupted. Clear suggestions of the perverse moral and sexual influence of the deceased adults upon the children are provided by an outside person, Mrs. Grose. Our governess then becomes an avenging angel fighting for the souls of her charges. Miles, exhausted by the ordeal, dies an expiatory death. This interpretation makes the short novel grandly universal in its overtones.

Appraisement of the novella in these two ways still leaves much to be decided. The profusion of details offered by the author to bolster either interpretation must be thoughtfully scrutinized; the subtle shadings of style which help create the mood of the piece must be fully discerned; and following upon the prologue, the unfaltering certainty of the point of view from which the account is narrated has to be warily explored: a narrator tells of a narrator who reads from a manuscript written in the first person.

The Turn of the Screw

THE STORY had held us, round the fire, sufficiently breathless, but except the obvious remark that it was gruesome, as, on Christmas eve in an old house, a strange tale should essentially be, I remember no comment uttered till somebody happened to say that it was the only case he had met in which such a visitation had fallen on a child. The case, I may mention, was that of an apparition in just such an old house as had gathered us for the occasion—an appearance, of a dreadful kind, to a little boy sleeping in the room with his mother and waking her up in the terror of it; waking her not to dissipate his dread and soothe him to sleep again, but to encounter also, herself, before she had succeeded in doing so, the same sight that had shaken him. It was this observation that drew from Douglas—not immediately, but later in the evening—a reply that had the interesting consequence to which I call attention. Someone else told a story not particularly effective, which I saw he was not following. This I took for a sign that he had himself something to produce and that we should only have to wait. We waited in fact till two nights later; but that same evening, before we scattered, he brought out what was in his mind.

"I quite agree—in regard to Griffin's ghost, or whatever it was—that its appearing first to the little boy, at so tender an age, adds a particular touch. But it's not the first occurrence of its charming kind that I know to have involved a child. If the child gives the effect another turn of the screw, what do you say to *two* children—?"

"We say, of course," somebody exclaimed, "that they give two turns! Also that we want to hear about them."

I can see Douglas there before the fire, to which he had got up to present his back, looking down at his interlocutor with his hands in his pockets. "Nobody but me, till now, has ever heard. It's quite too horrible." This, naturally, was declared by several voices to give the thing the utmost price, and our friend, with quiet art, prepared his triumph by turning his eyes over the rest of us and going on: "It's beyond everything. Nothing at all that I know touches it."

"For sheer terror?" I remember asking.

He seemed to say it was not so simple as that; to be really at a loss how to qualify it. He passed his hand over his eyes, made a little wincing grimace. "For dreadful—dreadfulness!"

"Oh, how delicious!" cried one of the women.

He took no notice of her; he looked at me, but as if, instead of me, he saw what he spoke of. "For general uncanny ugliness and horror and pain."

"Well then," I said, "just sit right down and begin."

He turned round to the fire, gave a kick to a log, watched it an instant. Then as he faced us again: "I can't begin. I shall have to send to town." There was a unanimous groan at this, and much reproach; after which, in his preoccupied way, he explained. "The story's written. It's in a locked drawer—it has not been out for years. I could write to my man and enclose the key; he could send down the packet as he finds it." It was to me in particular that he appeared to propound this—appeared almost to appeal for aid not to hesitate. He had broken a thickness of ice, the formation of many a winter; had had his reasons for a long silence. The others resented postponement, but it was just his scruples that charmed me. I adjured him to write by the first post and to agree with us for an early hearing; then I asked him if the experience in question had been his own. To this his answer was prompt. "Oh, thank God, no!"

"And is the record yours? You took the thing down?"

"Nothing but the impression. I took that *here*"—he tapped his heart. "I've never lost it."

"Then your manuscript—?"

"Is in old, faded ink, and in the most beautiful hand." He hung fire again. "A woman's. She has been dead these twenty years. She sent me the pages in question before she died." They were all listening now, and of course there was somebody to be arch, or at any rate to draw the inference. But if he put the inference by without a smile it was also with irritation. "She was a most charming person, but she was ten years older than I. She was my sister's governess," he quietly said. "She was the most agreeable woman I've ever known in her position; she would have been worthy of any whatever. It was long ago, and this episode was long before. I was at Trinity, and I found her at home on my coming down the second summer. I was much there that year—it was a beautiful one; and we had, in her off-hours, some strolls and talks in the garden—talks in which she struck me as awfully clever and nice. Oh yes; don't grin: I liked her extremely and am glad to this day to think she liked me too. If she hadn't she wouldn't have told me. She had never told anyone. It wasn't simply that she said so, but that I knew she hadn't. I was sure; I could see. You'll easily judge why when you hear."

"Because the thing had been such a scare?"

He continued to fix me. "You'll easily judge," he repeated: "*you* will."

I fixed him too. "I see. She was in love."

He laughed for the first time. "You *are* acute. Yes, she was in love. That is, she had been. That came out—she couldn't tell her story without its coming out. I saw it, and she saw I saw it; but neither of us spoke of it. I remember the time and the place—the corner of the lawn, the shade of the great beeches and the long, hot summer afternoon. It wasn't a scene for a shudder; but oh—!" He quitted the fire and dropped back into his chair.

"You'll receive the packet Thursday morning?" I inquired.

"Probably not till the second post."

"Well then; after dinner—"

"You'll all meet me here?" He looked us round again. Isn't anybody going?" It was almost the tone of hope.

"Everybody will stay!"

"*I* will—and *I* will!" cried the ladies whose departure had been fixed. Mrs. Griffin, however, expressed the need for a little more light. "Who was it she was in love with?"

"The story will tell," I took upon myself to reply.

"Oh, I can't wait for thc story!"

"The story *won't* tell," said Douglas; "not in any literal, vulgar way."

"More's the pity, then. That's the only way I ever understand."

"Won't *you* tell, Douglas?" somebody else inquired.

He sprang to his feet again. "Yes—tomorrow. Now I must go to bed. Good-night." And quickly catching up a candlestick, he left us slightly bewildered. From our end of the great brown hall we heard his step on the stair; whereupon Mrs. Griffin spoke. 'Well, if I don't know who she was in love with, I know who *he* was."

"She was ten years older," said her husband.

"*Raison de plus*—at that age! But it's rather nice, his long reticence."

"Forty years!" Griffin put in.

"With this outbreak at last."

"The outbreak," I returned, "will make a tremendous occasion of Thursday night"; and everyone so agreed with me that, in the light of it, we lost all attention for everything else. The last story, however incomplete and like the mere opening of a serial, had been told; we hand-shook and "candlestuck," as somebody said, and went to bed.

I knew the next day that a letter containing the key had, by the first post, gone off to his London apartments; but in spite of—or perhaps just on account of—the eventual diffusion of this knowledge we quite let him alone till after dinner, till such an hour of the evening, in fact, as might best accord with the kind of emotion on which our hopes were fixed. Then he became as communicative as we could desire and indeed gave us his best reason for being so. We had it from him again before the fire in the hall, as we had had our mild wonders of the previous night. It appeared that the narrative he had promised to read us really required for a proper intelligence a few words of prologue. Let me say here distinctly, to have done with it, that this narrative, from an exact transcript of my own made much later, is what I shall presently give. Poor Douglas, before his death—when it was in sight—committed to me the manuscript that reached him on the third of these days and that, on the same spot, with immense effect, he began to read to our hushed little circle on the night of the fourth. The departing ladies who had said they would stay didn't, of course, thank heaven, stay:

they departed, in consequence of arrangements made, in a rage of curiosity, as they professed, produced by the touches with which he had already worked us up. But that only made his little final auditory more compact and select, kept it, round the hearth, subject to a common thrill.

The first of these touches conveyed that the written statement took up the talc at a point after it had, in a manner, begun. The fact to be in possession of was therefore that his old friend, the youngest of several daughters of a poor country parson, had, at the age of twenty, on taking service for the first time in the schoolroom, come up to London, in trepidation, to answer in person an advertisement that had already placed her in brief correspondence with the advertiser. This person proved, on her presenting herself, for judgment, at a house in Harley Street, that impressed her as vast and imposing—this prospective patron proved a gentleman, a bachelor in the prime of life, such a figure as had never risen, save in a dream or an old novel, before a fluttered, anxious girl out of a Hampshire vicarage. One could easily fix his type; it never, happily, dies out. He was handsome and bold and pleasant, offhand and gay and kind. He struck her, inevitably, as gallant and splendid, but what took her most of all and gave her the courage she afterwards showed was that he put the whole thing to her as a kind of favour, an obligation he should gratefully incur. She conceived him as rich, but as fearfully extravagant—saw him all in a glow of high fashion, of good looks, of expensive habits, of charming ways with women. He had for his own town residence a big house filled with the spoils of travel and the trophies of the chase; but it was to his country home, an old family place in Essex, that he wished her immediately to proceed.

He had been left, by the death of their parents in India, guardian to a small nephew and a small niece, children of a younger, a military brother, whom he had lost two years before. These children were, by the strangest of chances for a man in his position,—a lone man without the right sort of experience or a grain of patience,—very heavily on his hands. It had all been a great worry and, on his own part doubtless, a series of blunders, but he immensely pitied the poor chicks and had done all he could: had in particular sent them down to his other house, the proper place for them being of course the country, and kept them there, from the first, with the best people he could find to look after them, parting even with his own servants to wait on them and going down himself, whenever he might, to see how they were doing. The awkward thing was that they had practically no other relations and that his own affairs took up all his time. He had put them in possession of Bly, which was healthy and secure, and had placed at the head of their little establishment—but below stairs only—an excellent woman, Mrs. Grose, whom he was sure his visitor would like and who had formerly been maid to his mother. She was now housekeeper and was also acting for the time as superintendent to the little girl, of whom, without children of her own, she was, by good luck, extremely fond. There were plenty of people to help, but of course the young lady who should go down as governess would be in supreme authority. She would also have, in holidays, to look after the small boy, who had been for a term at school—young as he was to be sent, but what else could be done? —and who, as the holidays were about to begin, would be back from one day to the other. There had been for the two children at first a young lady whom they had had the misfortune to lose. She had done for them quite beautifully—she was a most respectable person—till her death, the great awkwardness of which had, precisely, left no alternative but the school for little Miles. Mrs. Grose, since then, in the way of manners and things, had done as she could for Flora; and there were, further, a cook, a housemaid, a dairywoman, an old pony, an old groom, and an old gardener, all likewise thoroughly respectable.

So far had Douglas presented his picture when someone put a question. "And what did the former governess die of?—of so much respectability?"

Our friend's answer was prompt. "That will come out. I don't anticipate."

"Excuse me—I thought that was just what you *are* doing."

"In her successor's place," I suggested, "I should have wished to learn if the office brought with it—"

"Necessary danger to life?" Douglas completed my thought. "She did wish to learn, and she did learn. You shall hear tomorrow what she learnt. Meanwhile, of course, the prospect struck her as slightly grim. She was young, untried, nervous: it was a vision of serious duties and little company, of really great loneliness. She hesitated—took a couple of days to consult and consider. But the salary offered much exceeded her modest measure, and on a second interview she faced the music, she engaged." And Douglas, with this, made a pause that, for the benefit of the company, moved me to throw in—

"The moral of which was of course the seduction exercised by the splendid young man. She succumbed to it."

He got up and, as he had done the night before, went to the fire, gave a stir to a log with his foot, then stood a moment with his back to us. "She saw him only twice."

"Yes, but that's just the beauty of her passion."

A little to my surprise, on this, Douglas turned round to me. "It *was* the beauty of it. There were others," he went on, "who hadn't succumbed. He told her frankly all his difficulty—that for several applicants the conditions had been prohibitive. They were, somehow, simply afraid. It sounded dull—it sounded strange; and all the more so because of his main condition."

"Which was—?"

"That she should never trouble him—but never, never: neither appeal nor complain nor write about anything; only meet all questions herself, receive all moneys from his solicitor, take the whole thing over and let him alone. She promised to do this, and she mentioned to me that when, for a moment, disburdened, delighted, he held her hand, thanking her for the sacrifice, she already felt rewarded."

"But was that all her reward?" one of the ladies asked.

"She never saw him again."

"Oh!" said the lady; which, as our friend immediately left us again, was the only other word of importance contributed to the subject till, the next night, by the corner of the hearth, in the best chair, he opened the faded red cover of a thin old-fashioned gilt-edged album. The whole thing took indeed more nights than one, but on the first occasion the same lady put another question. "What is your title?"

"I haven't one."

"Oh, *I* have!" I said. But Douglas, without heeding me, had begun to read with a fine clearness that was like a rendering to the ear of the beauty of his author's hand.

I

I REMEMBER the whole beginning as a succession of flights and drops, a little see-saw of the right throbs and the wrong. After rising, in town, to meet his appeal, I had at all events a couple of very bad days—found myself doubtful again, felt indeed sure I had made a mistake. In this state of mind I spent the long hours of bumping, swinging coach that carried me to the stopping-place at which I was to be met by a vehicle from the house. This convenience, I was told, had been ordered, and I found, toward the close of the June afternoon, a commodious fly in waiting for me. Driving at that hour, on a lovely day, through a country to which the summer sweetness seemed to offer me a friendly welcome, my fortitude mounted afresh and, as we turned into the avenue, encountered a reprieve that was probably but a proof of the point to which it had sunk. I suppose I had expected, or had dreaded, something so melancholy that what greeted me was a good surprise. I remember as a most pleasant impression the broad, clear front, its open windows and fresh curtains and the pair of maids looking out; I remember the lawn and the bright flowers and the crunch of my wheels on the gravel and the clustered treetops over which the rooks circled and cawed in the golden sky. The scene had a greatness that made it a different affair from my own scant home, and there immediately appeared at the door, with a little girl in her hand, a civil person who dropped me as decent a curtsey as if I had been the mistress or a distinguished visitor. I had received in Harley Street a narrower notion of the place, and that, as I recalled it, made me think the proprietor still more of a gentleman, suggested that what I was to enjoy might be something beyond this promise.

I had no drop again till the next day, for I was carried triumphantly through the following hours by my introduction to the younger of my pupils. The little girl who accompanied Mrs. Grose appeared to me on the spot a creature so charming as to make it a great fortune to have to do with her. She was the most beautiful child I had ever seen, and I afterwards wondered that my employer had not told me more of her. I slept little that night—I was too much excited; and this astonished me too, I recollect, remained with me, adding to my sense of the liberality with which I was treated. The large, impressive room, one of the best in the house, the great state bed, as I almost felt it, the full, figured draperies, the long glasses in which, for the first time, I could see myself from head to foot, all struck me—like the extraordinary charm of my small charge—as so many things thrown in. It was thrown in as well, from the first moment, that I should get on with Mrs. Grose in a relation over which, on my way, in the coach, I fear I had rather brooded. The only thing indeed that in this early outlook might have made me shrink again was the clear circumstance of her being so glad to see me. I perceived within half an hour that she was so glad—stout, simple, plain, clean, wholesome woman—as to be positively on her guard against showing it too much. I wondered even then a little why she should wish not to show it, and that, with reflection, with suspicion, might of course have made me uneasy.

But it was a comfort that there could be no uneasiness in a connection with anything so beatific as the radiant image of my little girl, the vision of whose angelic beauty had probably more than anything else to do with the restlessness that, before morning, made me several times rise and wander about my room to take in the whole picture and prospect; to watch, from my open window, the faint summer dawn, to look at such portions of the rest of the house as I could catch, and to listen, while, in the fading dusk, the first birds began to twitter, for the possible recurrence of a sound or two, less natural and not without, but within, that I had fancied I heard. There had been a moment when I believed I recognised, faint and far, the cry of a child; there had been another when I found myself just consciously starting as at the passage, before my door, of a light footstep. But these fancies were not marked enough not to be thrown off, and it is only in the light, or the gloom, I should rather say, of other and subsequent matters that they now come back to me. To watch, teach, "form" little Flora would too evidently be the making of a happy and a useful life. It had been agreed between us downstairs that after this first occasion I should have her as a matter of course at night, her small white bed being already arranged, to that end, in my room. What I had undertaken was the whole care of her, and she had remained, just this last time, with Mrs. Grose only as an effect of our consideration for my inevitable strangeness and her natural timidity. In spite of this timidity—which the child herself, in the oddest way in the world, had been perfectly frank and brave about, allowing it, without a sign of uncomfortable consciousness, with the deep, sweet serenity indeed of one of Raphael's holy infants, to be discussed, to be imputed to her and to determine us—I felt quite sure she would presently like me. It was part of what I already liked Mrs. Grose herself for, the pleasure I could see her feel in my admiration and wonder as I sat at supper with four tall candles and with my pupil, in a high chair and a bib, brightly facing me, between them, over bread and milk. There were naturally things that in Flora's presence could pass between us only as prodigious and gratified looks, obscure and roundabout allusions.

"And the little boy—does he look like her? Is he too so very remarkable?"

One wouldn't flatter a child. "Oh, Miss, *most* remarkable. If you think well of this one!"—and she stood there with a plate in her hand, beaming at our companion, who looked from one of us to the other with placid heavenly eyes that contained nothing to check us.

"Yes; if I do—?"

"You *will* be carried away by the little gentleman!"

"Well, that, I think, is what I came for—to be carried away. I'm afraid, however," I remember feeling the impulse to add, "I'm rather easily carried away. I was carried away in London!"

I can still see Mrs. Grose's broad face as she took this in. "In Harley Street?"

"In Harley Street."

"Well, Miss, you're not the first—and you won't be the last."

"Oh, I've no pretension," I could laugh, "to being the only one. My other pupil, at any rate, as I understand, comes back tomorrow?"

"Not tomorrow—Friday, Miss. He arrives, as you did, by the coach, under care of the guard, and is to be met by the same carriage."

I forthwith expressed that the proper as well as the pleasant and friendly thing would be therefore that on the arrival of the public conveyance I should be in waiting for him with his little sister; an idea in which Mrs. Grose concurred so heartily that I somehow took her manner as a kind of comforting pledge—never falsified, thank heaven!—that we should on every question be quite at one. Oh, she was glad I was there!

What I felt the next day was, I suppose, nothing that could be fairly called a reaction from the cheer of my arrival; it was probably at the most only a slight oppression produced by a fuller measure of the scale, as I walked round them, gazed up at them, took them in, of my new circumstances. They had, as it were, an extent and mass for which I had not been prepared and in the presence of which I found myself, freshly, a little scared as well as a little proud. Lessons, in this agitation, certainly suffered some delay; I reflected that my first duty was, by the gentlest arts I could contrive, to win the child into the sense of knowing me. I spent the day with her out of doors; I arranged with her, to her great satisfaction, that it should be she, she only, who might show me the place. She showed it step by step and room by room and secret by secret, with droll, delightful, childish talk about it and with the result, in half an hour, of our becoming immense friends. Young as she was, I was struck, throughout our little tour, with her confidence and courage with the way, in empty chambers and dull corridors, on crooked staircases that made me pause and even on the summit of an old machicolated square tower that made me dizzy, her morning music, her disposition to tell me so many more things than she asked, rang out and led me on. I have not seen Bly since the day I left it, and I dare say that to my older and more informed eyes it would now appear sufficiently contracted. But as my little conductress, with her hair of gold and her frock of blue, danced before me round corners and pattered down passages, I had the view of a castle of romance inhabited by a rosy sprite, such a place as would somehow, for diversion of the young idea, take all colour out of storybooks and fairy-tales. Wasn't it just a storybook over which I had fallen a-doze and a-dream? No; it was a big, ugly, antique, but convenient house, embodying a few features of a building still older, half replaced and half utilised, in which I had the fancy of our being almost as lost as a handful of passengers in a great drifting ship. Well, I was, strangely, at the helm!

II

THIS CAME home to me when, two days later, I drove over with Flora to meet, as Mrs. Grose said, the little gentleman; and all the more for an incident that, presenting itself the second evening, had deeply disconcerted me. The first day had been, on the whole, as I have expressed, reassuring; but I was to see it wind up in keen apprehension. The postbag, that evening,—it came late,—contained a letter for me, which, however, in the hand of my employer, I found to be composed but of a few words enclosing another, addressed to himself, with a seal still unbroken. "This, I recognise, is from the head-master, and the head-master's an awful bore. Read him, please; deal with him; but mind you don't report. Not a word. I'm off!" I broke the seal with a great effort—so great a one that I was a long time coming to it; took the unopened missive at last up to my room and only attacked it just before going to bed. I had better have let it wait till morning, for it gave me a second sleepless night. With no counsel to take, the next day, I was full of distress; and it finally got so the better of me that I determined to open myself at least to Mrs. Grose.

"What does it mean? The child's dismissed his school."

She gave me a look that I remarked at the moment; then, visibly, with a quick blankness, seemed to try to take it back. "But aren't they all—?"

"Sent home—yes. But only for the holidays. Miles may never go back at all."

Consciously, under my attention, she reddened. "They won't take him?"

"They absolutely decline."

At this she raised her eyes, which she had turned from me; I saw them fill with good tears. "What has he done?"

I hesitated; then I judged best simply to hand her my letter—which, however, had the effect of making her, without taking it, simply put her hands behind her. She shook her head sadly. "Such things are not for me, Miss."

My counsellor couldn't read! I winced at my mistake, which I attenuated as I could, and opened my letter again to repeat it to her; then, faltering in the act and folding it up once more, I put it back in my pocket. "Is he really *bad?*"

The tears were still in her eyes. "Do the gentlemen say so?"

"They go into no particulars. They simply express their regret that it should be impossible to keep him. That can have only one meaning." Mrs. Grose listened with dumb emotion; she forebore to ask me what this meaning might be; so that, presently, to put the thing with some coherence and with the mere aid of her presence to my own mind, I went on: "That he's an injury to the others."

At this, with one of the quick turns of simple folk, she suddenly flamed up. "Master Miles! *him* an injury?"

There was such a flood of good faith in it that, though I had not yet seen the child, my very fears made me jump to the absurdity of the idea. I found myself, to meet my friend the better, offering it, on the spot, sarcastically. "To his poor little innocent mates!"

"It's too dreadful," cried Mrs. Grose, "to say such cruel things! Why, he's scarce ten years old."

"Yes, yes; it would be incredible."

She was evidently grateful for such a profession. "See him, Miss, first. *Then* believe it!" I felt forthwith a new impatience to see him; it was the beginning of a curiosity that, for all the next hours, was to deepen almost to pain. Mrs. Grose was aware, I could judge, of what she had produced in me, and she followed it up with assurance. "You might as well believe it of the little lady. Bless her," she added the next moment—"*look* at her!"

I turned and saw that Flora, whom, ten minutes before, I had established in the schoolroom with a sheet of white paper, a pencil, and a copy of nice "round O's," now presented herself to view at the open door. She expressed in her little way an extraordinary detachment from disagreeable duties, looking to me, however, with a great childish light that seemed to offer it as a mere result of the affection she had conceived for my person, which had rendered necessary that she should follow me. I needed nothing more than this to feel the full force of Mrs. Grose's comparison, and, catching my pupil in my arms, covered her with kisses in which there was a sob of atonement.

None the less, the rest of the day, I watched for further occasion to approach my colleague, especially as, toward evening, I began to fancy she rather sought to avoid me. I overtook her, I remember, on the staircase; we went down together, and at the bottom I detained her, holding her there with a hand on her arm. "I take what you said to me at noon as a declaration that *you've* never known him to be bad."

She threw back her head; she had clearly, by this time, and very honestly, adopted an attitude. "Oh, never known him—I don't pretend *that!*"

I was upset again. "Then you *have* known him—?"

"Yes indeed, Miss, thank God!"

On reflection I accepted this. "You mean that a boy who never is—?"

"Is no boy for *me!*"

I held her tighter. "You like them with the spirit to be naughty?" Then, keeping pace with her answer, "So do I!" I eagerly brought out. "But not to the degree to contaminate—"

"To contaminate?"—my big word left her at a loss. I explained it. "To corrupt."

She stared, taking my meaning in; but it produced in her an odd laugh. "Are you afraid he'll corrupt *you*?" She put the question with such a fine bold humour that, with a laugh, a little silly doubtless, to match her own, I gave way for the time to the apprehension of ridicule.

But the next day, as the hour for my drive approached, I cropped up in another place. "What was the lady who was here before?"

"The last governess? She was also young and pretty—almost as young and almost as pretty, Miss, even as you."

"Ah, then, I hope her youth and her beauty helped her!" I recollect throwing off. "He seems to like us young and pretty!"

"Oh, he *did*," Mrs. Grose assented: "it was the way he liked everyone!" She had no sooner spoken indeed than she caught herself up. "I mean that's *his* way—the master's."

I was struck. "But of whom did you speak first?"

She looked blank, but she coloured. "Why, of *him*."

"Of the master?"

"Of who else?"

There was so obviously no one else that the next moment I had lost my impression of her having accidentally said more than she meant; and I merely asked what I wanted to know. "Did *she* see anything in the boy—?"

"That wasn't right? She never told me."

I had a scruple, but I overcame it. "Was she careful—particular?"

Mrs. Grose appeared to try to be conscientious. "About some things—yes."

"But not about all?"

Again she considered. "Well, Miss—she's gone. I won't tell tales."

"I quite understand your feeling," I hastened to reply; but I thought it, after an instant, not opposed to this concession to pursue: "Did she die here?"

"No—she went off."

I don't know what there was in this brevity of Mrs. Grose's that struck me as ambiguous. "Went off to die?" Mrs. Grose looked straight out of the window, but I felt that, hypothetically, I had a right to know what young persons engaged for Bly were expected to do. "She was taken ill, you mean, and went home?"

"She was not taken ill, so far as appeared, in this house. She left it, at the end of the year, to go home, as she said, for a short holiday, to which the time she had put in had certainly given her a right. We had then a young woman—a nursemaid who had stayed on and who was a good girl and clever; and *she* took the children altogether for the interval. But our young lady never came back, and at the very moment I was expecting her I heard from the master that she was dead."

I turned this over. "But of what?"

"He never told me! But please, Miss," said Mrs. Grose, "I must get to my work."

III

HER THUS turning her back on me was fortunately not, for my just preoccupations, a snub that could check the growth of our mutual esteem. We met, after I had brought home little Miles, more intimately than ever on the ground of my stupefaction, my general emotion: so monstrous was I then ready to pronounce it that such a child as had now been revealed to me should be under an interdict. I was a little late on the scene, and I felt, as he stood wistfully looking out for me before the door of the inn at which the coach had put him down, that I had seen him, on the instant, without and within, in the great glow of freshness, the same positive fragrance of purity, in which I had, from the first moment, seen his little sister. He was incredibly beautiful, and Mrs. Grose had put her finger on it: everything but a sort of passion of tenderness for him was swept away by his presence. What I then and there took him to my heart for was something divine that I have never found to the same degree in any child—his indescribable little air of knowing nothing in the world but love. It would have been impossible to carry a bad name with a greater sweetness of innocence, and by the time I had got back to Bly with him I remained merely bewildered—so far, that is, as I was not outraged—by the sense of the horrible letter locked up in my room, in a drawer. As soon as I could compass a private word with Mrs. Grose I declared to her that it was grotesque.

She promptly understood me. "You mean the cruel charge—?"

"It doesn't live an instant. My dear woman, *look* at him!"

She smiled at my pretension to have discovered his charm. "I assure you, Miss, I do nothing else! What will you say, then?" she immediately added.

"In answer to the letter?" I had made up my mind. "Nothing."

"And to his uncle?"

I was incisive. "Nothing."

"And to the boy himself?"

I was wonderful. "Nothing."

She gave with her apron a great wipe to her mouth. "Then I'll stand by you. We'll see it out."

"We'll see it out!" I ardently echoed, giving her my hand to make it a vow.

She held me there a moment, then whisked up her apron again with her detached hand. "Would you mind, Miss, if I used the freedom—"

"To kiss me? No!" I took the good creature in my arms and, after we had embraced like sisters, felt still more fortified and indignant.

This, at all events, was for the time: a time so full that, as I recall the way it went, it reminds me of all the art I now need to make it a little distinct. What I look back at with amazement is the situation I accepted. I had undertaken, with my companion, to see it out, and I was under a charm, apparently, that could smooth away the extent and the far and difficult connections of such an effort. I was lifted aloft on a great wave of infatuation and pity. I found it simple, in my ignorance, my confusion, and perhaps my conceit, to assume that I could deal with a boy whose education for the world was all on the point of beginning. I am unable even to remember at this day what proposal I framed for the end of his holidays and the resumption of his studies. Lessons with me, indeed, that charming summer, we all had a theory that he was to have; but I now feel that, for weeks, the lessons must have been rather my own. I learnt something—at first certainly—that had not been one of the teachings of my small, smothered life; learnt to be amused, and even amusing, and not to think for the morrow. It was the first time, in a manner, that I had known space and air and freedom, all the music of summer and all the mystery of nature. And then there was consideration—and consideration was sweet. Oh, it was a trap—not designed, but deep—to my imagination, to my delicacy, perhaps to my vanity; to whatever, in me, was most excitable. The best way to picture it all is to say that I was off my guard. They gave me so little trouble—they were of a gentleness so extraordinary. I used to speculate—but even this with a dim disconnectedness—as to how the rough future (for all futures are rough!) would handle them and might bruise them. They had the bloom of health and happiness; and yet, as if I had been in charge of a pair of little grandees, of princes of the blood, for whom everything, to be right, would have to be enclosed and protected, the only form that, in my fancy, the after-years could take for them was that of a romantic, a really royal extension of the garden and the park. It may be, of course, above all, that what suddenly broke into this gives the previous time a charm of stillness—that hush in which something gathers or crouches. The change was actually like the spring of a beast.

In the first weeks the days were long; they often, at their finest, gave me what I used to call my own hour, the hour when, for my pupils, tea-time and bed-time having come and gone, I had, before my final retirement, a small interval alone. Much as I liked my companions, this hour was the thing in the day I liked most; and I liked it best of all when, as the light faded—or rather, I should say, the day lingered and the last calls of the last birds sounded, in a flushed sky, from the old trees—I could take a turn into the grounds and enjoy, almost with a sense of property that amused and flattered me, the beauty and dignity of the place. It was a pleasure at these moments to feel myself tranquil and justified; doubtless, perhaps, also to reflect that by my discretion, my quiet good sense and general high propriety, I was giving pleasure —if he ever thought of it!—to the person to whose pressure I had responded. What I was doing was what he had earnestly hoped and directly asked of me, and that I *could*, after all, do it proved even a greater joy than I had expected. I dare say I fancied myself, in short, a remarkable young woman and took comfort in the faith that this would more publicly appear. Well, I needed to be remarkable to offer a front to the remarkable things that presently gave their first sign.

It was plump, one afternoon, in the middle of my very hour: the children were tucked away and I had come out for my stroll. One of the thoughts that, as I don't in the least shrink now from noting, used to be with me in these wanderings was that it would be as charming as a charming story suddenly to meet someone. Someone would appear there at the turn of a path and would stand before me and smile and approve. I didn't ask more than that—I only asked that he should *know*; and the only way to be sure he knew would be to see it, and the kind light of it, in his handsome face. That was exactly present to me— by which I mean the face was—when, on the first of these occasions, at the end of a long June day, I stopped short on emerging from one of the plantations and coming into view of the house. What arrested me on the spot—and with a shock much greater than any vision had allowed for—was the sense that my imagination had, in a flash, turned real. He did stand there!—but high up, beyond the lawn and at the very top of the tower to which, on that first morning, little Flora had conducted me. This tower was one of a pair—square, incongruous, crenelated structures—that were distinguished, for some reason, though I could see little difference, as the new and the old. They flanked opposite ends of the house and were probably architectural absurdities, redeemed in a measure indeed by not being wholly disengaged nor of a height too pretentious, dating, in their gingerbread antiquity, from a romantic revival that was already a respectable past. I admired them, had fancies about them, for we could all profit in a degree, especially when they loomed through the dusk, by the grandeur of their actual battlements; yet it was not at such an elevation that the figure I had so often invoked seemed most in place.

It produced in me, this figure, in the clear twilight, I remember, two distinct gasps of emotion, which were, sharply, the shock of my first and that of my second surprise. My second was a violent perception of the mistake of my first: the man who met my eyes was not the person I had precipitately supposed. There came to me thus a bewilderment of vision of which, after these years, there is no living view that I can hope to give. An unknown man in a lonely place is a permitted object of fear to a young woman privately bred; and the figure that faced me was—a few more seconds assured me—as little anyone else I knew as it was the image that had been in my mind. I had not seen it in Harley Street—I had not seen it anywhere. The place, moreover, in the strangest way in the world, had, on the instant, and by the very fact of its appearance, become a solitude. To me at least, making my statement here with a deliberation with which I have never made it, the whole feeling of the moment returns. It was as if, while I took in—what I did take in—all the rest of the scene had been stricken with death. I can hear again, as I write, the intense hush in which the sounds of evening dropped. The rooks stopped cawing in the golden sky and the friendly hour lost, for the minute, all its voice. But there was no other change in nature, unless indeed it were a change that I saw with a stranger sharpness. The gold was still in the sky, the clearness in the air, and the man who looked at me over the battlements was as definite as a picture in a frame. That's how I thought, with extraordinary quickness, of each person that he might have been and that he was not. We were confronted across our distance quite long enough for me to ask myself with intensity who then he was and to feel, as an effect of my inability to say, a wonder that in a few instants more became intense.

The great question, or one of these, is, afterwards, I know, with regard to certain matters, the question of how long they have lasted. Well, this matter of mine, think what you will of it, lasted while I caught at a dozen possibilities, none of which made a difference for the better, that I could see, in there having been in the house—and for how long, above all?—a person of whom I was in ignorance. It lasted while I just bridled a little with the sense that my office demanded that there should be no such ignorance and no such person. It lasted while this visitant, at all events,—and there was a touch of the strange freedom, as I remember, in the sign of familiarity of his wearing no hat,—seemed to fix me, from his position, with just the question, just the scrutiny through the fading light, that his own presence provoked. We were too far apart to call to each other, but there was a moment at which, at shorter range, some challenge between us, breaking the hush, would have been the right result of our straight mutual stare. He was in one of the angles, the one away from the house, very erect, as it struck me, and with both hands on the ledge. So I saw him as I see the letters I form on this page; then, exactly, after a minute, as if to add to the spectacle, he slowly changed his place—passed, looking at me hard all the while, to the opposite corner of the platform. Yes, I had the sharpest sense that during this transit he never took his eyes from me, and I can see at this moment the way his hand, as he went, passed from one of the crenelations to the next. He stopped at the other corner, but less long, and even as he turned away still markedly fixed me. He turned away; that was all I knew.

IV

IT WAS not that I didn't wait, on this occasion, for more, for I was rooted as deeply as I was shaken. Was there a "secret" at Bly —a mystery of Udolpho or an insane, an unmentionable relative kept in unsuspected confinement? I can't say how long I turned it over, or how long, in a confusion of curiosity and dread, I remained where I had my collision; I only recall that when I re-entered the house darkness had quite closed in. Agitation, in the interval, certainly had held me and driven me, for I must, in circling about the place, have walked three miles; but I was to be, later on, so much more overwhelmed that this mere dawn of alarm was a comparatively human chill. The most singular part of it in fact—singular as the rest had been—was the part I became, in the hall, aware of in meeting Mrs. Grose. This picture comes back to me in the general train—the impression, as I received it on my return, of the wide white panelled space, bright in the lamplight and with its portraits and red carpet, and of the good surprised look of my friend, which immediately told me she had missed me. It came to me straightway, under her contact, that, with plain heartiness, mere relieved anxiety at my appearance, she knew nothing whatever that could bear upon the incident I had there ready for her. I had not suspected in advance that her comfortable face would pull me up, and I somehow measured the importance of what I had seen by my thus finding myself hesitate to mention it. Scarce anything in the whole history seems to me so odd as this fact that my real beginning of fear was one, as I may say, with the instinct of sparing my companion. On the spot, accordingly, in the pleasant hall and with her eyes on me, I, for a reason that I couldn't then have phrased, achieved an inward revolution—offered a vague pretext for my lateness and, with the plea of the beauty of the night and of the heavy dew and wet feet, went as soon as possible to my room.

Here it was another affair; here, for many days after, it was a queer affair enough. There were hours, from day to day,—or at least there were moments, snatched even from clear duties,—when I had to shut myself up to think. It was not so much yet that I was more nervous than I could bear to be as that I was remarkably afraid of becoming so; for the truth I had now to turn over was, simply and clearly, the truth that I could arrive at no account whatever of the visitor with whom I had been so inexplicably and yet, as it seemed to me, so intimately concerned. It took little time to see that I could sound without forms of inquiry and without exciting remark any domestic complication. The shock I had suffered must have sharpened all my senses; I felt sure, at the end of three days and as the result of mere closer attention, that I had not been practised upon by the servants nor made the object of any "game." Of whatever it was that I knew nothing was known around me. There was but one sane inference: someone had taken a liberty rather gross. That was what, repeatedly, I dipped into my room and locked the door to say to myself. We had been, collectively, subject to an intrusion; some unscrupulous traveller, curious in old houses, had made his way in unobserved, enjoyed the prospect from the best point of view, and then stolen out as he came. If he had given me such a bold hard stare, that was but a part of his indiscretion. The good thing, after all, was that we should surely see no more of him.

This was not so good a thing, I admit, as not to leave me to judge that what, essentially, made nothing else much signify was simply my charming work. My charming work was just my life with Miles and Flora, and through nothing could I so like it as through feeling that I could throw myself into it in trouble. The attraction of my small charges was a constant joy, leading me to wonder afresh at the vanity of my original fears, the distaste I had begun by entertaining for the probable grey prose of my office. There was to be no grey prose, it appeared, and no long grind; so how could work not be charming that presented itself as daily beauty? It was all the romance of the nursery and the poetry of the schoolroom. I don't mean by this, of course, that we studied only fiction and verse; I mean I can express no otherwise the sort of interest my companions inspired. How can I describe that except by saying that instead of growing used to them—and it's a marvel for a governess: I call the sisterhood to witness!—I made constant fresh discoveries. There was one direction, assuredly, in which these discoveries stopped: deep obscurity continued to cover the region of the boy's conduct at school. It had been promptly given me, I have noted, to face that mystery without a pang. Perhaps even it would be nearer the truth to say that—without a word—he himself had cleared it up. He had made the whole charge absurd. My conclusion bloomed there with the real rose-flush of his innocence: he was only too fine and fair for the little horrid, unclean school-world, and he had paid a price for it. I reflected acutely that the sense of such differences, such superiorities of quality, always, on the part of the majority—which could include even stupid, sordid head-masters—turns infallibly to the vindictive.

Both the children had a gentleness (it was their only fault, and it never made Miles a muff) that kept them—how shall I express it?—almost impersonal and certainly quite unpunishable. They were like the cherubs of the anecdote, who had—morally, at any rate—nothing to whack! I remember feeling with Miles in especial as if he had had, as it were, no history. We expect of a small child a scant one, but there was in this beautiful little boy something extraordinarily sensitive, yet extraordinarily happy, that, more than in any creature of his age I have seen, struck me as beginning anew each day. He had never for a second suffered. I took this as a direct disproof of his having really been chastised. If he had been wicked he would have "caught" it, and I should have caught it by the rebound—I should have found the trace. I found nothing at all, and he was therefore angel. He never spoke of his school, never mentioned a comrade or a master; and I, for my part, was quite too much disgusted to allude to them. Of course I was under the spell, and the wonderful part is that, even at the time, I perfectly knew I was. But I gave myself up to it; it was an antidote to any pain, and I had more pains than one. I was in receipt in these days of disturbing letters from home, where things were not going well. But with my children, what things in the world mattered? That was the question I used to put to my scrappy retirements. I was dazzled by their loveliness.

There was a Sunday—to get on—when it rained with such force and for so many hours that there could be no procession to church;

in consequence of which, as the day declined, I had arranged with Mrs. Grose that, should the evening show improvement, we would attend together the late service. The rain happily stopped, and I prepared for our walk, which, through the park and by the good road to the village, would be a matter of twenty minutes. Coming downstairs to meet my colleague in the hall, I remembered a pair of gloves that had required three stitches and that had received them—with a publicity perhaps not edifying—while I sat with the children at their tea, served on Sundays, by exception, in that cold, clean temple of mahogany and brass, the "grown-up" dining-room. The gloves had been dropped there, and I turned in to recover them. The day was grey enough, but the afternoon light still lingered, and it enabled me, on crossing the threshold, not only to recognise, on a chair near the wide window, then closed, the articles I wanted, but to become aware of a person on the other side of the window and looking straight in. One step into the room had sufficed; my vision was instantaneous; it was all there. The person looking straight in was the person who had already appeared to me. He appeared thus again with I won't say greater distinctness, for that was impossible, but with a nearness that represented a forward stride in our intercourse and made me, as I met him, catch my breath and turn cold. He was the same—he was the same, and seen, this time, as he had been seen before, from the waist up, the window, though the dining-room was on the ground-floor, not going down to the terrace on which he stood. His face was close to the glass, yet the effect of this better view was, strangely, only to show me how intense the former had been. He remained but a few seconds—long enough to convince me he also saw and recognised; but it was as if I had been looking at him for years and had known him always. Something, however, happened this time that had not happened before; his stare into my face, through the glass and across the room, was as deep and hard as then, but it quitted me for a moment during which I could still watch it, see it fix successively several other things. On the spot there came to me the added shock of a certitude that it was not for me he had come there. He had come for someone else.

The flash of this knowledge—for it was knowledge in the midst of dread—produced in me the most extraordinary effect, started, as I stood there, a sudden vibration of duty and courage. I say courage because I was beyond all doubt already far gone. I bounded straight out of the door again, reached that of the house, got, in an instant, upon the drive, and, passing along the terrace as fast as I could rush, turned a corner and came full in sight. But it was in sight of nothing now—my visitor had vanished. I stopped, I almost dropped, with the real relief of this; but I took in the whole scene—I gave him time to reappear. I call it time, but how long was it? I can't speak to the purpose today of the duration of these things. That kind of measure must have left me: they couldn't have lasted as they actually appeared to me to last. The terrace and the whole place, the lawn and the garden beyond it, all I could see of the park, were empty with a great emptiness. There were shrubberies and big trees, but I remember the clear assurance I felt that none of them concealed him. He was there or was not there: not there if I didn't see him. I got hold of this; then, instinctively, instead of returning as I had come, went to the window. It was confusedly present to me that I ought to place myself where he had stood. I did so; I applied my face to the pane and looked, as he had looked, into the room. As if, at this moment, to show me exactly what his range had been, Mrs. Grose, as I had done for himself just before, came in from the hall. With this I had the full image of a repetition of what had already occurred. She saw me as I had seen my own visitant; she pulled up short as I had done; I gave her something of the shock that I had received. She turned white, and this made me ask myself if I had blanched as much. She stared, in short, and retreated on just *my* lines, and I knew she had then passed out and come round to me and that I should presently meet her. I remained where I was, and while I waited I thought of more things than one. But there's only one I take space to mention. I wondered why *she* should be scared.

V

OH, SHE let me know as soon as, round the corner of the house, she loomed again into view. "What in the name of goodness is the matter—?" She was now flushed and out of breath.

I said nothing till she came quite near. "With me?" I must have made a wonderful face. "Do I show it?"

"You're as white as a sheet. You look awful."

I considered; I could meet on this, without scruple, any innocence. My need to respect the bloom of Mrs. Grose's had dropped, without a rustle, from my shoulders, and if I wavered for the instant it was not with what I kept back. I put out my hand to her and she took it; I held her hard a little, liking to feel her close to me. There was a kind of support in the shy heave of her surprise. "You came for me for church, of course, but I can't go."

"Has anything happened?"

"Yes. You must know now. Did I look very queer?"

"Through this window? Dreadful!"

"Well," I said, "I've been frightened." Mrs. Grose's eyes expressed plainly that *she* had no wish to be, yet also that she knew too well her place not to be ready to share with me any marked inconvenience. Oh, it was quite settled that she *must* share! "Just what you saw from the dining-room a minute ago was the effect of that. What *I* saw—just before—was much worse."

Her hand tightened. "What was it?"

"An extraordinary man. Looking in."

"What extraordinary man?"

"I haven't the least idea."

Mrs. Grose gazed round us in vain. "Then where is he gone?"

"I know still less."

"Have you seen him before?"

"Yes—once. On the old tower."

She could only look at me harder. "Do you mean he's a stranger?"

"Oh, very much!"

"Yet you didn't tell me?"

"No—for reasons. But now that you've guessed—"

Mrs. Grose round eyes encountered this change. "Ah, I haven't guessed!" she said very simply. "How can I if *you* don't imagine?"

"I don't in the very least."

"You've seen him nowhere but on the tower?"

"And on this spot just now."

Mrs. Grose looked round again. "What was he doing on the tower?"

"Only standing there and looking down at me."

She thought a minute. "Was he a gentleman?"

I found I had no need to think. "No." She gazed in deeper wonder. "No."

"Then nobody about the place? Nobody from the village?"

"Nobody—nobody. I didn't tell you, but I made sure."

She breathed a vague relief: this was, oddly, so much to the good. It only went indeed a little way. "But if he isn't a gentleman—"

"What *is* he? He's a horror."

"A horror?"

"He's—God help me if I know *what* he is!"

Mrs. Grose looked round once more; she fixed her eyes on the duskier distance, then, pulling herself together, turned to me with abrupt inconsequence. "It's time we should be at church."

"Oh, I'm not fit for church!"

"Won't it do you good?"

"It won't do *them*—!" I nodded at the house.

"The children?"

"I can't leave them now."

"You're afraid—?"

I spoke boldly. "I'm afraid of *him*."

Mrs. Grose's large face showed me, at this, for the first time, the faraway faint glimmer of a consciousness more acute: I somehow made out in it the delayed dawn of an idea I myself had not given her and that was as yet quite obscure to me. It comes back to me that I thought instantly of this as something I could get from her; and I felt it to be connected with the desire she presently showed to know more. "When was it—on the tower?"

"About the middle of the month. At this same hour."

"Almost at dark?" said Mrs. Grose.

"Oh, no, not nearly. I saw him as I see you."

"Then how did he get in?"

"And how did he get out?" I laughed. "I had no opportunity to ask him! This evening, you see," I pursued, "he has not been able to get in."

"He only peeps?"

"I hope it will be confined to that!" She had now let go of my hand; she turned away a little. I waited an instant; then I brought out: "Go to church. Good-bye. I must watch."

Slowly she faced me again. "Do you fear for them?"

We met in another long look. "Don't *you*?" Instead of answering she came nearer to the window and, for a minute, applied her face to the glass. "You see how he could see," I meanwhile went on.

She didn't move. "How long was he here?"

"Till I came out. I came to meet him."

Mrs. Grose at last turned round, and there was still more in her face. "*I* couldn't have come out."

"Neither could I!" I laughed again. "But I did come. I have my duty."

"So have I mine," she replied; after which she added: "What is he like?"

"I've been dying to tell you. But he's like nobody."

"Nobody?" she echoed.

"He has no hat." Then seeing in her face that she already, in this, with a deeper dismay, found a touch of picture, I quickly added stroke to stroke. "He has red hair, very red, close-curling, and a pale face, long in shape, with straight, good features and little, rather queer whiskers that are as red as his hair. His eyebrows are, somehow, darker; they look particularly arched and as if they might move a good deal. His eyes are sharp, strange—awfully; but I only know clearly that they're rather small and very fixed. His mouth's wide, and his lips are thin, and except for his little whiskers he's quite clean-shaven. He gives me a sort of sense of looking like an actor."

"An actor!" It was impossible to resemble one less, at least, than Mrs. Grose at that moment.

"I've never seen one, but so I suppose them. He's tall, active, erect," I continued, "but never—no, never!—a gentleman."

My companion's face had blanched as I went on; her round eyes started and her mild mouth gaped. "A gentleman?" she gasped, confounded, stupefied: "a gentleman *he*?"

"You know him then?"

She visibly tried to hold herself. "But he *is* handsome?"

I saw the way to help her. "Remarkably!"

"And dressed—?"

"In somebody's clothes. They're smart, but they're not his own."

She broke into a breathless affirmative groan. "They're the master's!"

I caught it up. "You *do* know him?"

She faltered but a second. "Quint!" she cried.

"Quint?"

"Peter Quint—his own man, his valet, when he was here!"

"When the master was?"

Gaping still, but meeting me, she pieced it all together. "He never wore his hat, but he did wear—well, there were waistcoats missed! They were both here—last year. Then the master went, and Quint was alone."

I followed, but halting a little. "Alone?"

"Alone with *us*." Then, as from a deeper depth, "In charge," she added.

"And what became of him?"

She hung fire so long that I was still more mystified. "He went too," she brought out at last.

"Went where?"

Her expression, at this, became extraordinary. "God knows where! He died."

"Died?" I almost shrieked.

She seemed fairly to square herself, plant herself more firmly to utter the wonder of it. "Yes. Mr. Quint is dead."

VI

IT TOOK of course more than that particular passage to place us together in presence of what we had now to live with as we could—my dreadful liability to impressions of the order so vividly exemplified, and my companion's knowledge, henceforth,—a knowledge half consternation and half compassion,—of that liability. There had been, this evening, after the revelation that left me, for an hour, so prostrate—there had been, for either of us, no attendance on any service but a little service of tears and vows, of prayers and promises, a climax to the series of mutual challenges and pledges that had straightway ensued on our retreating together to the schoolroom and shutting ourselves up there to have everything out. The result of our having everything out was simply to reduce our situation to the last rigour of its elements. She herself had seen nothing, not the shadow of a shadow, and nobody in the house but the governess was in the governess's plight; yet she accepted without directly impugning my sanity the truth as I gave it to her, and ended by showing me, on this ground, an awe-stricken tenderness, an expression of the sense of my more than questionable privilege, of which the very breath has remained with me as that of the sweetest of human charitieis.

What was settled between us, accordingly, that night, was that we thought we might bear things together; and I was not even sure that, in spite of her exemption, it was she who had the best of the burden. I knew at this hour, I think, as well as I knew later what I was capable of meeting to shelter my pupils; but it took me some time to be wholly sure of what my honest ally was prepared for to keep terms with so compromising a contract. I was queer company enough—quite as queer as the company I received; but as I trace over what we went through I see how much common ground we must have found in the one idea that, by good fortune, *could* steady us. It was the idea, the second movement, that led me straight out, as I may say, of the inner chamber of my dread. I could take the air in the court, at least, and there Mrs. Grose could join me. Perfectly can I recall now the particular way strength came to me before we separated for the night. We had gone over and over every feature of what I had seen.

"He was looking for someone else, you say—someone who was not you?"

"He was looking for little Miles." A portentous clearness now possessed me. "*That's* whom he was looking for."

"But how did you know?"

"I know, I know, I know!" My exaltation grew. "And *you* know, my dear!"

She didn't deny this, but I required, I felt, not even so much telling as that. She resumed in a moment, at any rate: "What if *he* should see him?"

"Little Miles? That's what he wants!"

She looked immensely scared again. "The child?"

"Heaven forbid! The man. He wants to appear to *them*." That he might was an awful conception, and yet, somehow, I could keep it at bay; which, moreover, as we lingered there, was what I succeeded in practically proving. I had an absolute certainty that I could see again what I had already seen, but something within me said that by offering myself bravely as the sole subject of such experience, by accepting, by inviting, by surmounting it all, I should serve as an expiatory victim and guard the tranquillity of my companions. The children, in especial, I should thus fence about and absolutely save. I recall one of the last things I said that night to Mrs. Grose.

"It does strike me that my pupils have never mentioned—"

She looked at me hard as I musingly pulled up. "His having been here and the time they were with him?"

"The time they were with him, and his name, his presence, his history, in any way."

"Oh, the little lady doesn't remember. She never heard or knew."

"The circumstances of his death?" I thought with some intensity. "Perhaps not. But Miles would remember—Miles would know."

"Ah, don't try him!" broke from Mrs. Grose.

I returned her the look she had given me. "Don't be afraid." I continued to think. "It *is* rather odd."

"That he has never spoken of him?"

"Never by the least allusion. And you tell me they were 'great friends'?"

"Oh, it wasn't *him*!" Mrs. Grose with emphasis declared. "It was Quint's own fancy. To play with him, I mean—to spoil him." She paused a moment; then added: "Quint was much too free."

This gavc mc, straight from my vision of his face—*such* a face—a sudden sickness of disgust. "Too free with *my* boy?"

"Too free with everyone!"

I forebore, for the moment, to analyse this description further than by the reflection that a part of it applied to several of the members of the household, or the half-dozen maids and men who were still of our small colony. But there was everything, for our apprehension, in the lucky fact that no discomfortable legend, no perturbation of scullions, had ever, within anyone's memory, attached to the kind old place. It had neither bad name nor ill fame, and Mrs. Grose, most apparently, only desired to cling to me and to quake in silence. I even put her, the very last thing of all, to the test. It was when, at midnight, she had her hand on the schoolroom door to take leave. "I have it from you then—for it's of great importance—that he was definitely and admittedly bad?"

"Oh, not admittedly. *I* knew it—but the master didn't."

"And you never told him?"

"Well, he didn't like tale-bearing—he hated complaints. He was terribly short with anything of that kind, and if people were all right to *him*—"

"He wouldn't be bothered with more?" This squared well enough with my impression of him: he was not a trouble-loving gentleman, nor so very particular perhaps about some of the company *he* kept. All the same, I pressed my interlocutress. "I promise you *I* would have told!"

She felt my discrimination. "I dare say I was wrong. But, really, I was afraid."

"Afraid of what?"

"Of things that man could do. Quint was so clever—he was so deep."

I took this in still more than, probably, I showed. "You weren't afraid of anything else? Not of his effect—?"

"His effect?" she repeated with a face of anguish and waiting while I faltered.

"On innocent little precious lives. They were in your charge."

"No, they were not in mine!" she roundly and distressfully returned.

"The master believed in him and placed him here because he was supposed not to be well and the country air so good for him. So he had everything to say. Yes"—she let me have it—"even about *them.*"

"Them—that creature?" I had to smother a kind of howl. "And you could bear it!"

"No, I couldn't—and I can't now!" And the poor woman burst into tears.

A rigid control, from the next day, was, as I have said, to follow them; yet how often and how passionately, for a week, we came back together to the subject! Much as we had discussed it that Sunday night, I was, in the immediate later hours in especial—for it may be imagined whether I slept—still haunted with the shadow of something she had not told me. I myself had kept back nothing, but there was a word Mrs. Grose had kept back. I was sure, moreover, by morning, that this was not from a failure of frankness, but because on every side there were fears. It seems to me indeed, in retrospect, that by the time the morrow's sun was high I had restlessly read into the facts before us almost all the meaning they were to receive from subsequent and more cruel occurrences. What they gave me above all was just the sinister figure of the living man—the dead one would keep awhile!—and of the months he had continuously passed at Bly, which, added up, made a formidable stretch. The limit of this evil time had arrived only when, on the dawn of a winter's morning, Peter Quint was found, by a labourer going to early work, stone dead on the road from the village: a catastrophe explained—superficially at least—by a visible wound to his head; such a wound as might have been produced—and as, on the final evidence, *had* been—by a fatal slip, in the dark and after leaving the public house, on the steepish icy slope, a wrong path, altogether, at the bottom of which he lay. The icy slope, the turn mistaken at night and in liquor, accounted for much—practically, in the end and after the inquest and boundless chatter, for everything; but there had been matters in his life—strange passages and perils, secret disorders, vices more than suspected—that would have accounted for a good deal more.

I scarce know how to put my story into words that shall be a credible picture of my state of mind; but I was in these days literally able to find a joy in the extraordinary flight of heroism the occasion demanded of me. I now saw that I had been asked for a service admirable and difficult; and there would be a greatness in letting it be seen—oh, in the right quarter!—that I could succeed where many another girl might have failed. It was an immense help to me—I confess I rather applaud myself as I look back!—that I saw my service so strongly and so simply. I was there to protect and defend the little creatures in the world the most bereaved and the most lovable, the appeal of whose helplessness had suddenly become only too explicit, a deep, constant ache of one's own committed heart. We were cut off, really, together; we were united in our danger. They had nothing but me, and I—well, I had *them*. It was in short a magnificent chance. This chance presented itself to me in an image richly material. I was a screen—I was to stand before them. The more I saw, the less they would. I began to watch them in a stifled suspense, a disguised excitement that might well, had it continued too long, have turned to something like madness. What saved me, as I now see, was that it turned to something else altogether. It didn't last as suspense—it was superseded by horrible proofs. Proofs, I say, yes—from the moment I really took hold.

This moment dated from an afternoon hour that I happened to spend in the grounds with the younger of my pupils alone. We had left Miles indoors, on the red cushion of a deep window-seat; he had wished to finish a book, and I had been glad to encourage a purpose so laudable in a young man whose only defect was an occasional excess of the restless. His sister, on the contrary, had been alert to come out, and I strolled with her half an hour, seeking the shade, for the sun was still high and the day exceptionally warm. I was aware afresh, with her, as we went, of how, like her brother, she contrived—it was the charming thing in both children—to let me alone without appearing to drop me and to accompany me without appearing to surround. They were never importunate and yet never listless. My attention to them all really went to seeing them amuse themselves immensely without me: this was a spectacle they seemed actively to prepare and that engaged me as an active admirer. I walked in a world of their invention—they had no occasion whatever to draw upon mine; so that my time was taken only with being, for them, some remarkable person or thing that the game of the moment required and that was merely, thanks to my superior, my exalted stamp, a happy and highly distinguished sinecure. I forget what I was on the present occasion; I only remember that I was something very important and very quiet and that Flora was playing very hard. We were on the edge of the lake, and, as we had lately begun geography, the lake was the Sea of Azof.

Suddenly, in these circumstances, I became aware that, on the other side of the Sea of Azof, we had an interested spectator. The way this knowledge gathered in me was the strangest thing in the world—the strangest, that is, except the very much stranger in which it quickly merged itself. I had sat down with a piece of work—for I was something or other that could sit—on the old stone bench which overlooked the pond; and in this position I began to take in with certitude, and yet without direct vision, the presence, at a distance, of a third person. The old trees, the thick shrubbery, made a great and pleasant shade, but it was all suffused with the brightness of the hot, still hour. There was no ambiguity in anything; none whatever, at least, in the conviction I from one moment to another found myself forming as to what I should see straight before me and across the lake as a consequence of raising my eyes. They were attached at this juncture to the stitching in which I was engaged, and I can feel once more the spasm of my effort not to move them till I should have steadied myself as to be able to make up my mind what to do. There was an alien object in view—a figure whose right of presence I instantly, passionately questioned. I recollect counting over perfectly the possibilities, reminding myself that nothing was more natural, for instance, than the appearance of one of the men about the place, or even of a messenger, a postman or a tradesman's boy, from the village. That reminder had as little effect on my practical certitude as I was conscious—still even without looking—of its having upon the character and attitude of our visitor. Nothing was more natural than that these things should be the other things that they absolutely were not.

Of the positive identity of the apparition I would assure myself as soon as the small clock of my courage should have ticked out the right second; meanwhile, with an effort that was already sharp enough, I transferred my eyes straight to little Flora, who, at the moment, was about ten yards away. My heart had stood still for an instant with the wonder and terror of the question whether she too would see; and I held my breath while I waited for what a cry from her, what some sudden innocent sign either of interest or of alarm, would tell me. I waited, but nothing came; then, in the first place—and there is something more dire in this, I feel, than in anything I have to relate—I was determined by a sense that, within a minute, all sounds from her had previously dropped; and, in the second, by the circumstance that, also within the minute, she had, in her play, turned her back to the water. This was her attitude when I at last looked at her—looked with the confirmed conviction that we were still, together, under direct personal notice. She had picked up a small flat piece of wood, which happened to have in it a little hole that had evidently suggested to her the idea of sticking in another fragment that might figure as a mast and make the thing a boat. This second morsel, as I watched her, she was very markedly and intently to tighten in its place. My apprehension of what she was doing sustained me so that after some seconds I felt I was ready for more. Then I again shifted my eyes—I faced what I had to face.

VII

I GOT hold of Mrs. Grose as soon after this as I could; and I can give no intelligible account of how I fought out the interval. Yet I still hear myself cry as I fairly threw myself into her arms: "They *know*—it's too monstrous: they know, they know!"

"And what on earth—?" I felt her incredulity as she held me.

"Why, all that *we* know—and heaven knows what else besides!" Then, as she released me, I made it out to her, made it out perhaps only now with full coherency even to myself. "Two hours ago, in the garden" —I could scarce articulate—"Flora *saw!*"

Mrs. Grose took it as she might have taken a blow in the stomach. "She has told you?" she panted.

"Not a word—that's the horror. She kept it to heself! The child of eight, *that* child!" Unutterable still, for me, was the stupefaction of it.

Mrs. Grose, of course, could only gape the wider. 'Then how do you know?"

"I was there—I saw with my eyes: saw that she was perfectly aware."

"Do you mean aware of *him*?"

"No—of *her*." I was conscious as I spoke that I looked prodigious things, for I got the slow reflection of them in my companion's face. "Another person—this time; but a figure of quite an unmistakable horror and evil: a woman in black, pale and dreadful—with such an air also, and such a face!—on the other side of the lake. I was there with the child—quiet for the hour; and in the midst of it she came."

"Came how—from where?"

"From where they come from! She just appeared and stood there—but not so near."

"And without coming nearer?"

"Oh, for the effect and the feeling, she might have been as close as you!"

My friend, with an odd impulse, fell back a step. "Was she someone you've never seen?"

"Yes. But someone the child has. Someone *you* have." Then, to show how I had thought it all out: "My predecessor—the one who died."

"Miss Jessel?"

"Miss Jessel. You don't believe me?" I pressed.

She turned right and left in her distress. "How can you be sure?"

This drew from me, in the state of my nerves, a flash of impatience. "Then ask Flora—*she's* sure!" But I had no sooner spoken than I caught myself up. "No, for God's sake, *don't!* She'll say she isn't—she'll lie!"

Mrs. Grose was not too bewildered instinctively to protest. "Ah, how *can* you?"

"Because I'm clear. Flora doesn't want me to know."

"It's only then to spare you."

"No, no—there are depths, depths! The more I go over it, the more I see in it, and the more I see in it the more I fear. I don't know what I *don't* see—what I *don't* fear!"

Mrs. Grose tried to keep up with me. "You mean you're afraid of seeing her again?"

"Oh, no; that's nothing—now!" Then I explained. "It's of *not* seeing her."

But my companion only looked wan. "I don't understand you."

"Why, it's that the child may keep it up—and that the child assuredly *will*—without my knowing it."

At the image of this possibility Mrs. Grose for a moment collapsed, yet presently to pull herself together again, as if from the positive force of the sense of what, should we yield an inch, there would really be to give way to. "Dear, dear—we must keep our heads! And after all, if she doesn't mind it—!" She even tried a grim joke. "Perhaps she likes it!"

"Likes *such* things—a scrap of an infant!"

"Isn't it just a proof of her blessed innocence?" my friend bravely inquired.

She brought me, for the instant, almost round. "Oh, we must clutch at *that*—we must cling to it! If it isn't a proof of what you say, it's a proof of—God knows what! For the woman's a horror of horrors."

Mrs. Grose, at this, fixed her eyes a minute on the ground; then at last raising them, "Tell me how you know," she said.

"Then you admit it's what she was?" I cried.

"Tell me how you know," my friend simply repeated.

"Know! By seeing her! By the way she looked."

"At you, do you mean—so wickedly?"

"Dear me, no—I could have borne that. She gave me never a glance. She only fixed the child."

Mrs. Grose tried to see it. "Fixed her?"

"Ah, with such awful eyes!"

She stared at mine as if they might really have resembled them. "Do you mean of dislike?"

"God help us, no. Of something much worse."

"Worse than dislike?"—this left her indeed at a loss.

"With a determination—indescribable. With a kind of fury of intention."

I made her turn pale. "Intention?"

"To get hold of her." Mrs. Grose—her eyes just lingering on mine —gave a shudder and walked to the window; and while she stood there looking out I completed my statement. "*That's* what Flora knows."

After a little she turned round. "The person was in black, you say?"

"In mourning—rather poor, almost shabby. But—yes—with extraordinary beauty." I now recognised to what I had at last, stroke by stroke, brought the victim of my confidence, for she quite visibly weighed this. "Oh, handsome—very, very," I insisted; "wonderfully handsome. But infamous."

She slowly came back to me. "Miss Jessel—*was* infamous." She once more took my hand in both her own, holding it as tight as if to fortify me against the increase of alarm I might draw from this disclosure. "They were both infamous," she finally said.

So, for a little, we faced it once more together; and I found absolutely a degree of help in seeing it now so straight. "I appreciate," I said, "the great decency of your not having hitherto spoken; but the time has certainly come to give me the whole thing." She appeared to assent to this, but still only in silence; seeing which I went on: "I must have it now. Of what did she die? Come, there was something between them."

"There was everything."

"In spite of the difference—?"

"Oh, of their rank, their condition"—she brought it woefully out. "*She* was a lady."

I turned it over; I again saw. "Yes—she was a lady."

"And he so dreadfully below," said Mrs. Grose.

I felt that I doubtless needn't press too hard, in such company, on the place of a servant in the scale; but there was nothing to prevent an acceptance of my companion's own measure of my predecessor's abasement. There was a way to deal with that, and I dealt; the more readily for my full vision—on the evidence—of our employer's late clever, good-looking "own" man; impudent, assured, spoiled, depraved. "The fellow was a hound."

Mrs. Grose considered as if it were perhaps a little a case for a sense of shades. "I've never seen one like him. He did what he wished."

"With *her*?"

"With them all."

It was as if now in my friend's own eyes Miss Jessel had again appeared. I seemed at any rate, for an instant, to see their evocation of her as distinctly as I had seen her by the pond; and I brought out with decision: "It must have been also what *she* wished!"

Mrs. Grose face signified that it had been indeed, but she said at the same time: "Poor woman—she paid for it!"

"Then you do know what she died of?" I asked.

"No—I know nothing. I wanted not to know; I was glad enough I didn't; and I thanked heaven she was well out of this!"

"Yet you had, then, your idea—"

"Of her real reason for leaving? Oh yes—as to that. She couldn't have stayed. Fancy it here—for a governess! And afterwards I imagined —and I still imagine. And what I imagine is dreadful."

"Not so dreadful as what *I* do," I replied; on which I must have shown her—as I was indeed but too conscious—a front of miserable defeat. It brought out again all her compassion for me, and at the renewed touch of her kindness my power to resist broke down. I burst, as I had, the other time, made her burst, into tears; she took me to her motherly breast, and my lamentation overflowed. "I don't do it!" I sobbed in despair; "I don't save or shield them. It's far worse than I dreamed—they're lost!"

VIII

WHAT I had said to Mrs. Grose was true enough: there were in the matter I had put before her depths and possibilities that I lacked resolution to sound; so that when we met once more in the wonder of it we were of a common mind about the duty of resistance to extravagant fancies. We were to keep our heads if we should keep nothing else—difficult indeed as that might be in the face of what, in our prodigious experience, was least to be questioned. Late that night, while the house slept, we had another talk in my room, when she went all the way with me as to its being beyond doubt that I had seen exactly what I had seen. To hold her perfectly in the pinch of that, I found I had only to ask her how, if I had "made it up," I came to to be able to give, of each of the persons appearing to me, a picture disclosing, to the last detail, their special marks—a portrait on the exhibition of which she had instantly recognised and named them. She wished, of course,—small blame to her!—to sink the whole subject; and I was quick to assure her that my own interest in it had now violently taken the form of a search for the way to escape from it. I encountered her on the ground of a probability that with recurrence—for recurrence we took for granted—I should get used to my danger, distinctly professing that my personal exposure had suddenly become the least of my discomforts. It was my new suspicion that was intolerable; and yet even to this complication the later hours of the day had brought a little ease.

On leaving her, after my first outbreak, I had of course returned to my pupils, associating the right remedy for my dismay with that sense of their charm which I had already found to be a thing I could positively cultivate and which had never failed me yet. I had simply, in other words, plunged afresh into Flora's special society and there become aware—it was almost a luxury!—that she could put her little conscious hand straight upon the spot that ached. She had looked at me in sweet speculation and then had accused me to my face of having "cried." I had supposed I had brushed away the ugly signs: but I could literally—for the time, at all events—rejoice, under this fathomless charity, that they had not entirely disappeared. To gaze into the depths of blue of the child's eyes and pronounce their loveliness a trick of premature cunning was to be guilty of cynicism in preference to which I naturally preferred to abjure my judgment and, so far as might be, my agitation. I couldn't abjure for merely wanting to, but I could repeat to Mrs. Grose—as I did there, over and over, in the small hours—that with their voices in the air, their pressure on one's heart and their fragrant faces against one's cheek, everything fell to the ground but their incapacity and their beauty. It was a pity that, somehow, to settle this once for all, I had equally to re-enumerate the signs of subtlety that, in the afternoon, by the lake, had made a miracle of my show of self-possession. It was a pity to be obliged to re-investigate the certitude of the moment itself and repeat how it had come to me as a revelation that the inconceivable communion I then surprised was a matter, for either party, of habit. It was a pity that I should have had to quaver out again the reasons for my not having, in my delusion, so much as questioned that the little girl saw our visitant even as I actually saw Mrs. Grose herself, and that she wanted, by just so much as she did thus see, to make me suppose she didn't, and at the same time, without showing anything, arrive at a guess as to whether I myself did! It was a pity that I needed once more to describe the portentous little activity by which she sought to divert my attention—the perceptible increase of movement, the greater intensity of play, the singing, the gabbling of nonsense, and the invitation to romp.

Yet if I had not indulged, to prove there was nothing in it, in this review, I should have missed the two or three dim elements of comfort that still remained to me. I should not for instance have been able to asseverate to my friend that I was certain—which was so much to the good—that I at least had not betrayed myself. I should not have been prompted, by stress of need, by desperation of mind,—I scarce know what to call it,—to invoke such further aid to intelligence as might spring from pushing my colleague fairly to the wall. She had told me, bit by bit, under pressure, a great deal; but a small shifty spot on the wrong side of it all still sometimes brushed my brow like the wing of a bat; and I remember how on this occasion—for the sleeping house and the concentration alike of our danger and our watch seemed to help—I felt the importance of giving the last jerk to the curtain. "I don't believe anything so horrible," I recollect saying; "no, let us put it definitely, my dear, that I don't. But if I did, you know, there's a thing I should require now, just without sparing you the least bit more—oh, not a scrap, come!—to get out of you. What was it you had in mind when, in our distress, before Miles came back, over the letter from his school, you said, under my insistence, that you didn't pretend for him that he had not literally *ever* been 'bad'? He was *not* literally 'ever,' in these weeks that I myself have lived with him and so closely watched him; he has been an imperturbable little prodigy of delightful, loveable goodness. Therefore you might perfectly have made the claim for him if you had not, as it happened, seen an exception to take. What was your exception, and to what passage in your personal observation of him did you refer?"

It was a dreadfully austere inquiry, but levity was not our note, and, at any rate, before the grey dawn admonished us to separate I had got my answer. What my friend had had in mind proved to be immensely to the purpose. It was neither more nor less than the circumstance that for a period of several months Quint and the boy had been perpetually together. It was in fact the very appropriate truth that she had ventured to criticise the propriety, to hint at the incongruity, of so close an alliance, and even to go so far on the subject as a frank overture to Miss Jessel. Miss Jessel had, with a most strange manner, requested her to mind her business, and the good woman had, on this, directly approached little Miles. What she had said to him, since I pressed, was that *she* liked to see young gentlemen not forget their station.

I pressed again, of course, at this. "You reminded him that Quint was only a base menial?"

"As you might say! And it was his answer, for one thing, that was bad."

"And for another thing?" I waited. "He repeated your words to Quint?"

"No, not that. It's just what he *wouldn't!*" she could still impress upon me. "I was sure, at any rate," she added, "that he didn't. But he denied certain occasions."

"What occasions?"

"When they had been about together quite as if Quint were his tutor—and a very grand one—and Miss Jessel only for the little lady. When he had gone off with the fellow, I mean, and spent hours with him."

"He then prevaricated about it—he said he hadn't?" Her assent was clear enough to cause me to add in a moment: "I see. He lied."

"Oh!" Mrs. Grose mumbled. This was a suggestion that it didn't matter; which indeed she backed up by a further remark. "You see, after all, Miss Jessel didn't mind. She didn't forbid him."

I considered. "Did he put that to you as a justification?"

At this she dropped again. "No, he never spoke of it."

"Never mentioned her in connection with Quint?"

She saw, visibly flushing, where I was coming out. "Well, he didn't show anything. He denied," she repeated; "he denied."

Lord, how I pressed her now! "So that you could see he knew what was between the two wretches?"

"I don't know—I don't know!" the poor woman groaned.

"You do know, you dear thing," I replied; "only you haven't my dreadful boldness of mind, and you keep back, out of timidity and modesty and delicacy, even the impression that, in the past, when you had, without my aid, to flounder about in silence, most of all made you miserable. But I shall get it out of you yet! There was something in the boy that suggested to you," I continued, "that he covered and concealed their relation."

"Oh, he couldn't prevent—"

"Your learning the truth? I dare say! But, heavens," I fell, with vehemence, a-thinking, "what it shows that they must, to that extent, have succeeded in making of him!"

"Ah, nothing that's not nice *now!*" Mrs. Grose lugubriously pleaded.

"I don't wonder you looked queer," I persisted, "when I mentioned to you the letter from his school!"

"I doubt if I looked as queer as you!" she retorted with homely force. "And if he was so bad then as that comes to, how is he such an angel now?"

"Yes, indeed—and if he was a fiend at school! How, how, how? Well," I said in my torment, "you must put it to me again, but I shall not be able to tell you for some days. Only, put it to me again!" I cried in a way that made my friend stare. "There are directions in which I must not for the present let myself go." Meanwhile I returned to her first example—the one to which she had just previously referred—of the boy's happy capacity for an occasional slip. "If Quint—on your remonstrance at the time you speak of—was a base menial, one of the things Miles said to you, I find myself guessing, was that you were another." Again her admission was so adequate that I continued: "And you forgave him that?"

"Wouldn't *you?*"

"Oh, yes!" And we exchanged there, in the stillness, a sound of the oddest amusement. Then I went on: "At all events, while he was with the man—"

"Miss Flora was with the woman. It suited them all!"

It suited me too, I felt, only too well; by which I mean that it suited exactly the particularly deadly view I was in the very act of forbidding myself to entertain. But I so far succeeded in checking the expression of this view that I will throw, just here, no further light on it than may be offered by the mention of my final observation to Mrs. Grose.

"His having lied and been impudent are, I confess, less engaging specimens than I had hoped to have from you of the outbreak in him of the little natural man. Still," I mused, "they must do, for they make me feel more than ever that I must watch."

It made me blush, the next minute, to see in my friend's face how much more unreservedly she had forgiven him than her anecdote struck me as presenting to my own tenderness an occasion for doing. This came out when, at the schoolroom door, she quitted me. "Surely you don't accuse *him*—"

"Of carrying on an intercourse that he conceals from me? Ah, remember that, until further evidence, I now accuse nobody." Then, before shutting her out to go, by another passage, to her own place, "I must just wait," I wound up.

IX

I WAITED and waited, and the days, as they elapsed, took something from my consternation. A very few of them, in fact, passing, in constant sight of my pupils, without a fresh incident, sufficed to give to grievous fancies and even to odious memories a kind of brush of the sponge. I have spoken of the surrender to their extraordinary childish grace as a thing I could actively cultivate, and it may be imagined if I neglected now to address myself to this source for whatever it would yield. Stranger than I can express, certainly, was the effort to struggle against my new lights; it would doubtless have been, however, a greater tension still had it not been so frequently successful. I used to wonder how my little charges could help guessing that I thought strange things about them; and the circumstance that these things only made them more interesting was not by itself a direct aid to keeping them in the dark. I trembled lest they should see that they *were* so immensely more interesting. Putting things at the worst, at all events, as in meditation I so often did, any clouding of their innocence could only be—blameless and foredoomed as they were—a reason the more for taking risks. There were moments when, by an irresistible impulse, I found myself catching them up and pressing them to my heart. As soon as I had done so I used to say to myself: "What will they think of that? Doesn't it betray too much?" It would have been easy to get into a sad, wild tangle about how much I might betray; but the real account, I feel, of the hours of peace that I could still enjoy was that the immediate charm of my companions was a beguilement still effective even under the shadow of the possibility that it was studied. For if it occurred to me that I might occasionally excite suspicion by the little outbreaks of my sharper passion for them, so too I remember wondering if I mightn't see a queerness in the traceable increase of their own demonstrations. They were at this period extravagantly and preternaturally fond of me; which, after all, I could reflect, was no more than a graceful response in children perpetually bowed over and hugged. The homage of which they were so lavish succeeded, in truth, for my nerves, quite as well as if I never appeared to myself, as I may say, literally to catch them at a purpose in it. They had never, I think, wanted to do so many things for their poor protectress; I mean—though they got their lessons better and better, which was naturally what would please her most—in the way of diverting, entertaining, surprising her; reading her passages, telling her stories, acting her charades, pouncing out at her, in disguises, as animals and historical characters, and above all astonishing her by the "pieces" they had secretly got by heart and could interminably recite. I should never get to the bottom—were I to let myself go even now—of the prodigious private commentary, all under still more private correction, with which, in these days, I overscored their full hours. They had shown me from the first a facility for everything, a general faculty which, taking a fresh start, achieved remarkable flights. They got their little tasks as if they loved them, and indulged, from the mere exuberance of the gift, in the most unimposed little miracles of memory. They not only popped out at me as tigers and as Romans, but as Shakespeareans, astronomers, and navigators. This was so singularly the case that it had presumably much to do with the fact as to which, at the present day, I am at a loss for a different explanation: I allude to my unnatural composure on the subject of another school for Miles. What I remember is that I was content not, for the time, to open the question, and that contentment must have sprung from the sense of his perpetually striking show of cleverness. He was too clever for a bad governess, for a parson's daughter, to spoil; and the strangest if not the brightest thread in the pensive embroidery I just spoke of was the impression I might have got, if I had dared to work it out, that he was under some influence operating in his small intellectual life as a tremendous incitement.

If it was easy to reflect, however, that such a boy could postpone school, it was at least as marked that for such a boy to have been "kicked out" by a school-master was a mystification without end. Let me add that in their company now—and I was careful almost never to be out of it—I could follow no scent very far. We lived in a cloud of music and love and success and private theatricals. The musical sense in each of the children was of the quickest, but the elder in especial had a marvellous knack of catching and repeating. The schoolroom piano broke into all gruesome fancies; and when that failed there were confabulations in corners, with a sequel of one of them going out in the highest spirits in order to "come in" as something new. I had had brothers myself, and it was no revelation to me that little girls could be slavish idolaters of little boys. What surpassed everything was that there was a little boy in the world who could have for the inferior age, sex, and intelligence so fine a consideration. They were extraordinarily at one, and to say that they never either quarrelled or complained is to make the note of praise coarse for their quality of sweetness. Sometimes, indeed, when I dropped into coarseness, I perhaps came across traces of little understandings between them by which one of them should keep me occupied while the other slipped away. There is a *naïf* side, I suppose, in all diplomacy; but if my pupils practised upon me, it was surely with the minimum of grossness. It was all in the other quarter that, after a lull, the grossness broke out.

I find that I really hang back; but I must take my plunge. In going on with the record of what was hideous at Bly, I not only challenge the most liberal faith—for which I little care; but—and this is another matter—I renew what I myself suffered, I again push my way through it to the end. There came suddenly an hour after which, as I look back, the affair seems to me to have been all pure suffering; but I have at least reached the heart of it, and the straightest road out is doubtless to advance. One evening—with nothing to lead up or to prepare it—I felt the cold touch of the impression that had breathed on me the night of my arrival and which, much lighter then, as I have mentioned, I should probably have made little of in memory had my subsequent sojourn been less agitated. I had not gone to bed; I sat reading by a couple of candles. There was a roomful of old books at Bly—last-century fiction, some of it, which, to the extent of a distinctly deprecated renown, but never to so much as that of a stray specimen, had reached the sequestered home and appealed to the unavowed curiosity of my youth. I remember that the book I had in my hand was Fielding's *Amelia*; also that I was wholly awake. I recall further both a general conviction that it was horribly late and a particular objection to looking at my watch. I figure, finally, that the white curtain draping, in the fashion of those days, the head of Flora's little bed, shrouded, as I had assured myself long before, the perfection of childish rest. I recollect in short that, though I was deeply interested in my author, I found myself, at the turn of a page and with his spell all scattered, looking straight up from him and hard at the door of my room. There was a moment during which I listened, reminded of the faint sense I had had, the first night, of there being something undefineably astir in the house, and noted the soft breath of the open casement just move the half-drawn blind. Then, with all the marks of a deliberation that must have seemed magnificent had there been anyone to admire it, I laid down my book, rose to my feet, and, taking a candle, went straight out of the room and, from the passage, on which my light made little impression, noiselessly closed and locked the door.

I can say now neither what determined nor what guided me, but I went straight along the lobby, holding my candle high, till I came within sight of the tall window that presided over the great turn of the staircase. At this point I precipitately found myself aware of three things. They were practically simultaneous, yet they had flashes of succession. My candle, under a bold flourish, went out, and I perceived, by the uncovered window, that the yielding dusk of earliest morning rendered it unnecessary. Without it, the next instant, I saw that there was someone on the stair. I speak of sequences, but I required no lapse of seconds to stiffen myself for a third encounter with Quint. The apparition had reached the landing halfway up and was therefore on the spot nearest the window, where at sight of me, it stopped short and fixed me exactly as it had fixed me from the tower and from the garden. He knew me as well as I knew him; and so, in the cold, faint twilight, with a glimmer in the high glass and another on the polish of the oak stair below, we faced each other in our common intensity. He was absolutely, on this occasion, a living, detestable, dangerous presence. But that was not the wonder of wonders; I reserve this distinction for quite another circumstance: the circumstance that dread had unmistakeably quitted me and that there was nothing in me there that didn't meet and measure him.

I had plenty of anguish after that extraordinary moment, but I had, thank God, no terror. And he knew I had not—I found myself at the end of an instant magnificently aware of this. I felt, in a fierce rigour of confidence, that if I stood my ground a minute I should cease—for the time, at least—to have him to reckon with; and during the minute, accordingly, the thing was as human and hideous as a real interview: hideous just because it *was* human, as human as to have met alone, in the small hours, in a sleeping house, some enemy, some adventurer, some criminal. It was the dead silence of our long gaze at such close quarters that gave the whole horror, huge as it was, its only note of the unnatural. If I had met a murderer in such a place and at such an hour, we still at least would have spoken. Something would have passed, in life, between us; if nothing had passed one of us would have moved. The moment was so prolonged that it would have taken but little more to make me doubt if even I were in life. I can't express what followed it save by saying that the silence itself—which was indeed in a manner an attestation of my strength—became the element into which I saw the figure disappear; in which I definitely saw it turn as I might have seen the low wretch to which it had once belonged turn on receipt of an order, and pass, with my eyes on the villainous back that no hunch could have more disfigured, straight down the staircase and into the darkness in which the next bend was lost.

X

I REMAINED awhile at the top of the stair, but with the effect presently of understanding that when my visitor had gone, he had gone: then I returned to my room. The foremost thing I saw there by the light of the candle I had left burning was that Flora's little bed was empty; and on this I caught my breath with all the terror that, five minutes before, I had been able to resist. I dashed at the place in which I had left her lying and over which (for the small silk counterpane and the sheets were disarranged) the white curtains had been deceivingly pulled forward; then my step, to my unutterable relief, produced an answering sound: I perceived an agitation of the window-blind, and the child, ducking down, emerged rosily from the other side of it. She stood there in so much of her candour and so little of her nightgown, with her pink bare feet and the golden glow of her curls. She looked intensely grave, and I had never had such a sense of losing an advantage acquired (the thrill of which had just been so prodigious) as on my consciousness that she addressed me with a reproach. "You naughty: where *have* you been?"—instead of challenging her own irregularity I found myself arraigned and explaining. She herself explained, for that matter, with the loveliest, eagerest simplicity. She had known suddenly, as she lay there, that I was out of the room, and had jumped up to see what had become of me. I had dropped, with the joy of her reappearance, back into my chair—feeling then, and then only, a little faint; and she had pattered straight over to me, thrown herself upon my knee, given herself to be held with the flame of the candle full in the wonderful little face that was still flushed with sleep. I remember closing my eyes an instant, yielding, consciously, as before the excess of something beautiful that shone out of the blue of her own. "You were looking for me out of the window?" I said. "You thought I might be walking in the grounds?"

"Well, you know, I thought someone was"—she never blanched as she smiled out that at me.

Oh, how I looked at her now! "And did you see anyone?"

"Ah, *no*!" she returned, almost with the full privilege of childish inconsequence, resentfully, though with a long sweetness in her little drawl of the negative.

At that moment, in the state of my nerves, I absolutely believed she lied; and if I once more closed my eyes it was before the dazzle of the three or four possible ways in which I might take this up. One of these, for a moment, tempted me with such singular intensity that, to withstand it, I must have gripped my little girl with a spasm that, wonderfully, she submitted to without a cry or a sign of fright. Why not break out at her on the spot and have it all over?—give it to her straight in her lovely little lighted face? "You see, you see, you *know* that you do and that you already quite suspect I believe it; therefore why not frankly confess it to me, so that we may at least live with it together and learn perhaps, in the strangeness of our fate, where we are and what it means?" This solicitation dropped, alas, as it came: if I could immediately have succumbed to it I might have spared myself—well you'll see what. Instead of succumbing I sprang again to my feet, looked at her bed, and took a helpless middle way. "Why did you pull the curtain over the place to make me think you were still there?"

Flora luminously considered; after which, with her little divine smile: "Because I don't like to frighten you!"

"But if I had, by your idea, gone out—?"

She absolutely declined to be puzzled; she turned her eyes to the flame of the candle as if the question were as irrelevant, or at any rate as impersonal, as Mrs. Marcet or nine-times-nine. "Oh, but you know," she quite adequately answered, "that you might come back, you dear, and that you *have*!" And after a little, when she had got into bed, I had, for a long time, by almost sitting on her to hold her hand, to prove that I recognised the pertinence of my return.

You may imagine the general complexion, from that moment, of my nights. I repeatedly sat up till I didn't know when; I selected moments when my room-mate unmistakeably slept, and, stealing out, took noiseless turns in the passage and even pushed as far as to where I had last met Quint. But I never met him there again; and I may as well say at once that I on no other occasion saw him in the house. I just missed, on the staircase, on the other hand, a different adventure. Looking down it from the top I once recognised the presence of a woman seated on one of the lower steps with her back presented to me, her body half bowed and her head, in an attitude of woe, in her hands. I had been there but an instant, however, when she vanished without looking round at me. I knew, none the less, exactly what dreadful face she had to show; and I wondered whether, if instead of being above I had been below, I should have had, for going up, the same nerve I had lately shown Quint. Well, there continued to be plenty of chance for nerve. On the eleventh night after my last encounter with that gentleman—they were all numbered now—I had an alarm that perilously skirted it and that indeed, from the particular quality of its unexpectedness, proved quite my sharpest shock. It was precisely the first night during this series that, weary with watching, I had felt that I might again without laxity lay myself down at my old hour. I slept immediately and, as I afterwards know, till about one o'clock; but when I woke it was to sit straight up, as completely roused as if a hand had shook me. I had left a light burning, but it was now out, and I felt an instant certainty that Flora had extinguished it. This brought me to my feet and straight, in the darkness, to her bed, which I found she had left. A glance at the window enlightened me further, and the striking of a match completed the picture.

The child had again got up—this time blowing out the taper, and had again, for some purpose of observation or response, squeezed in behind the blind and was peering out into the night. That she now saw—as she had not, I had satisfied myself, the previous time—was proved to me by the fact that she was disturbed neither by my re-illumination nor by the haste I made to get into slippers and into a wrap. Hidden, protected, absorbed, she evidently rested on the sill—the casement opened forward—and gave herself up. There was a great still moon to help her, and this fact had counted in my quick decision. She was face to face with the apparition we had met at the lake, and could now communicate with it as she had not then been able to do. What I, on my side, had to care for was, without disturbing her, to reach, from the corridor, some other window in the same quarter. I got to the door without her hearing me; I got out of it, closed it and listened, from the other side, for some sound from her. While I stood in the passage I had my eyes on her brother's door, which was but ten steps off and which, indescribably, produced in me a renewal of the strange impulse that I lately spoke of as my temptation. What if I should go straight in and march to *his* window?—what if, by risking to his boyish bewilderment a revelation of my motive, I should throw across the rest of the mystery the long halter of my boldness?

This thought held me sufficiently to make me cross to his threshold and pause again. I preternaturally listened; I figured to myself what might portentously be; I wondered if his bed were also empty and he too were secretly at watch. It was a deep, soundless minute, at the end of which my impulse failed. He was quiet; he might be innocent; the risk was hideous; I turned away. There was a figure in the grounds—a figure prowling for a sight, the visitor with whom Flora was engaged; but it was not the visitor most concerned with my boy. I hesitated afresh, but on other grounds and only a few seconds; then I had made my choice. There were empty rooms at Bly, and it was only a question of choosing the right one. The right one suddenly presented itself to me as the lower one—though high above the gardens—in the solid corner of the house that I have spoken of as the old tower. This was a large, square chamber, arranged with some state as a bedroom, the extravagant size of which made it so inconvenient that it had not for years, though kept by Mrs. Grose in exemplary order, been occupied. I had often admired it and I knew my way about in it; I had only after just faltering at the first chill gloom of its disuse, to pass across it and unbolt as quietly as I could one of the shutters. Achieving this transit, I uncovered the glass without a sound and, applying my face to the pane, was able, the darkness without being much less than within, to see that I commanded the right direction. Then I saw something more. The moon made the night extraordinarily penetrable and showed me on the lawn a person, diminished by distance, who stood there motionless and as if fascinated, looking up to where I had appeared—looking, that is, not so much straight at me as at something that was apparently above me. There was clearly another person above me—there was a person on the tower; but the presence on the lawn was not in the least what I had conceived and had confidently hurried to meet. The presence on the lawn—I felt sick as I made it out—was poor little Miles himself.

XI

IT WAS not till late next day that I spoke to Mrs. Grose; the rigour with which I kept my pupils in sight making it often difficult to meet her privately, and the more as we each felt the importance of not provoking—on the part of the servants quite as much as on that of the children—any suspicion of a secret flurry or of a discussion of mysteries. I drew a great security in this particular from her mere smooth aspect. There was nothing in her fresh face to pass on to others my horrible confidences. She believed me, I was sure, absolutely: if she hadn't I don't know what would have become of me, for I couldn't have borne the business alone. But she was a magnificent monument to the blessing of a want of imagination, and if she could see in our little charges nothing but their beauty and amiability, their happiness and cleverness, she had no direct communication with the sources of my trouble. If they had been at all visibly blighted or battered, she would doubtless have grown, on tracing it back, haggard enough to match them; as matters stood, however, I could feel her, when she surveyed them, with her large white arms folded and the habit of serenity in all her look, thank the Lord's mercy that if they were ruined the pieces would still serve. Flights of fancy gave place, in her mind, to a steady fireside glow, and I had already begun to perceive how, with the development of the conviction that—as time went on without a public accident—our young things could, after all, look out for themselves, she addressed her greatest solicitude to the sad case presented by their instructress. That, for myself, was a sound simplification: I could engage that, to the world, my face should tell no tales, but it would have been, in the conditions, an immense added strain to find myself anxious about hers.

At the hour I now speak of she had joined me, under pressure, on the terrace, where, with the lapse of the season, the afternoon sun was now agreeable; and we sat there together while, before us, at a distance, but within call if we wished, the children strolled to and fro in one of their most manageable moods. They moved slowly, in unison, below us, over the lawn, the boy, as they went, reading aloud from a story-book and passing his arm round his sister to keep her quite in touch. Mrs. Grose watched them with positive placidity; then I caught the suppressed intellectual creak with which she conscientiously turned to take from me a view of the back of the tapestry. I had made her a receptacle of lurid things, but there was an odd recognition of my superiority—my accomplishments and my function—in her patience under my pain. She offered her mind to my disclosures as, had I wished to mix a witch's broth and proposed it with assurance, she would have held out a large clean saucepan. This had become thoroughly her attitude by the time that, in my recital of the events of the night, I reached the point of what Miles had said to me when, after seeing him, at such a monstrous hour, almost on the very spot where he happened now to be, I had gone down to bring him in; choosing then, at the window, with a concentrated need of not alarming the house, rather that method than a signal more resonant. I had left her meanwhile in little doubt of my small hope of representing with success even to her actual sympathy my sense of the real splendour of the little inspiration with which, after I had got him into the house, the boy met my final articulate challenge. As soon as I appeared in the moonlight on the terrace, he had come to me as straight as possible; on which I had taken his hand without a word and led him, through the dark spaces, up the staircase where Quint had so hungrily hovered for him, along the lobby where I had listened and trembled, and so to his forsaken room.

Not a sound, on the way, had passed between us, and I had wondered—oh, *how* I had wondered!—if he were groping about in his little mind for something plausible and not too grotesque. It would tax his invention, certainly, and I felt, this time, over his real embarrassment, a curious thrill of triumph. It was a sharp trap for the inscrutable! He couldn't play any longer at innocence; so how the deuce would he get out of it? There beat in me indeed, with the passionate throb of this question, an equal dumb appeal as to how the deuce *I* should. I was confronted at last, as never yet, with all the risk attached even now to sounding my own hurried note. I remember in fact that as we pushed into his little chamber, where the bed had not been slept in at all and the window, uncovered to the moonlight, made the place so clear that there was no need of striking a match—I remember how I suddenly dropped, sank upon the edge of the bed from the force of the idea that he must know how he really, as they say, "had" me. He could do what he liked, with all his cleverness to help him, so long as I should continue to defer to the old tradition of the criminality of those caretakers of the young who minister to superstitions and fears. He "had" me indeed, and in a cleft stick; for who would ever absolve me, who would consent that I should go unhung, if, by the faintest tremor of an overture, I were the first to introduce into our perfect intercourse an element so dire? No, no: it was useless to attempt to convey to Mrs. Grose, just as it is scarcely less so to attempt to suggest here, how, in our short, stiff brush in the dark, he fairly shook me with admiration. I was of course thoroughly kind and merciful; never, never yet had I placed on his little shoulders hands of such tenderness as those with which, while I rested against the bed, I held him there well under fire. I had no alternative but, in form at least, to put it to him.

"You must tell me now—and all the truth. What did you go out for? What were you doing there?"

I can still see his wonderful smile, the whites of his beautiful eyes, and the uncovering of his little teeth shine to me in the dusk. "If I tell you why, will you understand?" My heart, at this, leaped into my mouth. *Would* he tell me why? I found no sound on my lips to press it, and I was aware of replying only with a vague, repeated, grimacing nod. He was gentleness itself, and while I wagged my head at him he stood there more than ever a little fairy prince. It was his brightness indeed that gave me a respite. Would it be so great if he were really going to tell me? "Well," he said at last, "just exactly in order that you should do this."

"Do what?"

"Think me—for a change—*bad!*" I shall never forget the sweetness and gaiety with which he brought out the word, nor how, on top of it, he bent forward and kissed me. It was practically the end of everything. I met his kiss and I had to make, while I folded him for a minute in my arms, the most stupendous effort not to cry. He had given exactly the account of himself that permitted least of my going behind it, and it was only with the effect of confirming my acceptance of it that, as I presently glanced about the room, I could say—

"Then you didn't undress at all?"

He fairly glittered in the gloom. "Not at all. I sat up and read."

"And when did you go down?"

"At midnight. When I'm bad I *am* bad!"

"I see, I see—it's charming. But how could you be sure I would know it?"

"Oh, I arranged that with Flora." His answers rang out with a readiness! "She was to get up and look out."

"Which is what she did do." It was I who fell into the trap!

"So she disturbed you, and, to see what she was looking at, you also looked—you saw."

"While you," I concurred, "caught your death in the night air!"

He literally bloomed so from this exploit that he could afford radiantly to assent. "How otherwise should I have been bad enough?" he asked. Then, after another embrace, the incident and our interview closed on my recognition of all the reserves of goodness that, for his joke, he had been able to draw upon.

XII

THE PARTICULAR impression I had received proved in the morning light, I repeat, not quite successfully presentable to Mrs. Grose, though I reinforced it with the mention of still another remark that he had made before we separated. "It all lies in half-a-dozen words," I said to her, "words that really settle the matter. 'Think, you know, what I *might* do!' He threw that off to show me how good he is. He knows down to the ground what he 'might' do. That's what he gave them a taste of at school."

"Lord, you do change!" cried my friend.

"I don't change—I simply make it out. The four, depend upon it, perpetually meet. If on either of these last nights you had been with either child, you would clearly have understood. The more I've watched and waited the more I've felt that if there were nothing else to make it sure it would be made so by the systematic silence of each. *Never*, by a slip of the tongue, have they so much as alluded to either of their old friends, any more than Miles has alluded to his expulsion. Oh yes, we may sit here and look at them, and they may show off to us there to their fill; but even while they pretend to be lost in their fairy-tale they're steeped in their vision of the dead restored. He's not reading to her," I declared; "they're talking of *them*—they're talking horrors! I go on, I know, as if I were crazy; and it's a wonder I'm not. What I've seen would have made *you* so; but it has only made me more lucid, made me get hold of still other things."

My lucidity must have seemed awful, but the charming creatures who were victims of it, passing and repassing in their interlocked sweetness, gave my colleague something to hold on by; and I felt how tight she held as, without stirring in the breath of my passion, she covered them still with her eyes. "Of what other things have you got hold?"

"Why, of the very things that have delighted, fascinated, and yet, at bottom, as I now so strangely see, mystified and troubled me. Their more than earthly beauty, their absolutely unnatural goodness. It's a game," I went on; "it's a policy and a fraud!"

"On the part of little darlings—?"

"As yet mere lovely babies? Yes, mad as that seems!" The very act of bringing it out really helped me to trace it—follow it all up and piece it all together. "They haven't been good—they've only been absent. It has been easy to live with them, because they're simply leading a life of their own. They're not mine—they're not ours. They're his and they're hers!"

"Quint's and that woman's?"

"Quint's and that woman's. They want to get to them."

Oh, how, at this, poor Mrs. Grose appeared to study them! "But for what?"

"For the love of all the evil that, in those dreadful days, the pair put into them. And to ply them with that evil still, to keep up the work of demons, is what brings the others back."

"Laws!" said my friend under her breath. The exclamation was homely, but it revealed a real acceptance of my further proof of what, in the bad time—for there had been a worse even than this!—must have occurred. There could have been no such justification for me as the plain assent of her experience to whatever depth of depravity I found credible in our brace of scoundrels. It was in obvious submission of memory that she brought out after a moment: "They *were* rascals! But what can they now do?" she pursued.

"Do?" I echoed so loud that Miles and Flora, as they passed at their distance, paused an instant in their walk and looked at us. "Don't they do enough?" I demanded in a lower tone, while the children, having smiled and nodded and kissed hands to us, resumed their exhibition. We were held by it a minute; then I answered: "They can destroy them!" At this my companion did turn, but the inquiry she launched was a silent one, the effect of which was to make me more explicit. "They don't know, as yet, quite how—but they're trying hard. They're seen only across, as it were, and beyond—in strange places and on high places, the top of towers, the roof of houses, the outside of windows, the further edge of pools; but there's a deep design, on either side, to shorten the distance and overcome the obstacle; and the success of the tempters is only a question of time. They've only to keep to their suggestions of danger."

"For the children to come?"

"And perish in the attempt!" Mrs. Grose slowly got up, and I scrupulously added: "Unless, of course, we can prevent!"

Standing there before me while I kept my seat, she visibly turned things over. "Their uncle must do the preventing. He must take them away."

"And who's to make him?"

She had been scanning the distance, but she now dropped on me a foolish face. "You, Miss."

"By writing to him that his house is poisoned and his little nephew and niece mad?"

"But if they *are*, Miss?"

"And if I am myself, you mean? That's charming news to be sent him by a governess whose prime undertaking was to give him no worry."

Mrs. Grose considered, following the children again. "Yes, he do hate worry. That was the great reason—"

"Why those fiends took him in so long? No doubt, though his indifference must have been awful. As I'm not a fiend, at any rate, I shouldn't take him in."

My companion, after an instant and for all answer, sat down again and grasped my arm. "Make him at any rate come to you."

I stared. "To *me*?" I had a sudden fear of what she might do. " 'Him'?"

"He ought to *be* here—he ought to help."

I quickly rose, and I think I must have shown her a queerer face than ever yet. "You see me asking him for a visit?" No, with her eyes on my face she evidently couldn't. Instead of it even—as a woman reads another—she could see what I myself saw: his derision, his amusement, his contempt for the break-down of my resignation at being left alone and for the fine machinery I had set in motion to attract his attention to my slighted charms. She didn't know—no one knew—how proud I had been to serve him and to stick to our terms; yet she none the less took the measure, I think, of the warning I now gave her. "If you should so lose your head as to appeal to him for me—"

She was really frightened. "Yes, Miss?"

"I would leave, on the spot, both him and you."

XIII

IT WAS all very well to join them, but speaking to them proved quite as much as ever an effort beyond my strength—offered, in close quarters, difficulties as insurmountable as before. This situation continued a month, and with new aggravations and particular notes, the note above all, sharper and sharper, of the small ironic consciousness on the part of my pupils. It was not, I am as sure today as I was sure then, my mere infernal imagination: it was absolutely traceable that they were aware of my predicament and that this strange relation made, in a manner, for a long time, the air in which we moved. I don't mean that they had their tongues in their cheeks or did anything vulgar, for that was not one of their dangers: I do mean, on the other hand, that the element of the unnamed and untouched became, between us, greater than any other, and that so much avoidance could not have been so successfully effected without a great deal of tacit arrangement. It was as if, at moments, we were perpetually coming into sight of subjects before which we must stop short, turning suddenly out of alleys that we perceived to be blind, closing with a little bang that made us look at each other—for, like all bangs, it was something louder than we had intended—the doors we had indiscreetly opened. All roads lead to Rome, and there were times when it might have struck us that almost every branch of study or subject of conversation skirted forbidden ground. Forbidden ground was the question of the return of the dead in general and of whatever, in especial, might survive, in memory, of the friends little children had lost. There were days when I could have sworn that one of them had, with a small invisible nudge, said to the other: "She thinks she'll do it this time—but she *won't!*" To "do it" would have been to indulge for instance—and for once in a way—in some direct reference to the lady who had prepared them for my discipline. They had a delightful endless appetite for passages in my own history, to which I had again and again treated them; they were in possession of everything that had ever happened to me, had had, with every circumstance, the story of my smallest adventures and of those of my brothers and sisters and of the cat and the dog at home, as well as many particulars of the eccentric nature of my father, of the furniture and arrangement of our house, and of the conversation of the old women of our village. There were things enough, taking one with another, to chatter about, if one went very fast and knew by instinct when to go round. They pulled with an art of their own the strings of my invention and my memory; and nothing else perhaps, when I thought of such occasions afterwards, gave me so the suspicion of being watched from under cover. It was in any case over *my* life, *my* past, and *my* friends alone that we could take anything like our ease—a state of affairs that led them sometimes without the least pertinence to break out into sociable reminders. I was invited —with no visible connection—to repeat afresh Goody Gosling's celebrated *mot* or to confirm the details already supplied as to the cleverness of the vicarage pony.

It was partly at such junctures as these and partly at quite different ones that, with the turn my matters had now taken, my predicament, as I have called it, grew most sensible. The fact that the days passed for me without another encounter ought, it would have appeared, to have done something toward soothing my nerves. Since the light brush, that second night on the upper landing, of the presence of a woman at the foot of the stair, I had seen nothing, whether in or out of the house, that one had better not have seen. There was many a corner round which I expected to come upon Quint, and many a situation that, in a merely sinister way, would have favoured the appearance of Miss Jessel. The summer had turned, the summer had gone; the autumn had dropped upon Bly and had blown out half our lights. The place, with its grey sky and withered garlands, its bared spaces and scattered dead leaves, was like a theatre after the performance—all strewn with crumpled playbills. There were exactly states of the air, conditions of sound and of stillness, unspeakable impressions of the *kind* of ministering moment, that brought back to me, long enough to catch it, the feeling of the medium in which, that June evening out-of-doors, I had had my first sight of Quint, and in which, too, at those other instants, I had, after seeing him through the window, looked for him in vain in the circle of shrubbery. I recognised the signs, the portents—I recognised the moment, the spot. But they remained unaccompanied and empty, and I continued unmolested; if unmolested one could call a young woman whose sensibility had, in the most extraordinary fashion, not declined but deepened. I had said in my talk with Mrs. Grose on that horrid scene of Flora's by the lake—and had perplexed her by so saying—that it would from that moment distress me much more to lose my power than to keep it. I had then expressed what was vividly in my mind: the truth that, whether the children really saw or not—since, that is, it was not yet definitely proved—I greatly preferred, as a safeguard, the fulness of my own exposure. I was ready to know the very worst that was to be known. What I had then had an ugly glimpse of was that my eyes might be sealed just while theirs were most opened. Well, my eyes *were* sealed, it appeared, at present—a consummation for which it seemed blasphemous not to thank God. There was, alas, a difficulty about that: I would have thanked him with all my soul had I not had in a proportionate measure this conviction of the secret of my pupils.

How can I retrace today the strange steps of my obsession? There were times of our being together when I would have been ready to swear that, literally, in my presence, but with my direct sense of it closed, they had visitors who were known and were welcome. Then it was that, had I not been deterred by the very chance that such an injury might prove greater than the injury to be averted, my exultation would have broken out. "They're here, they're here, you little wretches," I would have cried, "and you can't deny it now!" The little wretches denied it with all the added volume of their sociability and their tenderness, in just the crystal depths of which—like the flash of a fish in a stream—the mockery of their advantage peeped up. The shock, in truth, had sunk into me still deeper than I knew on the night when, looking out to see either Quint or Miss Jessel under the stars, I had beheld the boy over whose rest I watched and who had immediately brought in with him—had straightway, there, turned it on me—the lovely upward look with which from the battlements above me, the hideous apparition of Quint had played. If it was a question of a scare, my discovery on this occasion had scared me more than any other, and it was in the condition of nerves produced by it that I made my actual inductions. They harassed me so that sometimes, at odd moments, I shut myself up audibly to rehearse—it was at once a fantastic relief and a renewed despair—the manner in which I might come to the point. I approached it from one side and the other while, in my room, I flung myself about, but I always broke down in the monstrous utterance of names. As they died away on my lips, I said to myself that I should indeed help them to represent something infamous if, by pronouncing them, I should violate as rare a little case of instinctive delicacy as any schoolroom, probably, had ever known. When I said to myself: "*They* have the manners to be silent, and you, trusted as you are, the baseness to speak!" I felt myself crimson and I covered my face with my hands. After these secret scenes I chattered more than ever, going on volubly enough 'til one of our prodigious, palpable hushes occurred—I can call them nothing else—the strange, dizzy lift or swim (I try for terms!) into a stillness, a pause of all life, that had nothing to do with the more or less noise that at the moment we might be engaged in making and that I could hear through any deepened exhilaration or quickened recitation or louder strum of the piano. Then it was that the others, the outsiders, were there. Though they were not angels, they "passed," as the French say, causing me, while they stayed, to tremble with the fear of their addressing to their younger victims some yet more infernal message or more vivid image than they had thought good enough for myself.

What it was most impossible to get rid of was the cruel idea that, whatever I had seen, Miles and Flora saw *more*—things terrible and unguessable and that sprang from dreadful passages of intercourse in the past. Such things naturally left on the surface, for the time, a chill which we vociferously denied that we felt; and we had, all three, with repetition, got into such splendid training that we went, each time, almost automatically, to mark the close of the incident, through the very same movements. It was striking of the children, at all events, to kiss me inveterately, with a kind of wild irrelevance and never to fail—one or the other—of the precious question that had helped us through many a peril. "When do you think he *will* come? Don't you think we *ought* to write?"—there was nothing like that inquiry, we found by experience, for carrying off an awkwardness. "He" of course was their uncle in Harley Street; and we lived in much profusion of theory that he might at any moment arrive to mingle in our circle. It was impossible to have given less encouragement than he had done to such a doctrine, but if we had not had the doctrine to fall back upon we should have deprived each other of some of our finest exhibitions. He never wrote to them—that may have been selfish, but it was a part of the flattery of his trust of me; for the way in which a man pays his highest tribute to a woman is apt to be but by the more festal celebration of one of the sacred laws of his comfort; and I held that I carried out the spirit of the pledge given not to appeal to him when I let my charges understand that their own letters were but charming literary exercises. They were too beautiful to be posted; I kept them myself; I have them all to this hour. This was a rule indeed which only added to the satiric effect of my being plied with the supposition that he might at any moment be among us. It was exactly as if my charges knew how almost more awkward than anything else that might be for me. There appears to me, moreover, as I look back, no note in all this more extraordinary than the mere fact that, in spite of my tension and of their triumph, I never lost patience with them. Adorable they must in truth have been, I now reflect, that I didn't in these days hate them! Would exasperation, however, if relief had longer been postponed, finally have betrayed me? It little matters, for relief arrived. I call it relief, though it was only the relief that a snap brings to a strain or the burst of a thunderstorm to a day of suffocation. It was at least change, and it came with a rush.

XIV

WALKING to church a certain Sunday morning, I had little Miles at my side and his sister, in advance of us and at Mrs. Grose's, well in sight. It was a crisp, clear day, the first of its order for some time; the night had brought a touch of frost, and the autumn air, bright and sharp, made the church-bells almost gay. It was an odd accident of thought that I should have happened at such a moment to be particularly and very gratefully struck with the obedience of my little charges. Why did they never resent my inexorable, my perpetual society? Something or other had brought nearer home to me that I had all but pinned the boy to my shawl and that, in the way our companions were marshalled before me, I might have appeared to provide against some danger of rebellion. I was like a gaoler with an eye to possible surprises and escapes. But all this belonged—I mean their magnificent little surrender—just to the special array of the facts that were most abysmal. Turned out for Sunday by his uncle's tailor, who had had a free hand and a notion of pretty waistcoats and of his grand little air, Miles's whole title to independence, the rights of his sex and situation, were so stamped upon him that if he had suddenly struck for freedom I should have had nothing to say. I was by the strangest of chances wondering how I should meet him when the revolution unmistakeably occurred. I call it a revolution because I now see how, with the word he spoke, the curtain rose on the last act of my dreadful drama and the catastrophe was precipitated. "Look here, my dear, you know," he charmingly said, "when in the world, please, am I going back to school?"

Transcribed here the speech sounds harmless enough, particularly as uttered in the sweet, high, casual pipe with which, at all interlocutors, but above all at his eternal governess, he threw off intonations as if he were tossing roses. There was something in them that always made one "catch," and I caught, at any rate, now so effectually that I stopped as short as if one of the trees of the park had fallen across the road. There was something new, on the spot, between us, and he was perfectly aware that I recognised it, though, to enable me to do so, he had no need to look a whit less candid and charming than usual. I could feel in him how he already, from my at first finding nothing to reply, perceived the advantage he had gained. I was so slow to find anything that he had plenty of time, after a minute, to continue with his suggestive but inconclusive smile: "You know, my dear, that for a fellow to be with a lady *always*—!" His "my dear" was constantly on his lips for me, and nothing could have expressed more the exact shade of the sentiment with which I desired to inspire my pupils than its fond familiarity. It was so respectfully easy.

But, oh, how I felt that at present I must pick my own phrases! I remember that, to gain time, I tried to laugh, and I seemed to see in the beautiful face with which he watched me how ugly and queer I looked. "And always with the same lady?" I returned.

He neither blenched nor winked. The whole thing was virtually out between us. "Ah, of course, she's a jolly, 'perfect' lady; but, after all, I'm a fellow, don't you see? that's—well, getting on."

I lingered there with him an instant ever so kindly. "Yes, you're getting on." Oh, but I felt helpless!

I have kept to this day the heartbreaking little idea of how he seemed to know that and to play with it. "And you can't say I've not been awfully good, can you?"

I laid my hand on his shoulder, for, though I felt how much better it would have been to walk on, I was not yet quite able. "No, I can't say that, Miles."

"Except just that one night, you know—!"

"That one night?" I couldn't look as straight as he.

"Why, when I went down—went out of the house."

"Oh, yes. But I forget what you did it for."

"You forget?"—he spoke with the sweet extravagance of childish reproach. "Why, it was to show you I could!"

"Oh, yes, you could."

"And I can again."

I felt that I might, perhaps, after all succeed in keeping my wits about me. "Certainly. But you won't."

"No, not *that* again. It was nothing."

"It was nothing," I said. "But we must go on."

He resumed our walk with me, passing his hand into my arm. "Then when *am* I going back?"

I wore, in turning it over, my most responsible air. "Were you very happy at school?"

He just considered. "Oh, I'm happy enough anywhere!"

"Well, then," I quavered, "if you're just as happy here—!"

"Ah, but that isn't everything! Of course *you* know a lot—"

"But you hint that you know almost as much?" I risked as he paused.

"Not half I want to!" Miles honestly professed. "But it isn't so much that."

"What is it, then?"

"Well—I want to see more life."

"I see; I see." We had arrived within sight of the church and of various persons, including several of the household of Bly, on their way to it and clustered about the door to see us go in. I quickened our step; I wanted to get there before the question between us opened up much further; I reflected hungrily that, for more than an hour, he would have to be silent; and I thought with envy of the comparative dusk of the pew and of the almost spiritual help of the hassock on which I might bend my knees. I seemed literally to be running a race with some confusion to which he was about to reduce me, but I felt that he had got in first when, before we had even entered the churchyard, he threw out—

"I want my own sort!"

It literally made me bound forward. "There are not many of your own sort, Miles!" I laughed. "Unless perhaps dear little Flora!"

"You really compare me to a baby girl?"

This found me singularly weak. "Don't you, then, *love* our sweet Flora?"

"If I didn't—and you too; if I didn't—!" he repeated as if retreating for a jump, yet leaving his thought so unfinished that, after we had come into the gate, another stop, which he imposed on me by the pressure of his arm, had become inevitable. Mrs. Grose and Flora had passed into the church, the other worshippers had followed, and we were, for the minute, alone among the old, thick graves. We had paused, on the path from the gate, by a low, oblong, table-like tomb.

"Yes. If you didn't—?"

He looked, while I waited, about at the graves. "Well, you know what!" But he didn't move, and he presently produced something that made me drop straight down on the stone slab, as if suddenly to rest. "Does my uncle think what *you* think?"

I markedly rested. "How do you know what I think?"

"Ah, well, of course I don't; for it strikes me you never tell me. But I mean does *he* know?"

"Know what, Miles?"

"Why, the way I'm going on."

I perceived quickly enough that I could make, to this inquiry, no answer that would not involve something of a sacrifice of my employer. Yet it appeared to me that we were all, at Bly, sufficiently sacrificed to make that venial. "I don't think your uncle much cares."

Miles, on this, stood looking at me. "Then don't you think he can be made to?"

"In what way?"

"Why, by his coming down."

"But who'll get him to come down?"

"*I* will!" the boy said with extraordinary brightness and emphasis. He gave me another look charged with that expression and then marched off alone into church.

XV

THE BUSINESS was practically settled from the moment I never followed him. It was a pitiful surrender to agitation, but my being aware of this had somehow no power to restore me. I only sat there on my tomb and read into what my little friend had said to me the fulness of its meaning; by the time I had grasped the whole of which I had also embraced, for absence, the pretext that I was ashamed to offer my pupils and the rest of the congregation such an example of delay. What I said to myself above all was that Miles had got something out of me and that the proof of it, for him, would be just this awkward collapse. He had got out of me that there was something I was much afraid of and that he should probably be able to make use of my fear to gain, for his own purpose, more freedom. My fear was of having to deal with the intolerable question of the grounds of his dismissal from school, for that was really but the question of the horrors gathered behind. That his uncle should arrive to treat with me of these things was a solution that, strictly speaking, I ought now to have desired to bring on; but I could so little face the ugliness and the pain of it that I simply procrastinated and lived from hand to mouth. The boy, to my deep discomposure, was immensely in the right, was in a position to say to me: "Either you clear up with my guardian the mystery of this interruption of my studies, or you cease to expect me to lead with you a life that's so unnatural for a boy." What was so unnatural for the particular boy I was concerned with was this sudden revelation of a consciousness and a plan.

That was what really overcame me, what prevented my going in. I walked round the church, hesitating, hovering; I reflected that I had already, with him, hurt myself beyond repair. Therefore I could patch up nothing, and it was too extreme an effort to squeeze beside him into the pew: he would be so much more sure than ever to pass his arm into mine and make me sit there for an hour in close, silent contact with his commentary on our talk. For the first minute since his arrival I wanted to get away from him. As I paused beneath the high east window and listened to the sounds of worship, I was taken with an impulse that might master me, I felt, completely should I give it the least encouragement. I might easily put an end to my predicament by getting away altogether. Here was my chance; there was no one to stop me; I could give the whole thing up—turn my back and retreat. It was only a question of hurrying again, for a few preparations, to the house which the attendance at church of so many of the servants would practically have left unoccupied. No one, in short, could blame me if I should just drive desperately off. What was it to get away if I got away only till dinner? That would be in a couple of hours, at the end of which—I had the acute prevision—my little pupils would play at innocent wonder about my non-appearance in their train.

"What *did* you do, you naughty, bad thing? Why in the world, to worry us so—and take our thoughts off too, don't you know?—did you desert us at the very door?" I couldn't meet such questions nor, as they asked them, their false little lovely eyes; yet it was all so exactly what I should have to meet that, as the prospect grew sharp to me, I at last let myself go.

I got, so far as the immediate moment was concerned, away; I came straight out of the churchyard and, thinking hard, retraced my steps through the park. It seemed to me that by the time I reached the house I had made up my mind I would fly. The Sunday stillness both of the approaches and of the interior, in which I met no one, fairly excited me with a sense of opportunity. Were I to get off quickly, this way, I should get off without a scene, without a word. My quickness would have to be remarkable, however, and the question of a conveyance was the great one to settle. Tormented, in the hall, with difficulties and obstacles, I remember sinking down at the foot of the staircase—suddenly collapsing there on the lowest step and then, with a revulsion, recalling that it was exactly where more than a month before, in the darkness of night and just so bowed with evil things, I had seen the spectre of the most horrible of women. At this I was able to straighten myself; I went the rest of the way up; I made, in my bewilderment, for the schoolroom, where there were objects belonging to me that I should have to take. But I opened the door to find again, in a flash, my eyes unsealed. In the presence of what I saw I reeled straight back upon my resistance.

Seated at my own table in clear noonday light I saw a person whom, without my previous experience, I should have taken at the first blush for some housemaid who might have stayed at home to look after the place and who, availing herself of rare relief from observation and of the schoolroom table and my pens, ink, and paper, had applied herself to the considerable effort of a letter to her sweetheart. There was an effort in the way that, while her arms rested on the table, her hands with evident weariness supported her head; but at the moment I took this in I had already become aware that, in spite of my entrance, her attitude strangely persisted. Then it was—with the very act of its announcing itself—that her identity flared up in a change of posture. She rose, not as if she had heard me, but with an indescribable grand melancholy of indifference and detachment, and, within a dozen feet of me, stood there as my vile predecessor. Dishonoured and tragic, she was all before me; but even as I fixed and, for memory, secured it, the awful image passed away. Dark as midnight in her black dress, her haggard beauty and her unutterable woe, she had looked at me long enough to appear to say that her right to sit at my table was as good as mine to sit at hers. While these instants lasted indeed I had the extraordinary chill of a feeling that it was I who was the intruder. It was as a wild protest against it that, actually addressing her—"You terrible, miserable woman!"—I heard myself break into a sound that, by the open door, rang through the long passage and the empty house. She looked at me as if she heard me, but I had recovered myself and cleared the air. There was nothing in the room the next minute but the sunshine and a sense that I must stay.

XVI

I HAD so perfectly expected that the return of my pupils would be marked by a demonstration that I was freshly upset at having to take into account that they were dumb about my absence. Instead of gaily denouncing and caressing me, they made no allusion to my having failed them, and I was left, for the time, on perceiving that she too said nothing, to study Mrs. Grose's odd face. I did this to such purpose that I made sure they had in some way bribed her to silence; a silence that, however, I would engage to break down on the first private opportunity. This opportunity came before tea: I secured five minutes with her in the housekeeper's room, where, in the twilight, amid a smell of lately-baked bread, but with the place all swept and garnished, I found her sitting in pained placidity before the fire. So I see her still, so I see her best: facing the flame from her straight chair in the dusky, shining room, a large clean image of the "put away"—of drawers closed and locked and rest without a remedy.

"Oh, yes, they asked me to say nothing; and to please them—so long as they were there—of course I promised. But what had happened to you?"

"I only went with you for the walk," I said. "I had then to come back to meet a friend."

She showed her surprise. "A friend—*you?*"

"Oh, yes, I have a couple!" I laughed. "But did the children give you a reason?"

"For not alluding to your leaving us? Yes; they said you would like it better. Do you like it better?"

My face had made her rueful. "No, I like it worse!" But after an instant I added: "Did they say why I should like it better?"

"No; Master Miles only said, 'We must do nothing but what she likes'!"

"I wish indeed he would! And what did Flora say?"

"Miss Flora was too sweet. She said, 'Oh, of course, of course!'—and I said the same."

I thought a moment. "You were too sweet too—I can hear you all. But none the less, between Miles and me, it's now all out."

"All out?" My companion stared. "But what, Miss?"

"Everything. It doesn't matter. I've made up my mind. I came home, my dear," I went on, "for a talk with Miss Jessel."

I had by this time formed the habit of having Mrs. Grose literally well in hand in advance of my sounding that note; so that even now, as she bravely blinked under the signal of my word, I could keep her comparatively firm. "A talk! Do you mean she spoke?"

"It came to that. I found her, on my return, in the schoolroom."

"And what did she say?" I can hear the good woman still, and the candour of her stupefaction.

"That she suffers the torments—!"

It was this, of a truth, that made her, as she filled out my picture, gape. "Do you mean," she faltered, "—of the lost?"

"Of the lost. Of the damned. And that's why, to share them—" I faltered myself with the horror of it.

But my companion, with less imagination, kept me up. "To share them—?"

"She wants Flora." Mrs. Grose might, as I gave it to her, fairly have fallen away from me had I not been prepared. I still held her there, to show I was. "As I've told you, however, it doesn't matter."

"Because you've made up your mind? But to what?"

"To everything."

"And what do you call 'everything'?"

"Why, sending for their uncle."

"Oh, Miss, in pity do," my friend broke out.

"Ah, but I will, I *will!* I see it's the only way. What's 'out,' as I told you, with Miles is that if he thinks I'm afraid to—and has ideas of what he gains by that—he shall see he's mistaken. Yes, yes; his uncle shall have it here from me on the spot (and before the boy himself if necessary) that if I'm to be reproached with having done nothing again about more school—"

"Yes, Miss—" my companion pressed me.

"Well, there's that awful reason."

There were now clearly so many of these for my poor colleague that she was excusable for being vague. "But—a—which?"

"Why, the letter from his old place."

"You'll show it to the master?"

"I ought to have done so on the instant."

"Oh, no!" said Mrs. Grose with decision.

"I'll put it before him," I went on inexorably, "that I can't undertake to work the question on behalf of a child who has been expelled—"

"For we've never in the least known what!" Mrs. Grose declared.

"For wickedness. For what else—when he's so clever and beautiful and perfect? Is he stupid? Is he untidy? Is he infirm? Is he ill-natured? He's exquisite—so it can be only *that;* and that would open up the whole thing. After all," I said, "it's their uncle's fault. If he left here such people—!"

"He didn't really in the least know them. The fault's mine." She had turned quite pale.

"Well, you shan't suffer," I answered.

I was silent awhile; we looked at each other. "Then what am I to tell him?"

"You needn't tell him anything. *I'll* tell him."

I measured this. "Do you mean you'll write—?" Remembering she couldn't, I caught myself up. "How do you communicate?"

"I tell the bailiff. *He* writes."

"And should you like him to write our story?"

My question had a sarcastic force that I had not fully intended, and it made her, after a moment, inconsequently break down. The tears were again in her eyes. "Ah, Miss, *you* write!"

"Well—tonight," I at last answered; and on this we separated.

XVII

I WENT so far, in the evening, as to make a beginning. The weather had changed back, a great wind was abroad, and beneath the lamp, in my room, with Flora at peace beside me, I sat for a long time before a blank sheet of paper and listened to the lash of the rain and the batter of the gusts. Finally I went out, taking a candle; I crossed the passage and listened a minute at Miles's door. What, under my endless obsession, I had been impelled to listen for was some betrayal of his not being at rest, and I presently caught one, but not in the form I had expected. His voice tinkled out. "I say, you there—come in." It was a gaiety in the gloom!

I went in with my light and found him, in bed, very wide awake, but very much at his ease. "Well, what are *you* up to?" he asked with a grace of sociability in which it occurred to me that Mrs. Grose, had she been present, might have looked in vain for proof that anything was "out."

I stood over him with my candle. "How did you know I was there?"

"Why, of course I heard you. Did you fancy you made no noise? You're like a troop of cavalry!" he beautifully laughed.

"Then you weren't asleep?"

"Not much! I lie awake and think."

I had put my candle, designedly, a short way off, and then, as he held out his friendly old hand to me, had sat down on the edge of his bed. "What is it," I asked, "that you think of?"

"What in the world, my dear, but *you*?"

"Ah, the pride I take in your appreciation doesn't insist on that! I had so far rather you slept."

"Well, I think also, you know, of this queer business of ours."

I marked the coolness of his firm little hand. "Of what queer business, Miles?"

"Why, the way you bring me up. And all the rest!"

I fairly held my breath a minute, and even from my glimmering taper there was light enough to show how he smiled up at me from his pillow. "What do you mean by all the rest?"

"Oh, you know, you know!"

I could say nothing for a minute, though I felt, as I held his hand and our eyes continued to meet, that my silence had all the air of admitting his charge and that nothing in the whole world of reality was perhaps at that moment so fabulous as our actual relation. "Certainly you shall go back to school," I said, "if it be that that troubles you. But not to the old place—we must find another, a better. How could I know it did trouble you, this question, when you never told me so, never spoke of it at all?" His clear, listening face, framed in its smooth whiteness, made him for the minute as appealing as some wistful patient in a children's hospital; and I would have given, as the resemblance came to me, all I possessed on earth really to be the nurse or the sister of charity who might have helped to cure him. Well, even as it was, I perhaps might help! "Do you know you've never said a word to me about your school—I mean the old one; never mentioned it in any way?"

He seemed to wonder; he smiled with the same loveliness. But he clearly gained time; he waited, he called for guidance. "Haven't I?" It wasn't for *me* to help him—it was for the thing I had met!

Something in his tone and the expression of his face, as I got this from him, set my heart aching with such a pang as it had never yet known; so unutterably touching was it to see his little brain puzzled and his little resources taxed to play, under the spell laid on him, a part of innocence and consistency. "No, never—from the hour you came back. You've never mentioned to me one of your masters, one of your comrades, nor the least little thing that ever happened to you at school. Never, little Miles—no, never—have you given me an inkling of anything that *may* have happened there. Therefore you can fancy how much I'm in the dark. Until you came out, that way, this morning, you had, since the first hour I saw you, scarce even made a reference to anything in your previous life. You seemed so perfectly to accept the present." It was extraordinary how my absolute conviction of his secret precocity (or whatever I might call the poison of an influence that I dared but half to phrase) made him, in spite of the faint breath of his inward trouble, appear as accessible as an older person—imposed him almost as an intellectual equal. "I thought you wanted to go on as you are."

It struck me that at this he just faintly coloured. He gave, at any rate, like a convalescent slightly fatigued, a languid shake of his head. "I don't—I don't. I want to get away."

"You're tired of Bly?"

"Oh, no, I like Bly."

"Well, then—?"

"Oh, *you* know what a boy wants!"

I felt that I didn't know as well as Miles, and I took temporary refuge. "You want to go to your uncle?"

Again, at this, with his sweet ironic face, he made a movement on the pillow. "Ah, you can't get off with that!"

I was silent a little, and it was I, now, I think, who changed colour. "My dear, I don't want to get off!"

"You can't, even if you do. You can't, you can't!"—he lay beautifully staring. "My uncle must come down, and you must completely settle things."

"If we do," I returned with some spirit, "you may be sure it will be to take you quite away."

"Well, don't you understand that that's exactly what I'm working for? You'll have to tell him—about the way you've let it all drop: you'll have to tell him a tremendous lot!"

The exultation with which he uttered this helped me somehow, for the instant, to meet him rather more. "And how much will *you*, Miles, have to tell him? There are things he'll ask you!"

He turned it over. "Very likely. But what things?"

"The things you've never told me. To make up his mind what to do with you. He can't send you back—"

"Oh, I don't want to go back!" he broke in. "I want a new field."

He said it with admirable serenity, with positive unimpeachable gaiety; and doubtless it was that very note that most evoked for me the poignancy, the unnatural childish tragedy, of his probable reappearance at the end of three months with all this bravado and still more dishonour. It overwhelmed me now that I should never be able to bear that, and it made me let myself go. I threw myself upon him and in the tenderness of my pity I embraced him. "Dear little Miles, dear little Miles—!"

My face was close to his, and he let me kiss him, simply taking it with indulgent good-humour. "Well, old lady?"

"Is there nothing—nothing at all that you want to tell me?"

He turned off a little, facing round toward the wall and holding up his hand to look at as one had seen sick children look. "I've told you—I told you this morning."

Oh, I was sorry for him! "That you just want me not to worry you?"

He looked round at me now, as if in recognition of my understanding him; then ever so gently, "To let me alone," he replied.

There was even a singular little dignity in it, something that made me release him, yet, when I had slowly risen, linger beside him. God knows I never wished to harass him, but I felt that merely, at this, to turn my back on him was to abandon or, to put it more truly, to lose him. "I've just begun a letter to your uncle," I said.

"Well, then, finish it!"

I waited a minute. "What happened before?"

He gazed up at me again. "Before what?"

"Before you came back. And before you went away."

For some time he was silent, but he continued to meet my eyes. "What happened?"

It made me, the sound of the words, in which it seemed to me that I caught for the very first time a small faint quaver of consenting consciousness—it made me drop on my knees beside the bed and seize once more the chance of possessing him. "Dear little Miles, dear little Miles, if you *knew* how I want to help you! It's only that, it's nothing but that, and I'd rather die than give you a pain or do you a wrong—I'd rather die than hurt a hair of you. Dear little Miles"—oh, I brought it out now even if I *should* go too far—"I just want you to help me to save you!" But I knew in a moment after this that I had gone too far. The answer to my appeal was instantaneous, but it came in the form of an extraordinary blast and chill, a gust of frozen air and a shake of the room as great as if, in the wild wind, the casement had crashed in. The boy gave a loud, high shriek, which, lost in the rest of the shock of sound, might have seemed, indistinctly, though I was so close to him, a note either of jubilation or of terror. I jumped to my feet again and was conscious of darkness. So for a moment we remained, while I stared about me and saw that the drawn curtains were unstirred and the window tight. "Why, the candle's out!" I then cried.

"It was I who blew it, dear!" said Miles.

XVIII

THE NEXT day, after lessons, Mrs. Grose found a moment to say to me quietly: "Have you written, Miss?"

"Yes—I've written." But I didn't add—for the hour—that my letter, sealed and directed, was still in my pocket. There would be time enough to send it before the messenger should go to the village. Meanwhile there had been, on the part of my pupils, no more brilliant, more exemplary morning. It was exactly as if they had both had at heart to gloss over any recent little friction. They performed the dizziest feats of arithmetic, soaring quite out of *my* feeble range, and perpetrated, in higher spirits than ever, geographical and historical jokes. It was conspicuous of course in Miles in particular that he appeared to wish to show how easily he could let me down. This child, to my memory, really lives in a setting of beauty and misery that no words can translate; there was a distinction all his own in every impulse he revealed; never was a small natural creature, to the uninitiated eye all frankness and freedom, a more ingenious, a more extraordinary little gentleman. I had perpetually to guard against the wonder of contemplation into which my initiated view betrayed me; to check the irrelevant gaze and discouraged sigh in which I constantly both attacked and renounced the enigma of what such a little gentleman could have done that deserved a penalty. Say that, by the dark prodigy I knew, the imagination of all evil *had* been opened up to him: all the justice within me ached for the proof that it could ever have flowered into an act.

He had never, at any rate, been such a little gentleman as when, after our early dinner on this dreadful day, he came round to me and asked if I shouldn't like him, for half an hour, to play to me. David playing to Saul could never have shown a finer sense of the occasion. It was literally a charming exhibition of tact, of magnanimity, and quite tantamount to his saying outright: "The true knights we love to read about never push an advantage too far. I know what you mean now: you mean that—to be let alone yourself and not followed up—you'll cease to worry and spy upon me, won't keep me so close to you, will let me go and come. Well, I 'come,' you see—but I don't go! There'll be plenty of time for that. I do really delight in your society, and I only want to show you that I contended for a principle." It may be imagined whether I resisted this appeal or failed to accompany him again, hand in hand, to the schoolroom. He sat down at the old piano and played as he had never played, and if there are those who think he had better have been kicking a football I can only say that I wholly agree with them. For at the end of a time that under his influence I had quite ceased to measure I started up with a strange sense of having literally slept at my post. It was after luncheon, and by the schoolroom fire, and yet I hadn't really, in the least, slept: I had only done something much worse—I had forgotten. Where, all this time, was Flora? When I put the question to Miles he played on a minute before answering, and then could only say: "Why, my dear, how do I know?"—breaking moreover into a happy laugh which, immediately after, as if it were a vocal accompaniment, he prolonged into incoherent, extravagant song.

I went straight to my room, but his sister was not there; then, before going downstairs, I looked into several others. As she was nowhere about she would surely be with Mrs. Grose, whom, in the comfort of that theory, I accordingly proceeded in quest of. I found her where I had found her the evening before, but she met my quick challenge with blank, scared ignorance. She had only supposed that, after the repast, I had carried off both the children; as to which she was quite in her right, for it was the very first time I had allowed the little girl out of my sight without some special provision. Of course now indeed she might be with the maids, so that the immediate thing was to look for her without an air of alarm. This we promptly arranged between us; but when, ten minutes later and in pursuance of our arrangement, we met in the hall, it was only to report on either side that after guarded inquiries we had altogether failed to trace her. For a minute there, apart from observation, we exchanged mute alarms, and I could feel with what high interest my friend returned me all those I had from the first given her.

"She'll be above," she presently said—"in one of the rooms you haven't searched."

"No; she's at a distance." I had made up my mind. "She has gone out."

Mrs. Grose stared. "Without a hat?"

I naturally also looked volumes. "Isn't that woman always without one?"

"She's with *her*?"

"She's with *her*!" I declared. "We must find them."

My hand was on my friend's arm, but she failed for the moment, confronted with such an account of the matter, to respond to my pressure. She communed, on the contrary, on the spot, with her uneasiness. "And where's Master Miles?"

"Oh, *he's* with Quint. They're in the schoolroom."

"Lord, Miss!" My view, I was myself aware—and therefore I suppose my tone—had never yet reached so calm an assurance.

"The trick's played," I went on; "they've successfully worked their plan. He found the most divine little way to keep me quiet while she went off."

" 'Divine'?" Mrs. Grose bewilderedly echoed.

"Infernal, then!" I almost cheerfully rejoined. "He has provided for himself as well. But come!"

She had helplessly gloomed at the upper regions. "You leave him—?"

"So long with Quint? Yes—I don't mind that now."

She always ended, at these moments, by getting possession of my hand, and in this manner she could at present still stay me. But after gasping an instant at my sudden resignation, "Because of your letter?" she eagerly brought out.

I quickly, by way of answer, felt for my letter, drew it forth, held it up, and then, freeing myself, went and laid it on the great hall-table. "Luke will take it," I said as I came back. I reached the house-door and opened it; I was already on the steps.

My companion still demurred: the storm of the night and the early morning had dropped, but the afternoon was damp and grey. I came down to the drive while she stood in the doorway. "You go with nothing on?"

"What do I care when the child has nothing? I can't wait to dress," I cried, "and if you must do so, I leave you. Try meanwhile, yourself, upstairs."

"With *them?*" Oh, on this, the poor woman promptly joined me!

XIX

WE WENT straight to the lake, as it was called at Bly, and I dare say rightly called, though I reflect that it may in fact have been a sheet of water less remarkable than it appeared to my untravelled eyes. My acquaintance with sheets of water was small, and the pool of Bly, at all events on the few occasions of my consenting, under the protection of my pupils, to affront its surface in the old flat-bottomed boat moored there for our use, had impressed me both with its extent and its agitation. The usual place of embarkation was half a mile from the house, but I had an intimate conviction that, wherever Flora might be, she was not near home. She had not given me the slip for any small adventure, and, since the day of the very great one that I had shared with her by the pond, I had been aware, in our walks, of the quarter to which she most inclined. This was why I had now given to Mrs. Grose's steps so marked a direction—a direction that made her, when she perceived it, oppose a resistance that showed me she was freshly mystified. "You're going to the water, Miss?—you think she's *in*—?"

"She may be, though the depth is, I believe, nowhere very great. But what I judge most likely is that she's on the spot from which, the other day, we saw together what I told you."

"When she pretended not to see—?"

"With that astounding self-possession! I've always been sure she wanted to go back alone. And now her brother has managed it for her."

Mrs. Grose still stood where she had stopped. "You suppose they really *talk* of them?"

I could meet this with a confidence! "They say things that, if we heard them, would simply appall us."

"And if she *is* there—?"

"Yes?"

"Then Miss Jessel is?"

"Beyond a doubt. You shall see."

"Oh, thank you!" my friend cried, planted so firm that, taking it in, I went straight on without her. By the time I reached the pool, however, she was close behind me, and I knew that, whatever, to her apprehension, might befall me, the exposure of my society struck her as her least danger. She exhaled a moan of relief as we at last came in sight of the greater part of the water without a sight of the child. There was no trace of Flora on that nearer side of the bank where my observation of her had been most startling, and none on the opposite edge, where, save for a margin of some twenty yards, a thick copse came down to the water. The pond, oblong in shape, had a width so scant compared to its length that, with its ends out of view, it might have been taken for a scant river. We looked at the empty expanse, and then I felt the suggestion of my friend's eyes. I knew what she meant and I replied with a negative headshake.

"No, no; wait! She has taken the boat."

My companion stared at the vacant mooring-place and then again across the lake. "Then where is it?"

"Our not seeing it is the strongest of proofs. She has used it to go over, and then has manged to hide it."

"All alone—that child?"

"She's not alone, and at such times she's not a child: she's an old, old woman." I scanned all the visible shore while Mrs. Grose took again, into the queer element I offered her, one of her plunges of submission; then I pointed out that the boat might perfectly be in a small refuge formed by one of the recesses of the pool, an indentation masked, for the hither side, by a projection of the bank and by a clump of trees growing close to the water.

"But if the boat's there, where on earth's *she?*" my colleague anxiously asked.

"That's exactly what we must learn." And I started to walk further.

"By going all the way round?"

"Certainly, far as it is. It will take us but ten minutes, but it's far enough to have made the child prefer not to walk. She went straight over."

"Laws!" cried my friend again; the chain of my logic was ever too much for her. It dragged her at my heels even now, and when we had got half-way round—a devious, tiresome process, on ground much broken and by a path choked with overgrowth—I paused to give her breath. I sustained her with a grateful arm, assuring her that she might hugely help me; and this started us afresh, so that in the course of but a few minutes more we reached a point from which we found the boat to be where I had supposed it. It had been intentionally left as much as possible out of sight and was tied to one of the stakes of a fence that came, just there, down to the brink and that had been an assistance to disembarking. I recognised, as I looked at the pair of short, thick oars, quite safely drawn up, the prodigious character of the feat for a little girl; but I had lived, by this time, too long among wonders and had panted to too many livelier measures. There was a gate in the fence, through which we passed, and that brought us, after a trifling interval, more into the open. Then, "There she is!" we both exclaimed at once.

Flora, a short way off, stood before us on the grass and smiled as if her performance was now complete. The next thing she did, however, was to stoop straight down and pluck—quite as if it were all she was there for—a big, ugly spray of withered fern. I instantly became sure she had just come out of the copse. She waited for us, not herself taking a step, and I was conscious of the rare solemnity with which we presently approached her. She smiled and smiled, and we met; but it was all done in a silence by this time flagrantly ominous. Mrs. Grose was the first to break the spell: she threw herself on her knees and, drawing the child to her breast, clasped in a long embrace the little tender, yielding body. While this dumb convulsion lasted I could only watch it —which I did the more intently when I saw Flora's face peep at me over our companion's shoulder. It was serious now—the flicker had left it; but it strengthened the pang with which I at that moment envied Mrs. Grose the simplicity of *her* relation. Still, all this while, nothing more passed between us save that Flora had let her foolish fern again drop to the ground. What she and I had virtually said to each other was that pretexts were useless now. When Mrs. Grose finally got up she kept the child's hand, so that the two were still before me; and the singular reticence of our communion was even more marked in the frank look she launched me. "I'll be hanged," it said, "if *I'll* speak!"

It was Flora who, gazing all over me in candid wonder, was the first. She was struck with our bareheaded aspect. "Why, where are your things?"

"Where yours are, my dear!" I promptly returned.

She had already got back her gaiety, and appeared to take this as an answer quite sufficient. "And where's Miles?" she went on.

There was something in the small valour of it that quite finished me: these three words from her were, in a flash like the glitter of a drawn blade, the jostle of the cup that my hand, for weeks and weeks, had held high and full to the brim and that now, even before speaking, I felt overflow in a deluge. "I'll tell you if you'll tell *me*—" I heard myself say, then heard the tremor in which it broke.

"Well, what?"

Mrs. Grose's suspense blazed at me, but it was too late now, and I brought the thing out handsomely. "Where, my pet, is Miss Jessel?"

XX

JUST AS in the churchyard with Miles, the whole thing was upon us. Much as I had made of the fact that this name had never once, between us, been sounded, the quick, smitten glare with which the child's face now received it fairly likened my breach of the silence to the smash of a pane of glass. It added to the interposing cry, as if to stay the blow, that Mrs. Grose, at the same instant, uttered over my violence—the shriek of a creature scared, or rather wounded, which, in turn, within a few seconds, was completed by a gasp of my own. I seized my colleague's arm. "She's there, she's there!"

Miss Jessel stood before us on the opposite bank exactly as she had stood the other time, and I remember, strangely, as the first feeling now produced in me, my thrill of joy at having brought on a proof. She was there, and I was justified; she was there, and I was neither cruel nor mad. She was there for poor scared Mrs. Grose, but she was there most for Flora; and no moment of my monstrous time was perhaps so extraordinary as that in which I consciously threw out to her—with the sense that, pale and ravenous demon as she was, she would catch and understand it—an inarticulate message of gratitude. She rose erect on the spot my friend and I had lately quitted, and there was not, in all the long reach of her desire, an inch of her evil that fell short. This first vividness of vision and emotion were things of a few seconds, during which Mrs. Grose's dazed blink across to where I pointed struck me as a sovereign sign that she too at last saw, just as it carried my own eyes precipitately to the child. The revelation then of the manner in which Flora was affected startled me, in truth, far more that it would have done to find her also merely agitated, for direct dismay was of course not what I had expected. Prepared and on her guard as our pursuit had actually made her, she would repress every betrayal; and I was therefore shaken, on the spot, by my first glimpse of the particular one for which I had not allowed. To see her, without a convulsion of her small pink face, not even feign to glance in the direction of the prodigy I announced, but only, instead of that, turn at *me* an expression of hard, still gravity, an expression absolutely new and unprecedented and that appeared to read and accuse and judge me—this was a stroke that somehow converted the little girl herself into the very presence that could make me quail. I quailed even though my certitude that she thoroughly saw was never greater than at that instant, and in the immediate need to defend myself I called it passionately to witness. "She's there, you little unhappy thing—there, there, *there*, and you see her as well as you see me!" I had said shortly before to Mrs. Grose that she was not at these times a child, but an old, old woman, and that description of her could not have been more strikingly confirmed than in the way in which, for all answer to this, she simply showed me, without a concession, an admission, of her eyes, a countenance of deeper and deeper, of indeed suddenly quite fixed, reprobation. I was by this time—if I can put the whole thing at all together—more appalled at what I may properly call her manner than at anything else, though it was simultaneously with this that I became aware of having Mrs. Grose also, and very formidably, to reckon with. My elder companion, the next moment, at any rate, blotted out everything but her own flushed face and her loud, shocked protest, a burst of high disapproval. "What a dreadful turn, to be sure, Miss! Where on earth do you see anything?"

I could only grasp her more quickly yet, for even while she spoke the hideous plain presence stood undimmed and undaunted. It had already lasted a minute, and it lasted while I continued, seizing my colleague, quite thrusting her at it and presenting her to it, to insist with my pointing hand. "You don't see her exactly as *we* see?—you mean to say you don't now—*now?* She's as big as a blazing fire! Only look, dearest woman, *look—!*" She looked, even as I did, and gave me, with her deep groan of negation, repulsion, compassion—the mixture with her pity of her relief at her exemption—a sense, touching to me even then, that she would have backed me up if she could. I might well have needed that, for with this hard blow of the proof that her eyes were hopelessly sealed I felt my own situation horribly crumble, I felt—I saw—my livid predecessor press, from her position, on my defeat, and I was conscious, more than all, of what I should have from this instant to deal with in the astounding little attitude of Flora. Into this attitude Mrs. Grose immediately and violently entered, breaking, even while there pierced through my sense of ruin a prodigious private triumph, into breathless reassurance.

"She isn't there, little lady, and nobody's there—and you never see nothing, my sweet! How can poor Miss Jessel? when poor Miss Jessel's dead and buried? *We* know, don't we, love?"—and she appealed, blundering in, to the child. "It's all a mere mistake and a worry and a joke—and we'll go home as fast as we can!"

Our companion, on this, had responded with a strange, quick primness of propriety, and they were again, with Mrs. Grose on her feet, united, as it were, in pained opposition to me. Flora continued to fix me with her small mask of reprobation, and even at that minute I prayed God to forgive me for seeming to see that, as she stood there holding tight to our friend's dress, her incomparable childish beauty had suddenly failed, had quite vanished. I've said it already—she was literally, she was hideously, hard; she had turned common and almost ugly. "I don't know what you mean. I see nobody. I see nothing. I never *have*. I think you're cruel. I don't like you!" Then, after this deliverance, which might have been that of a vulgarly pert little girl in the street, she hugged Mrs. Grose more closely and buried in her skirts the dreadful little face. In this position she produced an almost furious wail. "Take me away, take me away—oh, take me away from *her!*"

"From *me?*" I panted.

"From you—from you!" she cried.

Even Mrs. Grose looked across at me dismayed, while I had nothing to do but communicate again with the figure that, on the opposite bank, without a movement, as rigidly still as if catching, beyond the interval, our voices, was as vividly there for my disaster as it was not there for my service. The wretched child had spoken exactly as if she had got from some outside source each of her stabbing little words, and I could therefore, in the full despair of all I had to accept, but sadly shake my head at her. "If I had ever doubted, all my doubt would at present have gone. I've been living with the miserable truth, and now it has only too much closed round me. Of course I've lost you: I've interfered, and you've seen—under her dictation"—with which I faced, over the pool again, our infernal witness—"the easy and perfect way to meet it. I've done my best, but I've lost you. Good-bye." For Mrs. Grose I had an imperative, an almost frantic "Go, go!" before which, in infinite distress, but mutely possessed of the little girl and clearly convinced, in spite of her blindness, that something awful had occurred and some collapse engulfed us, she retreated, by the way we had come, as fast as she could move.

Of what first happened when I was left alone I had no subsequent memory. I only knew that at the end of, I suppose, a quarter of an hour, an odorous dampness and roughness, chilling and piercing my trouble, had made me understand that I must have thrown myself, on my face, on the ground and given way to a wildness of grief. I must have lain there long and cried and sobbed, for when I raised my head the day was almost done. I got up and looked a moment, through the twilight, at the grey pool and its blank, haunted edge, and then I took, back to the house, my dreary and difficult course. When I reached the gate in the fence the boat, to my surprise, was gone, so that I had a fresh reflection to make on Flora's extraordinary command of the situation. She passed that night, by the most tacit, and I should add, were not the word so grotesque a false note, the happiest of arrangements, with Mrs. Grose. I saw neither of them on my return, but, on the other hand as by an ambiguous compensation, I saw a great deal of Miles. I saw—I can use no other phrase—so much of him that it was as if it were more than it had ever been. No evening I had passed at Bly had the portentous quality of this one; in spite of which—and in spite also of the deeper depths of consternation that had opened beneath my feet—there was literally, in the ebbing actual, an extraordinarily sweet sadness. On reaching the house I had never so much as looked for the boy; I had simply gone straight to my room to change what I was wearing and to take in, at a glance, much material testimony to Flora's rupture. Her little belongings had all been removed. When later, by the schoolroom fire, I was served with tea by the usual maid, I indulged, on the article of my other pupil, in no inquiry whatever. He had his freedom now—he might have it to the end! Well, he did have it; and it consisted—in part at least—of his coming in at about eight o'clock and sitting down with me in silence. On the removal of the tea-things I had blown out the candles and drawn my chair closer: I was conscious of a mortal coldness and felt as if I should never again be warm. So, when he appeared, I was sitting in the glow with my thoughts. He paused a moment by the door as if to look at me; then—as if to share them—came to the other side of the hearth and sank into a chair. We sat there in absolute stillness; yet he wanted, I felt, to be with me.

XXI

BEFORE a new day, in my room, had fully broken, my eyes opened to Mrs. Grose, who had come to my bedside with worse news. Flora was so markedly feverish that an illness was perhaps at hand; she had passed a night of extreme unrest, a night agitated above all by fears that had for their subject not in the least her former, but wholly her present, governess. It was not against the possible re-entrance of Miss Jessel on the scene that she protested—it was conspicuously and passionately against mine. I was promptly on my feet of course, and with an immense deal to ask; the more that my friend had discernibly now girded her loins to meet me once more. This I felt as soon as I had put to her the question of her sense of the child's sincerity as against my own. "She persists in denying to you that she saw, or has ever seen, anything?"

My visitor's trouble, truly, was great. "Ah, Miss, it isn't a matter on which I can push her! Yet it isn't either, I must say, as if I much needed to. It has made her, every inch of her, quite old."

"Oh, I see her perfectly from here. She resents, for all the world like some high little personage, the imputation on her truthfulness and, as it were, her respectability. 'Miss Jessel indeed—*she!*' Ah, she's 'respectable,' the chit! The impression she gave me there yesterday was, I assure you, the very strangest of all; it was quite beyond any of the others. I *did* put my foot in it! She'll never speak to me again."

Hideous and obscure as it all was, it held Mrs. Grose briefly silent; then she granted my point with a frankness which, I made sure, had more behind it. "I think indeed, Miss, she never will. She do have a grand manner about it!"

"And that manner"—I summed it up—"is practically what's the matter with her now!"

Oh, that manner, I could see in my visitor's face, and not a little else besides! "She asks me every three minutes if I think you're coming in."

"I see—I see." I too, on my side, had so much more than worked it out. "Has she said to you since yesterday—except to repudiate her familiarity with anything so dreadful—a single other word about Miss Jessel?"

"Not one, Miss. And of course you know," my friend added, "I took it from her, by the lake, that, just then and there at least, there *was* nobody."

"Rather! And, naturally, you take it from her still."

"I don't contradict her. What else can I do?"

"Nothing in the world! You've the cleverest little person to deal with. They've made them—their two friends, I mean—still cleverer even than nature did; for it was wondrous material to play on! Flora has now her grievance, and she'll work it to the end."

"Yes, Miss; but to *what* end?"

"Why, that of dealing with me to her uncle. She'll make me out to him the lowest creature—!"

I winced at the fair show of the scene in Mrs. Grose's face; she looked for a minute as if she sharply saw them together. "And him who thinks so well of you!"

"He has an odd way—it comes over me now." I laughed, "—of proving it! But that doesn't matter. What Flora wants, of course, is to get rid of me."

My companion bravely concurred. "Never again to so much as look at you."

"So that what you've come to me now for," I asked, "is to speed me on my way?" Before she had time to reply, however, I had her in check. "I've a better idea—the result of my reflections. My going *would* seem the right thing, and on Sunday I was terribly near it. Yet that won't do. It's *you* who must go. You must take Flora."

My visitor, at this, did speculate. "But where in the world—?"

"Away from here. Away from *them*. Away, even most of all, now, from me. Straight to her uncle."

"Only to tell on you—?"

"No, not 'only'! To leave me, in addition, with my remedy."

She was still vague. "And what *is* your remedy?"

"Your loyalty, to begin with. And then Miles's."

She looked at me hard. "Do you think he—?"

"Won't if he has the chance, turn on me? Yes, I venture still to think it. At all events, I want to try. Get off with his sister as soon as possible and leave me with him alone." I was amazed, myself, at the spirit I had still in reserve, and therefore perhaps a trifle the more disconcerted at the way in which, in spite of this fine example of it, she hesitated. "There's one thing, of course," I went on: "they mustn't, before she goes, see each other for three seconds."

Then it came over me that, in spite of Flora's presumable sequestration from the instant of her return from the pool, it might already be too late. "Do you mean," I anxiously asked, "that they *have* met?"

At this she quite flushed. "Ah, Miss, I'm not such a fool as that! If I've been obliged to leave her three or four times, it has been each time with one of the maids, and at present, though she's alone, she's locked in safe. And yet—and yet!" There were too many things.

"And yet what?"

"Well, are you so sure of the little gentleman?"

"I'm not sure of anything but *you*. But I have, since last evening, a new hope. I think he wants to give me an opening. I do believe that —poor little exquisite wretch!—he wants to speak. Last evening, in the firelight and the silence, he sat with me for two hours as if it were just coming."

Mrs. Grose looked hard, through the window, at the grey, gathering day. "And did it come?"

"No, though I waited and waited, I confess it didn't, and it was without a breach of the silence or so much as a faint allusion to his sister's condition and absence that we at last kissed for good-night. All the same," I continued, "I can't, if her uncle sees her, consent to his seeing her brother without my having given the boy—and most of all because things have got so bad—a little more time."

My friend appeared on this ground more reluctant than I could quite understand. "What do you mean by more time?"

"Well, a day or two—really to bring it out. He'll then be on *my* side —of which you see the importance. If nothing comes, I shall only fail, and you will, at the worst, have helped me by doing, on your arrival in town, whatever you may have found possible." So I put it before her, but she continued for a little so inscrutably embarrassed that I came again to her aid. "Unless, indeed," I wound up, "you really want *not* to go."

I could see it, in her face, at last clear itself; she put out her hand to me as a pledge. "I'll go—I'll go. I'll go this morning."

I wanted to be very just. "If you *should* wish still to wait, I would engage she shouldn't see me."

"No, no: it's the place itself. She must leave it." She held me a moment with heavy eyes, then brought out the rest. "Your idea's the right one. I myself, Miss—"

"Well?"

"I can't stay."

The look she gave me with it made me jump at possibilities. "You mean that, since yesterday, you *have* seen—?"

She shook her head with dignity. "I've *heard—!*"

"Heard?"

"From that child—horrors! There!" she sighed with tragic relief. "On my honour, Miss, she says things—!" But at this evocation she broke down; she dropped, with a sudden sob, upon my sofa and, as I had seen her do before, gave way to all the grief of it.

It was quite in another manner that I, for my part, let myself go. "Oh, thank God!"

She sprang up again at this, drying her eyes with a groan. " 'Thank God'?"

"It so justifies me!"

"It does that, Miss!"

I couldn't have desired more emphasis, but I just hesitated. "She's so horrible?"

I saw my colleague scarce knew how to put it. "Really shocking."

"And about me?"

"About you, Miss—since you must have it. It's beyond everything, for a young lady; and I can't think wherever she must have picked up—"

"The appalling language she applied to me? I can, then!" I broke in with a laugh that was doubtless significant enough.

It only, in truth, left my friend still more grave. "Well, perhaps I ought to also—since I've heard some of it before! Yet I can't bear it," the poor woman went on while, with the same movement, she glanced, on my dressing-table, at the face of my watch. "But I must go back."

I kept her, however. "Ah, if you can't bear it—!"

"How can I stop with her, you mean? Why, just *for* that: to get her away. Far from this," she pursued, "far from *them*—"

"She may be different? she may be free?" I seized her almost with joy. "Then, in spite of yesterday, you *believe*—"

"In such doings?" Her simple description of them required, in the light of her expression, to be carried no further, and she gave me the whole thing as she had never done. "I believe."

Yes, it was a joy, and we were still shoulder to shoulder: if I might continue sure of that I should care but little what else happened. My support in the presence of disaster would be the same as it had been in my early need of confidence, and if my friend would answer for my honesty, I would answer for all the rest. On the point of taking leave of her, none the less, I was to some extent embarrassed. "There's one thing of course—it occurs to me—to remember. My letter, giving the alarm, will have reached town before you."

I now perceived still more how she had been beating about the bush and how weary at last it had made her. "Your letter won't have got there. Your letter never went."

"What then became of it?"

"Goodness knows! Master Miles—"

"Do you mean *he* took it?" I gasped.

She hung fire, but she overcame her reluctance. "I mean that I saw yesterday, when I came back with Miss Flora, that it wasn't where you had put it. Later in the evening I had the chance to question Luke, and he declared that he had neither noticed nor touched it." We could only exchange, on this, one of our deeper mutual soundings, and it was Mrs. Grose who first brought up the plumb with an almost elate "You see!"

"Yes, I see that if Miles took it instead he probably will have read it and destroyed it."

"And don't you see anything else?"

I faced her a moment with a sad smile. "It strikes me that by this time your eyes are open even wider than mine."

They proved to be so indeed, but she could still blush, almost, to show it. "I make out now what he must have done at school." And she gave, in her simple sharpness, an almost droll disillusioned nod. "He stole!"

I turned it over—I tried to be more judicial. "Well—perhaps."

She looked as if she found me unexpectedly calm. "He stole *letters!*"

She couldn't know my reasons for a calmness after all pretty shallow; so I showed them off as I might. "I hope then it was to more purpose than in this case! The note, at any rate, that I put on the table yesterday," I pursued, "will have given him so scant an advantage—for it contained only the bare demand for an interview—that he is already much ashamed of having gone so far for so little, and that what he had on his mind last evening was precisely the need of confession." I seemed to myself, for the instant, to have mastered it, to see it all. "Leave us, leave us"—I was already, at the door, hurrying her off. "I'll get it out of him. He'll meet me—he'll confess. If he confesses, he's saved. And if he's saved—"

"Then *you* are?" The dear woman kissed me on this, and I took her farewell. "I'll save you without him!" she cried as she went.

XXII

YET IT was when she had got off—and I missed her on the spot—that the great pinch really came. If I had counted on what it would give me to find myself alone with Miles, I speedily perceived, at least, that it would give me a measure. No hour of my stay in fact was so assailed with apprehensions as that of my coming down to learn that the carriage containing Mrs. Grose and my younger pupil had already rolled out of the gates. Now I *was*, I said to myself, face to face with the elements, and for much of the rest of the day, while I fought my weakness, I could consider that I had been supremely rash. It was a tighter place still than I had yet turned round in; all the more that, for the first time, I could see in the aspect of others a confused reflection of the crisis. What had happened naturally caused them all to stare; there was too little of the explained, throw out whatever we might, in the suddenness of my colleague's act. The maids and the men looked blank; the effect of which on my nerves was an aggravation until I saw the necessity of making it a positive aid. It was precisely, in short, by just clutching the helm that I avoided total wreck; and I dare say that, to bear up at all, I became, that morning, very grand and very dry. I welcomed the consciousness that I was charged with much to do, and I caused it to be known as well that, left thus to myself, I was quite remarkably firm. I wandered with that manner, for the next hour or two, all over the place and looked, I have no doubt, as if I were ready for any onset. So, for the benefit of whom it might concern, I paraded with a sick heart.

The person it appeared least to concern proved to be, till dinner, little Miles himself. My perambulations had given me meanwhile, no glimpse of him, but they had tended to make more public the change taking place in our relation as a consequence of his having at the piano, the day before, kept me, in Flora's interest, so beguiled and befooled. The stamp of publicity had of course been fully given by her confinement and departure, and the change itself was now ushered in by our non-observance of the regular custom of the schoolroom. He had already disappeared when, on my way down, I pushed open his door, and I learned below that he had breakfasted—in the presence of a couple of the maids—with Mrs. Grose and his sister. He had then gone out, as he said, for a stroll; than which nothing, I reflected, could better have expressed his frank view of the abrupt transformation of my office. What he would now permit this office to consist of was yet to be settled: there was a queer relief, at all events—I mean for myself in especial—in the renouncement of one pretension. If so much had sprung to the surface, I scarce put it too strongly in saying that what had perhaps sprung highest was the absurdity of our prolonging the fiction that I had anything more to teach him. It sufficiently stuck out that, by tacit little tricks in which even more than myself he carried out the care for my dignity, I had had to appeal to him to let me off straining to meet him on the ground of his true capacity. He had at any rate his freedom now; I was never to touch it again; as I had amply shown, moreover, when, on his joining me in the schoolroom the previous night, I had uttered, on the subject of the interval just concluded, neither challenge nor hint. I had too much, from this moment, my other ideas. Yet when he at last arrived the difficulty of applying them, the accumulations of my problem, were brought straight home to me by the beautiful little presence on which what had occurred had as yet, for the eye, dropped neither stain nor shadow.

To mark, for the house, the high state I cultivated I decreed that my meals with the boy should be served, as we called it, downstairs; so that I had been awaiting him in the ponderous pomp of the room outside of the window of which I had had from Mrs. Grose, that first scared Sunday, my flash of something it would scarce have done to call light. Here at present I felt afresh—for I had felt it again and again—how my equilibrium depended on the success of my rigid will, the will to shut my eyes as tight as possible to the truth that what I had to deal with was, revoltingly, against nature. I could only get on at all by taking "nature" into my confidence and my account, by treating my monstrous ordeal as a push in a direction unusual, of course, and unpleasant, but demanding, after all, for a fair front, only another turn of the screw of ordinary human virtue. No attempt, none the less, could well require more tact than just this attempt to supply, one's self, *all* the nature. How could I put even a little of that article into a suppression of reference to what had occurred? How, on the other hand, could I make a reference without a new plunge into the hideous obscure? Well, a sort of answer, after a time, had come to me, and it was so far confirmed as that I was met, incontestably, by the quickened vision of what was rare in my little companion. It was indeed as if he had found even now—as he had so often found at lessons—still some other delicate way to ease me off. Wasn't there light in the fact which, as we shared our solitude, broke out with a specious glitter it had never yet quite worn?—the fact that (opportunity aiding, precious opportunity which had now come) it would be preposterous, with a child so endowed, to forgo the help one might wrest from absolute intelligence? What had his intelligence been given him for but to save him? Mightn't one, to reach his mind, risk the stretch of an angular arm over his character? It was as if, when we were face to face in the dining-room, he had literally shown me the way. The roast mutton was on the table, and I had dispensed with attendance. Miles, before he sat down, stood a moment with his hands in his pockets and looked at the joint, on which he seemed on the point of passing some humorous judgment. But what he presently produced was: "I say, my dear, is she really very awfully ill?"

"Little Flora? Not so bad but that she'll presently be better. London will set her up. Bly had ceased to agree with her. Come here and take your mutton."

He alertly obeyed me, carried the plate carefully to his seat, and, when he was established, went on. "Did Bly disagree with her so terribly suddenly?"

"Not so suddenly as you might think. One had seen it coming on."

"Then why didn't you get her off before?"

"Before what?"

"Before she became too ill to travel."

I found myself prompt. "She's *not* too ill to travel: she only might have become so if she had stayed. This was just the moment to seize.

The journey will dissipate the influence"—oh, I was grand!—"and carry it off."

"I see, I see"—Miles, for that matter, was grand too. He settled to his repast with the charming little "table manner" that, from the day of his arrival, had relieved me of all grossness of admonition. Whatever he had been driven from school for, it was not for ugly feeding. He was irreproachable, as always, today; but he was unmistakeably more conscious. He was discernibly trying to take for granted more things than he found, without assistance, quite easy; and he dropped into peaceful silence while he felt his situation. Our meal was of the briefest—mine a vain pretence, and I had the things immediately removed. While this was done Miles stood again with his hands in his little pockets and his back to me—stood and looked out of the wide window through which, that other day, I had seen what pulled me up. We continued silent while the maid was with us—as silent, it whimsically occurred to me, as some young couple who, on their wedding-journey, at the inn, feel shy in the presence of the waiter. He turned round only when the waiter had left us. "Well—so we're alone!"

XXIII

"OH, MORE or less." I fancy my smile was pale. "Not absolutely. We shouldn't like that!" I went on.

"No—I suppose we shouldn't. Of course we have the others."

"We have the others—we have indeed the others," I concurred.

"Yet even though we have them," he returned, still with his hands in his pockets and planted there in front of me, "they don't much count, do they?"

I made the best of it, but I felt wan. "It depends on what you call 'much'!"

"Yes"—with all accommodation—"everything depends!" On this, however, he faced to the window again and presently reached it with his vague, restless, cogitating step. He remained there awhile, with his forehead against the glass, in contemplation of the stupid shrubs I knew and the dull things of November. I had always my hypocrisy of "work," behind which, now, I gained the sofa. Steadying myself with it there as I had repeatedly done at those moments of torment that I have described as the moments of my knowing the children to be given to something from which I was barred, I sufficiently obeyed my habit of being prepared for the worst. But an extraordinary impression dropped on me as I extracted a meaning from the boy's embarrassed back—none other than the impression that I was not barred now. This inference grew in a few minutes to sharp intensity and seemed bound up with the direct perception that it was positively *he* who was. The frames and squares of the great window were a kind of image, for him, of a kind of failure. I felt that I saw him, at any rate, shut in or shut out. He was admirable, but not comfortable: I took it in with a throb of hope. Wasn't he looking, through the haunted pane, for something he couldn't see?—and wasn't it the first time in the whole business that he had known such a lapse? The first, the very first: I found it a splendid portent. It made him anxious, though he watched himself; he had been anxious all day and, even while in his usual sweet little manner he sat at table, had needed all his small strange genius to give it a gloss. When he at last turned round to meet me, it was almost as if this genius had succumbed. "Well, I think I'm glad Bly agrees with *me!*"

"You would certainly seem to have seen, these twenty-four hours, a good deal more of it than for some time before. I hope," I went on bravely, "that you've been enjoying yourself."

"Oh, yes, I've been ever so far; all round about—miles and miles away. I've never been so free."

He had really a manner of his own, and I could only try to keep up with him. "Well, do you like it?"

He stood there smiling; then at last he put into two words—"Do *you?*"—more discrimination than I had ever heard two words contain. Before I had time to deal with that, however, he continued as if with the sense that this was an impertinence to be softened. "Nothing could be more charming than the way you take it, for of course if we're alone together now it's you that are alone most. But I hope," he threw in, "you don't particularly mind!"

"Having to do with you?" I asked. "My dear child, how can I help minding? Though I've renounced all claim to your company,—you're so beyond me,—I at least greatly enjoy it. What else should I stay on for?"

He looked at me more directly, and the expression of his face, graver now, struck me as the most beautiful I had ever found in it. "You stay on just for *that?*"

"Certainly. I stay on as your friend and from the tremendous interest

I take in you till something can be done for you that may be more worth your while. That needn't surprise you." My voice trembled so that I felt it impossible to suppress the shake. "Don't you remember how I told you, when I came and sat on your bed the night of the storm, that there was nothing in the world I wouldn't do for you?"

"Yes, yes!" He, on his side, more and more visibly nervous, had a tone to master; but he was so much more successful than I that, laughing out through his gravity, he could pretend we were pleasantly jesting. "Only that, I think, was to get me to do something for *you!*"

"It was partly to get you to do something," I conceded. "But, you know, you didn't do it."

"Oh, yes," he said with the brightest superficial eagerness, "you wanted me to tell you something."

"That's it. Out, straight out. What you have on your mind, you know."

"Ah, then, is *that* what you've stayed over for?"

He spoke with a gaiety through which I could still catch the finest little quiver of resentful passion; but I can't begin to express the effect upon me of an implication of surrender even so faint. It was as if what I had yearned for had come at last only to astonish me. "Well, yes—I may as well make a clean breast of it. It was precisely for that."

He waited so long that I supposed it for the purpose of repudiating the assumption on which my action had been founded; but what he finally said was: "Do you mean now—here?"

"There couldn't be a better place or time." He looked round him uneasily, and I had the rare—oh, the queer!—impression of the very first symptom I had seen in him of the approach of immediate fear. It was as if he were suddenly afraid of me—which struck me indeed as perhaps the best thing to make him. Yet in the very pang of the effort I felt it vain to try sternness, and I heard myself the next instant so gentle as to be almost grotesque. "You want so to go out again?"

"Awfully!" He smiled at me heroically, and the touching little bravery of it was enhanced by his actually flushing with pain. He had picked up his hat, which he had brought in, and stood twirling it in a way that gave me, even as I was just nearly reaching port, a perverse horror of what I was doing. To do it in *any* way as an act of violence, for what did it consist of but the obtrusion of the idea of grossness and guilt on a small helpless creature who had been for me a revelation of the possibilities of beautiful intercourse? Wasn't it base to create for a being so exquisite a mere alien awkwardness? I suppose I now read into our situation a clearness it couldn't have had at the time, for I seem to see our poor eyes already lighted with some spark of a prevision of the anguish that was to come. So we circled about, with terrors and scruples, like fighters not daring to close. But it was for each other we feared! That kept us a little longer suspended and unbruised. "I'll tell you everything," Miles said—"I mean I'll tell you anything you like. You'll stay on with me, and we shall both be all right and I *will* tell you —I *will*. But not now."

"Why not now?"

My insistence turned him from me and kept him once more at his window in a silence during which, between us, you might have heard a pin drop. Then he was before me again with the air of a person for whom, outside, someone who had frankly to be reckoned with was waiting. "I have to see Luke."

I had not yet reduced him to quite so vulgar a lie, and I felt proportionately ashamed. But, horrible as it was, his lies made up my truth. I achieved thoughtfully a few loops of my knitting. "Well, then, go to Luke, and I'll wait for what you promise. Only, in return for that, satisfy, before you leave me, one very much smaller request."

He looked as if he felt he had succeeded enough to be able still a little to bargain. "Very much smaller—?"

"Yes, a mere fraction of the whole. Tell me"—oh, my work preoccupied me, and I was off-hand!—"if, yesterday afternoon, from the table in the hall, you took, you know, my letter."

XXIV

MY SENSE of how he received this suffered for a minute from something that I can describe only as a fierce split of my attention —a stroke that at first, as I sprang straight up, reduced me to the mere blind movement of getting hold of him, drawing him close, and, while I just fell for support against the nearest piece of furniture, instinctively keeping him with his back to the window. The appearance was full upon us that I had already had to deal with here: Peter Quint had come into view like a sentinel before a prison. The next thing I saw was that, from outside, he had reached the window, and then I knew that, close to the glass and glaring in through it, he offered once more to the room his white face of damnation. It represents but grossly what took place within me at the sight to say that on the second my decision was made; yet I believe that no woman so overwhelmed ever in so short a time recovered her grasp of the *act*. It came to me in the very horror of the immediate presence that the act would be, seeing and facing what I saw and faced, to keep the boy himself unaware. The inspiration—I can call it by no other name—was that I felt how voluntarily, how transcendently, I *might*. It was like fighting with a demon for a human soul, and when I had fairly so appraised it I saw how the human soul—held out, in the tremor of my hands, at arm's length—had a perfect dew of sweat on a lovely childish forehead. The face that was close to mine was as white as the face against the glass, and out of it presently came a sound, not low nor weak, but as if from much further away, that I drank like a waft of fragrance.

"Yes—I took it."

At this, with a moan of joy, I enfolded, I drew him close; and while I held him to my breast, where I could feel in the sudden fever of his little body the tremendous pulse of his little heart, I kept my eyes on the thing at the window and saw it move and shift its posture. I have likened it to a sentinel, but its slow wheel, for a moment, was rather the prowl of a baffled beast. My present quickened courage, however, was such that, not too much to let it through, I had to shade, as it were, my flame. Meanwhile the glare of the face was again at the window, the scoundrel fixed as if to watch and wait. It was the very confidence that I might now defy him, as well as the positive certitude, by this time, of the child's unconsciousness, that made me go on. "What did you take it for?"

"To see what you said about me."

"You opened the letter?"

"I opened it."

My eyes were now, as I held him off a little again, on Miles's own face, in which the collapse of mockery showed me how complete was the ravage of uneasiness. What was prodigious was that at last, by my success, his sense was sealed and his communication stopped: he knew that he was in presence, but knew not of what, and knew still less that I also was and that I did know. And what did this strain of trouble matter when my eyes went back to the window only to see that the air was clear again and—by my personal triumph—the influence quenched? There was nothing there. I felt that the cause was mine and that I should surely get *all*. "And you found nothing!"—I let my elation out.

He gave the most mournful, thoughtful little headshake. "Nothing."

"Nothing, nothing!" I almost shouted in my joy.

"Nothing, nothing," he sadly repeated.

I kissed his forehead; it was drenched. "So what have you done with it?"

"I've burnt it."

"Burnt it?" It was now or never. "Is that what you did at school?"

Oh, what this brought up! "At school?"

"Did you take letters?—or other things?"

"Other things?" He appeared now to be thinking of something far off and that reached him only through the pressure of his anxiety. Yet it did reach him. "Did I *steal?*"

I felt myself redden to the roots of my hair as well as wonder if it were more strange to put to a gentleman such a question or to see him take it with allowances that gave the very distance of his fall in the world. "Was it for that you mightn't go back?"

The only thing he felt was rather a dreary little surprise. "Did you know I mightn't go back?"

"I know everything."

He gave me at this the longest and strangest look. "Everything?"

"Everything. Therefore *did* you—?" But I couldn't say it again.

Miles could, very simply. "No. I didn't steal."

My face must have shown him I believed him utterly; yet my hands—but it was for pure tenderness—shook him as if to ask him why, if it was all for nothing, he had condemned me to months of torment. "What then did you do?"

He looked in vague pain all round the top of the room and drew his breath, two or three times over, as if with difficulty. He might have been standing at the bottom of the sea and raising his eyes to some faint green twilight. "Well—I said things."

"Only that?"

"They thought it was enough!"

"To turn you out for?"

Never, truly, had a person "turned out" shown so little to explain it as this little person! He appeared to weigh my question, but in a manner quite detached and almost helpless. "Well, I suppose I oughtn't."

"But to whom did you say them?"

He evidently tried to remember, but it dropped—he had lost it. "I don't know!"

He almost smiled at me in the desolation of his surrender, which was indeed practically, by this time, so complete that I ought to have left it there. But I was infatuated—I was blind with victory, though even then the very effect that was to have brought him so much nearer was already that of added separation. "Was it to everyone?" I asked.

"No; it was only to—" But he gave a sick little headshake. "I don't remember their names."

"Were they then so many?"

"No—only a few. Those I liked."

Those he liked? I seemed to float not into clearness, but into a darker obscure, and within a minute there had come to me out of my very pity the appalling alarm of his being perhaps innocent. It was for the instant confounding and bottomless, for if he *were* innocent, what then on earth was *I*? Paralysed, while it lasted, by the mere brush of the question, I let him go a little, so that, with a deep-drawn sigh, he turned away from me again; which, as he faced toward the clear window, I suffered, feeling that I had nothing now there to keep him from. "And did they repeat what you said?" I went on after a moment.

He was soon at some distance from me, still breathing hard and again with the air, though now without anger for it, of being confined against his will. Once more, as he had done before, he looked up at the dim day as if, of what had hitherto sustained him, nothing was left but an unspeakable anxiety. "Oh, yes," he nevertheless replied—"they must have repeated them. To those *they* liked," he added.

There was, somehow, less of it than I had expected; but I turned it over. "And these things came round—?"

"To the masters? Oh, yes!" he answered very simply. "But I didn't know they'd tell."

"The masters? They didn't—they've never told. That's why I ask you."

He turned to me again his little beautiful fevered face. "Yes, it was too bad."

"Too bad?"

"What I suppose I sometimes said. To write home."

I can't name the exquisite pathos of the contradiction given to such a speech by such a speaker; I only know that the next instant I heard myself throw off with homely force: "Stuff and nonsense!" But the next after that I must have sounded stern enough. "What were these things?"

My sternness was all for his judge, his executioner; yet it made him avert himself again, and that movement made *me*, with a single bound and an irrepressible cry, spring straight upon him. For there again, against the glass, as if to blight his confession and stay his answer, was the hideous author of our woe—the white face of damnation. I felt a sick swim at the drop of my victory and all the return of my battle, so that the wildness of my veritable leap only served as a great betrayal. I saw him, from the midst of my act, meet it with a divination, and on the perception that even now he only guessed, and that the window was still to his own eyes free, I let the impulse flame up to convert the climax of his dismay into the very proof of his liberation. "No more, no more, no more!" I shrieked, as I tried to press him against me, to my visitant.

"Is she *here*?" Miles panted as he caught with his sealed eyes the direction of my words. Then as his strange "she" staggered me and, with a gasp, I echoed it, "Miss Jessel, Miss Jessel!" he with a sudden fury gave me back.

I seized, stupefied, his supposition—some sequel to what we had done to Flora, but this made me only want to show him that it was better still than that. "It's not Miss Jessel! But it's at the window—straight before us. It's *there*—the coward horror, there for the last time!"

At this, after a second in which his head made the movement of a baffled dog's on a scent and then gave a frantic little shake for air and light, he was at me in a white rage, bewildered, glaring vainly over the place and missing wholly, though it now, to my sense, filled the room like the taste of poison, the wide, overwhelming presence. "It's *he*?"

I was so determined to have all my proof that I flashed into ice to challenge him. "Whom do you mean by 'he'?"

"Peter Quint—you devil!" His face gave again, round the room, its convulsed supplication. "*Where?*"

They are in my ears still, his supreme surrender of the name and his tribute to my devotion. "What does he matter now, my own?—what will he *ever* matter? *I* have you," I launched at the beast, "but he has lost you for ever!" Then, for the demonstration of my work, "There, *there!*" I said to Miles.

But he had already jerked straight round, stared, glared again, and seen but the quiet day. With the stroke of the loss I was so proud of he uttered the cry of a creature hurled over an abyss, and the grasp with which I recovered him might have been that of catching him in his fall. I caught him, yes, I held him—it may be imagined with what a passion; but at the end of a minute I began to feel what it truly was that I held. We were alone with the quiet day, and his little heart, dispossessed, had stopped.

SELECTED DATES

IN THE LIVES AND WORKS OF THE AUTHORS

Leo Tolstoy (1828–1910)

1828	Born at Tula, Russia—Count Leolyev Nikolayevich Tolstoy; family of gentry dating back to the 16th century.
1844	Attended Kazan University, after private early education by French tutors.
1847	Returned to family estate, Yasnaya Polyana, to take up farming and management of serfs. Began his renowned diary.
1851–1857	Army life—participation in the Crimean War. Years of violent dissipation and immoral living, punctuated by equally tempestuous self-searching and criticism.
1861	Back at family home, served as a local magistrate in connection with the Emancipation Act of 1861. Illicit love affair with a serf's wife, followed by marriage to a woman of his own class.
1864	Completed *War and Peace,* his epic novel.
1877	*Anna Karenina,* another great novel. Plagued by an unbearable obsession of approaching death, he began his tortured conversion to Christianity, but could not accept the full teaching of the Orthodox Church.
1884	A *Confession,* an account of his conversion, considered by some to rank with *Confessions* of St. Augustine. *The Death of Ivan Ilych,* the greatest imaginative work of his maturity. Began to become the most venerated man in the world; he renounced his title and his worldly fortunes and led a mendicant life, preaching his personal version of Christianity among the peasants.
1896	*What is Art,* important book of aesthetics.
1901	Excommunicated by the Russian Orthodox Church.
1910	Having left his wife and taken their daughter Alexandra with him, as he started out on one of his teaching pilgrimages, he died; interred at ancestral home, without a Christian burial.

Herman Melville (1819–1891)

1819–1836	Born in New York City. Attended New York Male High School. Father, at first a prosperous importer, suffered severe losses in 1830; family moved to Albany and then to Lansingburgh (Troy), New York, in serious financial straits.
1837	Became a country school teacher. Studied surveying and engineering at Lansingburgh Academy.
1839	Went to sea as a cabin boy aboard a Liverpool trading ship; upon return, again taught school in New York country towns and journeyed as far west as Illinois.
1841	As an ordinary seaman, sailed from New Bedford, Massachusetts, bound for the South Seas on the whaler *Acushnet*.
1842	Jumped ship in the Marquesas Islands, and lived among the natives for a month.
1843	Shipped as ordinary seaman on the frigate, *United States*.
1846	*Typee*, a novel.
1847	*Omoo*, a novel. Married Elizabeth Shaw and settled in New York City.
1850	Purchased farm near Pittsfield, Massachusetts; became friend of Hawthorne.
1851	*Moby Dick*, a novel, his masterpiece.
1856	*The Piazza Tales*, including the novella *Benito Cereno*, which at first had been anonymously published in a magazine.
1857	Journey to Europe and the Holy Land in effort to regain his mental and physical health which had been waning for some years.
1863	Sold farm and returned to residence in New York.
1866–1885	Was District Inspector of Customs in New York; lived quietly and obscurely, publishing little.
1891	Died in New York; his passing was hardly noted.

André Gide (1869–1951)

1869 Born in Paris. Strict Protestant religious upbringing. Educated at École Alsacienne and Lycée Henri IV.

1893 Journey to North Africa, resulting in great personal liberation from his background and in lyric praise of the life of the senses.

1895 Married his cousin, whom he had loved since childhood; honeymoon voyage to Algeria.

1897 *Les Nourritures Terrestres*, a novel recounting his new joy in life.

1902 *L'Immoraliste*, a novel reflecting the author's pagan enthusiasm for life and, at the same time, his aspiration toward holiness.

1909 Helped found the *Nouvelle Revue Française*, a magazine which gradually became the most influential in the French literary world.

1914 Aided Franco-Belgian war refugees. *Les Caves du Vatican*, a novel.

1919 *La Symphonie Pastorale*, a novella.

1924 Viciously attacked on all sides as a corrupter of public morals through his art. *Si le grain ne meurt*, an autobiographical account.

1925 *Les Faux-Monnayeurs*, his novel masterpiece. Gide had disposed of his French properties and was traveling in French Equatorial Africa when the book appeared.

1936 *Le Retour de l'U.R.S.S.*, essays reflecting disillusionment with the Soviet experiment.

1939 He remained an unappeasable anti-Fascist on the eve of and throughout World War II. During Spanish Civil War, he had ardently supported the Loyalist cause.

1947 Received the Nobel Prize for Literature for "his extensive and artistically important authorship in which he exposed the problems and conditions of mankind."

1951 Died in Paris.

James Joyce (1882–1941)

1882–1900 Born in Dublin. Educated at Belvedere College, a Jesuit school, and University College, Dublin. Brilliant student of languages, literature, and philosophy.

1902 Following graduation, went to Paris to pursue private literary study; financial struggle to remain alive and independent.

1904–1912 Brought back to Dublin by illness and then death of his mother. Married. Returned to Continent—Zurich, Pola, Trieste—where he struggled to support a wife and two children by teaching and translating. *Chamber Music* (1907), a collection of lyric poems. Made his final visit to Dublin, in an unsuccessful effort to gain publication for his stories.

1914–1918 During war, lived under free arrest in Trieste. *Dubliners* (1914), his short stories, including *The Dead. Portrait of the Artist As a Young Man* (1917), autobiographical novel. *Exiles* (1918), a play. Given small financial grants by wealthy patronesses of arts to insure his monetary and artistic independence. Eyesight began to fail.

1920 Moved to Paris. Continued failure of eyesight; later underwent many operations for cataract.

1922 Publication in Paris of *Ulysses,* the *magnum opus* on which he had been laboring for seven years. The book was banned in Britain and America and was then pirated, vilified, and clandestinely circulated until admitted for publication in the U.S.A. (1934) and Great Britain (1936).

1939 *Finnegan's Wake*, a novel on which he had worked for seventeen years and which he considered his greatest work, although its almost insuperable difficulty has prevented widespread acceptance.

1941 Died in Zurich to which, during World War II, he had migrated from Paris.

Katherine Mansfield (1888–1923)

1888 Born in Wellington, New Zealand—Kathleen Beauchamp.

1889–1902 Childhood in small town of Karori, near Wellington. Gained prize for English composition at the village school.

1902–1905 Sent to England to be educated at Queen's College, London; edited college magazine. Developed interest in music.

1906–1909 Returned to New Zealand but, dissatisfied with intellectual isolation, longed for England.

1909–1911 Back to London. Unhappy first marriage. Contributed selections to *The New Age*. *In a German Pension*, short stories; gained immediate recognition.

1911–1917 Met J. Middleton Murry. Published stories in *Rhythm*, *The Blue Review*, and *The Signature*, literary magazines edited by Murry, the last with the collaboration of D. H. Lawrence. Greatly saddened in 1915 by her beloved brother's death in World War I. Began work on long story of her New Zealand childhood days (*Prelude*).

1918–1919 *Prelude* published but completely ignored by critics and public. Marriage to Murry. Attacks of pleurisy and then of consumption. Contributed critical reviews to *The Athenaeum*, then under her husband's editorship. Wrote *At the Bay* as a continuation of *Prelude*.

1920 Unhappy journey to the south of France in an attempt to regain health. *Je ne parle pas français*, short story.

1921 *Bliss, and Other Stories*.

1922 *The Garden Party, and Other Stories*. Established her reputation as one of the most remarkable writers of her generation.

1923 Struggled against disease. Died, suddenly, at Fontainebleau, France.

Katherine Anne Porter (1894–)

1894 Born in Indian Creek, Texas; great-great-great-grandaughter of Daniel Boone. Brought up in Texas and Louisiana; educated in small Southern convent schools.

1931 From earliest childhood, wrote "and destroyed manuscripts quite literally by the trunkful." Isolated from other artists, she rigorously pursued private reading and her individual artistic formation. Guggenheim Fellowship for study abroad.

1933 Marriage followed by divorce.

1935 *Flowering Judas and Other Stories*.

1938 Second marriage followed by divorce.

1939	*Pale Horse, Pale Rider,* stories and short novels.
1940	Received medal of the Society for Libraries.
1943	Elected to membership in the National Institute of Arts and Letters.
1944	*The Leaning Tower and Other Stories,* her third collection. Fellow in Regional American Literature of the Library of Congress.
1949	Honorary Litt. D. from the Women's College of the University of North Carolina; guest lecturer in literature, Stanford University.
1952	Delegate to the International Festival of Twentieth Century Arts, Paris. *The Days Before,* critical pieces and sketches.
1962	*Ship of Fools,* a novel awaited many years by the literary world.

Joseph Conrad (1857–1924)

1857–1871	Born in Poland—Teodor Jozef Konrad Korzeniowski. His father was leader of a Polish revolutionary group seeking freedom from Russia; parents died prematurely due to privations of exile. Orphaned boy was reared by an uncle.
1872	Toured Continent with a tutor.
1873	Shipped on French vessel from Marseilles.
1878	Arrived at Lowestoft, Suffolk, England, speaking virtually no English.
1886–1893	After conquering his new language, received master mariner's ticket by examination; became naturalized British subject. He served as first or second in command on merchant ships, mostly in the Far Eastern trade. Also made one memorable voyage up the Congo. Abandoned the sea and settled in England.
1895	*Almayer's Folly,* novel.
1896	*An Outcast of the Islands,* novel; marriage to an Englishwoman. Granted small Civil List pension.
1897	*Nigger of the "Narcissus,"* novel.
1900	*Lord Jim,* novel.

1902	*Heart of Darkness.*
1923	His only visit to the United States.
1924	Death at Bishopsbourne, Kent, England.

Henry James (1843–1916)

1843–1861	Born in New York City, into a wealthy family. Childhood in Albany and boyhood in New York City; schooled by tutors and governesses; traveled in Europe; read voraciously. Excused from Civil War Army service because of minor back injury.
1862	Family settled in Cambridge, Massachusetts. He briefly attended Harvard. Contributed stories to the *Atlantic Monthly*; became friend of William Dean Howells.
1869–1870	First adult trip to Europe.
1872–1874	Renewed stay in Europe. *Roderick Hudson,* a novel.
1875	Paris. Met Turgenev, Flaubert, Zola, Goncourt, Daudet, and Maupassant.
1878	*Daisy Miller,* a novella; he began to be internationally recognized.
1881	*The Portrait of a Lady,* a novel. Revisited America before returning to Europe, chiefly to England, for an uninterrupted stay of twenty years.
1898	Purchased home in Rye, Sussex; *The Turn of the Screw.*
1902	*The Wings of the Dove,* a novel.
1903	*The Ambassadors,* a novel.
1904	*The Golden Bowl,* a novel. Revisited America.
1907–1909	Wrote famous prefaces for the twenty-four-volume New York Edition of his collected works. Prefaces later gathered into *The Art of the Novel.*
1915	To demonstrate sympathy for England in World War I, became a British subject.
1916	Died in England; his ashes were returned to America for burial.